AF352927

Viruses and Nuclear Egress

Viruses and Nuclear Egress

Editor

Donald M. Coen

MDPI • Basel • Beijing • Wuhan • Barcelona • Belgrade • Manchester • Tokyo • Cluj • Tianjin

Editor
Donald M. Coen
Department of Biological
Chemistry and Molecular
Pharmacology
Blavatnik Institute
Harvard Medical School
Boston
United States

Editorial Office
MDPI
St. Alban-Anlage 66
4052 Basel, Switzerland

This is a reprint of articles from the Special Issue published online in the open access journal *Viruses* (ISSN 1999-4915) (available at: www.mdpi.com/journal/viruses/special_issues/Nuclear_Egress).

For citation purposes, cite each article independently as indicated on the article page online and as indicated below:

LastName, A.A.; LastName, B.B.; LastName, C.C. Article Title. *Journal Name* **Year**, *Volume Number*, Page Range.

ISBN 978-3-0365-5696-3 (Hbk)
ISBN 978-3-0365-5695-6 (PDF)

Contents

Review

Host and Viral Factors Involved in Nuclear Egress of Herpes Simplex Virus 1

Jun Arii

Division of Clinical Virology, Center for Infectious Diseases, Kobe University Graduate School of Medicine, Kobe, Hyogo 650-0017, Japan; jarii@med.kobe-u.ac.jp; Tel.: +81-78-382-6272

Abstract: Herpes simplex virus 1 (HSV-1) replicates its genome and packages it into capsids within the nucleus. HSV-1 has evolved a complex mechanism of nuclear egress whereby nascent capsids bud on the inner nuclear membrane to form perinuclear virions that subsequently fuse with the outer nuclear membrane, releasing capsids into the cytosol. The viral-encoded nuclear egress complex (NEC) plays a crucial role in this vesicle-mediated nucleocytoplasmic transport. Nevertheless, similar system mediates the movement of other cellular macromolecular complexes in normal cells. Therefore, HSV-1 may utilize viral proteins to hijack the cellular machinery in order to facilitate capsid transport. However, little is known about the molecular mechanisms underlying this phenomenon. This review summarizes our current understanding of the cellular and viral factors involved in the nuclear egress of HSV-1 capsids.

Keywords: herpesviruses; nuclear egress; primary envelopment; de-envelopment

Citation: Arii, J. Host and Viral Factors Involved in Nuclear Egress of Herpes Simplex Virus 1. *Viruses* **2021**, *13*, 754. https://doi.org/10.3390/v13050754

Academic Editor: Donald M. Coen

Received: 10 March 2021
Accepted: 23 April 2021
Published: 25 April 2021

Publisher's Note: MDPI stays neutral with regard to jurisdictional claims in published maps and institutional affiliations.

1. Introduction

The nuclear membrane (NM) consists of an inner nuclear membrane (INM) and an outer nuclear membrane (ONM), which separate nuclear and cytoplasmic activities in the eukaryotic cell (Figure 1A). The segregation of the nucleoplasm from the cytoplasm separates translation from transcription and maintains genome integrity. The INM and ONM are biochemically distinct [1]. While the ONM is continuous and functionally interacts with the endoplasmic reticulum (ER), the INM contains its own set of integral membrane proteins [2]. The ONM and the INM are separated by the perinuclear space and connected at annular junctions, where nuclear pore complexes (NPCs) are found. NPCs are large macromolecular assemblies that form transport channels and regulate trafficking between the cytoplasm and the nucleoplasm [3]. It had long been assumed that the NPC was the only route out of the nucleus except during the early phases of mitosis when the nuclear envelope breaks down.

Recently, however, an alternative and NPC-independent pathway for nuclear export has been described in cells infected with viruses. Vesicle-mediated nucleocytoplasmic transport is a mechanism for the nuclear export of macromolecular complexes [4]. In this system, a macromolecular complex in the nucleus buds through the INM to form a vesicle in the perinuclear space ('primary envelopment'). This vesicle then fuses with the ONM to release the complex into the cytoplasm in a process called 'de-envelopment' [4] (Figure 1B). This type of transport is observed in herpesvirus-infected mammalian cells and is required for the nuclear export of viral capsids that assemble in the nucleus [5]. Nucleocytoplasmic transport (or 'nuclear egress') of capsids is essential for the life cycle of all herpesviruses, as final maturation of their virions occurs in the cytoplasm. Although vesicle-mediated nucleocytoplasmic transport is not common in uninfected mammalian cells, it has been reported that *Drosophila* cellular large ribonucleoprotein complexes (RNPs) utilize this nuclear export mechanism in the absence of viral infection [6].

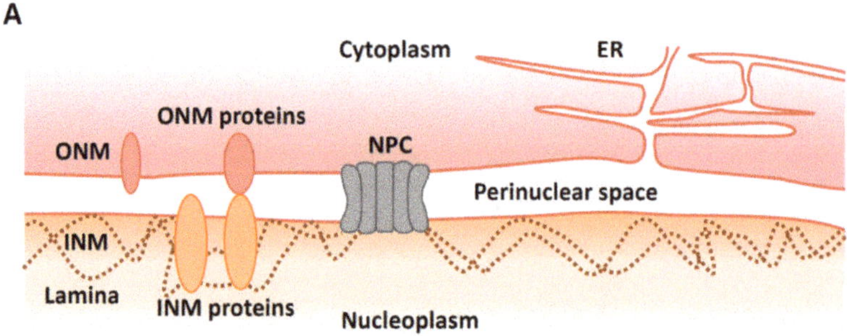

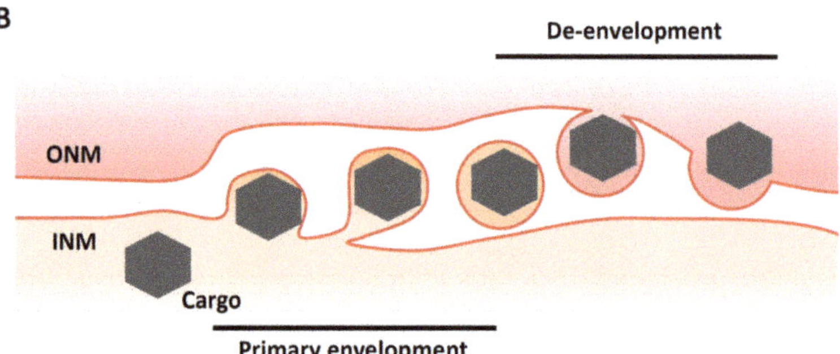

Figure 1. Overview of nuclear egress. (**A**) Illustration of a 'normal' nuclear envelope. INM, inner nuclear membrane; ONM, outer nuclear membrane; NPC, nuclear pore complex; ER, endoplasmic reticulum. (**B**) Illustration of vesicle-mediated nucleocytoplasmic transport. A cargo (e.g., viral capsid) in the nucleus buds through the INM to form a vesicle in the perinuclear space that then fuses with the ONM to release the complex into the cytoplasm (de-envelopment).

The goal of this review is to present the current knowledge of the cellular factors that are involved in vesicle-mediated nucleocytoplasmic transport of herpesvirus capsids. In particular, the review focuses on herpes simplex virus 1 (HSV-1), which has been extensively studied and for which the mechanisms of capsid nuclear egress are relatively well understood. The author hopes that the review will stimulate researchers in the field to engage in new research projects that will further elucidate the many mysteries of this transport system.

2. Overview of Nuclear Egress of Herpesviruses

Herpesviruses are enveloped double-stranded DNA viruses that establish lifelong latent infections in their natural hosts [7]. The *Herpesviridae* family is subdivided into the *Alphaherpesvirinae*, *Betaherpesvirinae*, and *Gammaherpesvirinae* subfamilies, based on their molecular and biological properties [7]. HSV-1 is the prototype of the alphaherpesvirus subfamily, which is comprised of the most common pathogenic agents in humans, and causes a variety of conditions such as mucocutaneous disease, keratitis, skin disease and encephalitis [8]. Herpesviruses share a common virion structure and similar proliferation strategies. A mature virion consists of an icosahedral capsid with a linear double-stranded DNA genome, a proteinaceous layer called tegument, and a host-membrane derived envelope spiked with viral glycoproteins.

Herpesviruses replicate their genomes and package them into capsids within the host cell nucleus. These capsids must traverse from the nucleus to the cytoplasm through a process called nuclear egress. In brief, the nascent nucleocapsids bud at the INM to form primary virions in the perinuclear space, and then the envelopes of primary virions fuse with the ONM, releasing the nucleocapsids into the cytoplasm for further maturation (Figure 1B). In the cytoplasm, nucleocapsids bud into vesicles derived from trans-Golgi networks or endosomes in a process called 'secondary envelopment'. Finally, enveloped virions are released from the cells through an exocytotic pathway [5].

This envelopment/de-envelopment model of herpesvirus egress is now generally accepted and appears to be the dominant mechanism employed by this virus. However, other routes have been proposed. For example, capsids may be transported concomitantly with nuclear envelope breakdown in certain situations [9]. It has also been reported that capsids may egress from the nucleus through defective nuclear pores [10,11].

3. The Nuclear Egress Complex (NEC)

The NEC is composed of the conserved viral proteins UL31 and UL34 and plays a crucial role in the nucleocytoplasmic transport of newly assembled nucleocapsids [5,12]. In the absence of either protein, capsids do not bud at the INM and instead accumulate inside the nucleus [13–18]. UL34 is a type II integral membrane protein that is targeted to both the INM and the ONM [19,20]. UL31 is a soluble nuclear protein that is held in close apposition to the inner and outer surfaces of the INM and ONM through its interaction with UL34 [20]. Conversely, INM localization of UL34 is enhanced in the presence of UL31 [16,20]. Crystal structures of NEC from HSV-1 and the other members of *Herpesviridae* reveal the highly conserved features of this complex [21–25].

4. Primary Envelopment

4.1. Lamina Dissociation

Nuclear shape and mechanical stability are maintained by the nuclear lamina, which is localized beneath the INM. This structure is primarily composed of lamins, which are members of the intermediate filament family of cytoskeletal proteins (Figure 1A). Lamins interact with a large number of proteins at the INM and in the nucleoplasm, thereby influencing structural stability of the nucleus, DNA replication, transcription, and chromatin remodeling [26].

Before HSV-1 capsids access the INM, the nuclear lamina meshwork has to be dissolved. Between prophase and metaphase of mitosis, the lamina disassembles before reassembling during interphase [1]. This lamina disassembly is tightly regulated by phosphorylation [1]. During herpesvirus infection, the lamina is disrupted locally [27–31] (Figure 2). It appears that the NEC component UL34 recruits protein kinase C (PKC) isoforms that phosphorylate lamin B during infection [27,32]. In addition, viral kinases are also required for lamina dissolution. Viral kinase Us3, which is conserved in alphaherpesviruses, phosphorylates the lamin A/C protein (LMNA) [33–35]. HSV-1 encodes the other kinase UL13, which is conserved in *Herpesviridae*. UL13 homologues found in *Betaherpesvirinae* and *Gammaherpesvirinae* have predominant roles in lamin dissociation [36–38]. However, this function of UL13 homologues is not conserved in HSV-1 UL13 [36].

Furthermore, the NEC binds directly to LMNA to modulate its conformation [29] and perturbs the interaction of lamins with emerin, a protein that links the lamina to the INM [39,40]. Taken together, the NEC directly and indirectly disrupts the fibrillar network of the nuclear lamina to ensure correct docking of capsids to the INM. Viral proteins also regulate cellular components beyond the NEC in order to facilitate capsid egress. For example, the HSV-1 neurovirulence factor ICP34.5 also contributes to lamin dissociation through recruitment of PKC to the NM via the cellular protein p32 [41,42] (Figure 2).

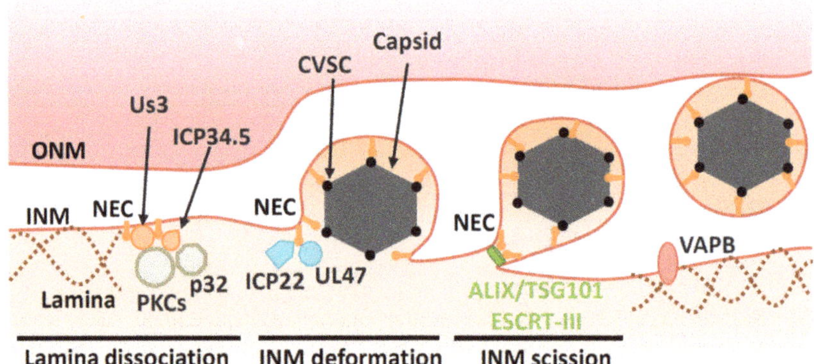

Figure 2. Multiple proteins are involved in primary envelopment of HSV-1 capsids. NEC and kinases (e.g., PKCs and Us3) induce lamina dissociation to allow capsid recruitment to the INM. ICP34.5 and p32 contribute to the recruitment of PKCs. The NEC is responsible for capsid recruitment and mediates budding by assembling into a hexagonal lattice. CVSC decorates vertices of DNA-filled nucleocapsid and promotes nuclear egress through interaction with the NEC. Viral proteins (e.g., UL47 and ICP22) and cellular protein VAPB regulate primary envelopment through as-yet unidentified mechanisms. The NEC recruits ESCRT-III via ALIX to pinch off primary virions from the INM.

New data have underscored the importance of the lamin meshwork as a barrier to viral replication. Lamin degradation is required for HSV-1 nuclear egress and growth in dendritic cells [43]. Tripartite Motif Containing (TRIM) 43 is a factor that restricts a broad range of herpesviruses and is upregulated during viral infection [44]. Interestingly, TRIM43 mediates degradation of the centrosomal protein, pericentrin, which subsequently leads to alterations of the nuclear lamina. This alteration suppresses transcriptionally active viral chromatin states and viral replication.

In general, lamins contribute to the regulation of chromatin organization by tethering peripheral heterochromatin and chromatin remodeling complexes to the nuclear envelope [26]. Evidence of LMNA's role in chromatin organization is provided by the finding that mutations in the human *LMNA* gene lead to premature aging and progressive loss of heterochromatin [45]. The same is true for the HSV-1 genome, as epigenetic regulation of viral DNA is modulated by its binding to LMNA [46–48]. These observations suggest that viral gene expression and nuclear egress must proceed in a coordinated manner.

4.2. Deformation of the INM

In addition to the disintegration of lamins, the NEC is responsible for recruitment of capsids to the INM [13,14,49,50]. Capsid vertex-specific component (CVSC), which consists of the UL17 and UL25 proteins, decorates vertices of DNA-filled nucleocapsids [51–54]. CVSC is proposed to promote nuclear egress of DNA-filled nucleocapsids through interaction with the NEC [55–59]. Once the capsid is recruited to the INM, primary envelopment occurs. This process involves membrane deformation around the capsid, followed by a 'pinching off' of the nascent bud; the resulting vesicle is formed in the perinuclear space and is referred to as the 'primary virion' (Figure 2). The NEC plays a direct role in nucleocapsid budding at the INM. Transient coexpression of NEC from pseudorabiesvirus (PRV, a porcine alphaherpesvirus) or Kaposi's sarcoma-associated herpesvirus (KSHV, a human gammaherpesvirus) induces formation of perinuclear vesicles [60–62]. Moreover, recombinant NEC from HSV-1 or PRV can promote vesiculation of synthetic lipid membranes in the absence of any other viral or cellular proteins [63,64]. These findings suggest that the NEC has an intrinsic ability to deform membranes. Disruption of the membrane-coupled hexagonal NEC lattice by mutation of the inter- or intrahexagonal contact site impairs vesiculation both in vitro and in infected cells [4,21,49,63,65]. Thus, it is likely that formation of the NEC lattice drives vesiculation [4,21,63].

Biochemical analysis of the NEC also shows the requirement for a specific lipid composition for vesicle formation in vitro. Although the role of lipid in infected cells has not been analyzed, a high concentration of phosphatidic acid (PA) is required for in vitro vesicle formation [63]. Recently, it has been reported that PA-rich membranes at nuclear envelope herniation sites recruit cellular membrane remodeling machinery to the NMs in budding yeast [66]. Cell membranes mainly consist of lipids and thus lipid composition must affect the viral entry and the process by which capsid vesicles form during the budding processes. Protein kinase D (PKD) regulates vesicle transport through phosphorylation of phosphatidylinositol 4-kinase β (PI4KIIIKβ), which converts phosphoinositide (PI) to phosphoinositol 4-phosphate (PI4P). Knockdown of PKD or its modulators impairs both primary envelopment and virion release [67], suggesting that lipid composition plays a role during primary envelopment. Phosphatidylethanolamine (PE) a phospholipid that plays significant roles in cellular processes, such as vesicle transport and membrane fusion. Inhibition of the PE synthesis pathway impairs envelopment of HSV-1 capsids at the cytoplasm, but has no effect on nuclear egress [68]. The exact composition of specific lipids and the requirement for these during primary envelopment in infected cells should be determined during future studies.

4.3. Factors Involved in INM Deformation

Although the NEC has ability to vesiculate membranes in transfected cells and in vitro [4,21,63], empty perinuclear vesicles are rarely observed during infection. Therefore, in infected cells, the intrinsic budding potential of the NEC has to be regulated tightly before capsids arrive to the INM. In other words, NEC activity alone may not be sufficient for primary envelopment in infected cells where the environment is more complex.

Indeed, several other viral factors are reported to contribute nuclear egress in infected cells (Table 1). Tegument is a layer comprising thousands of densely packaged proteins that are localized between the virion envelope and capsid. The assembly of tegument on capsids occurs predominantly in the cytoplasm following nuclear egress. However, accumulating evidence suggests that capsids acquire some of the tegument proteins in the nucleus. In particular, tegument proteins UL36 (VP1/2), UL37, UL41 (vhs), UL47 (VP13/14) UL48 (VP16), UL49 (VP22), Us3, ICP0 and ICP4 have been reported to localize in the nucleus or associate with capsids [69–73]. While UL36 is not essential for nuclear egress of HSV-1 or PRV, a nuclear-specific isoform comprising the C-terminal region of UL36 is recruited to PRV capsids in the nucleus and may enhance their nuclear egress [74,75]. Cryo-EM analyses reveal that UL36 forms part of the HSV-1 CVSC, which was originally considered to be a complex of the UL17 and UL25 proteins [76,77]. Contributions of other HSV-1 proteins on primary envelopment have been described. UL47, the major tegument protein of HSV-1, interacts with the NEC and promotes primary envelopment [78]. The immediate early protein ICP22, which is conserved in the subfamily *Alphaherpesvirinae*, interacts with the NEC and regulates NEC localization and HSV-1 primary envelopment [79] (Figure 2).

Table 1. Viral factors involved in nuclear egress of HSV capsid.

Protein	Conservation	Function
NEC (UL31/UL34)	*Herpesviridae*	Vesicle formation via hexagonal lattice [63]. Incorporation of capsid [13,14]. Dissociation of lamins [29]. Recruitment of ESCRT-III [80]. De-envelopment [81].
UL13	*Herpesviridae*	Dissociation of lamins [82]. (Weak in *Alphaherpesvirinae*).
Us3	*Alphaherpesvirinae*	Dissociation of lamins [33]. Promote de-envelopment [83]. Phosphorylate UL34 [84]. Phosphorylate gB [85,86]. Phosphorylate UL31 [87].
ICP34.5	HSV-1 and HSV-2	Dissociation of lamins via PKC [42].
CVSC (UL17/UL25)	*Herpesviridae*	Link between NEC and capsid [55,56].
UL47	*Alphaherpesvirinae*	Promote primary envelopment [78]. (Not observed in PRV [88,89]).
ICP22	*Alphaherpesvirinae*	Promote primary envelopment [79].
gB and gH/gL	*Herpesviridae*	Redundantly promote de-envelopment [90]. (Not observed in PRV [91]).
UL51	*Herpesviridae*	Promote de-envelopment [92]. (Not observed in PRV [93]).

In addition to the viral factors described above, host cell factors also contribute to vesicle formation at the INM during primary envelopment through as-yet undefined mechanisms. Vesicle-associated membrane protein-associated protein B (VAPB), which mediates cytoplasmic vesicle transport at the ER, facilitates HSV-1 nuclear egress [94]. Viral and cellular factors involved in HSV-1 primary envelopment are listed in Tables 1 and 2, respectively.

4.4. Scission at the INM during Primary Envelopment

Although NECs have the intrinsic capability to remodel membranes, they are unable to mediate primary envelopment by themselves in infected cells, as described above. It appears that host endosomal sorting complexes required for transport (ESCRT)-III machinery is responsible for the scission step during HSV-1 primary envelopment [80]. Viral budding can be divided into two stages: membrane deformation (the membrane is wrapped around the assembling virion) and membrane scission (the bud neck is severed) (Figure 2). A remarkable variety of enveloped viruses appear to use host ESCRT-III machinery at the scission stage of budding [95]. The ESCRT-III proteins were initially discovered as factors that are required for the biogenesis of multivesicular bodies (MVBs). It is now clear that the functions of the ESCRT-III proteins extend far beyond their role in MVB formation. Indeed, ESCRT complexes also regulate cytokinesis, the biogenesis of microvesicles and exosomes, plasma membrane repair, neuron pruning, the quality control of NPC, nuclear envelope reformation, and autophagy [96]. ESCRT-III proteins exist as soluble monomers in solution. When activated by membrane-associated upstream factors such as ESCRT-I/II, ALG-2-interacting protein X (ALIX) or charged multivesicular body protein (CHMP)7, ESCRT-III assembles into filamentous structures on membranes. These ESCRT-III assemblies are thought to be the main drivers of membrane remodeling and scission. The hexameric AAA ATPase Vps4 drives remodeling and disassembly of ESCRT-III filaments, and inactivation of Vps4 severely impairs ESCRT-III-mediated membrane remodeling [96].

Herpesvirus replication appears to rely on the ESCRT-III machinery [97–100]. At first, HSV-1 nuclear egress was thought to be independent of the ESCRT-III machinery in HEK293 cells as it is insensitive to a dominant negative mutation of Vps4 [97]. However, depletion of an ESCRT-III protein or its adaptor, ALIX in HeLa cells, reveals the role of ESCRT-III machinery during nuclear egress of HSV-1 [80]. In these cells, viral budding is arrested at the lollipop stage and nearly complete immature virions remained tethered to the INM. These nearly complete virions were morphologically similar to the virions of other viruses that are arrested during budding in the absence of ESCRT factors [95]. Furthermore, ESCRT-III is recruited by the NEC to the INM during the nuclear egress of HSV-1 capsids [80]. ESCRT-II, ALIX and CHMP7 have been identified as upstream factors of ESCRT-III that recruit ESCRT-III proteins and initiate their assembly [96]. In these contexts, CHMP7, but not ESCRT-II or ALIX, is required for NM reformation [101,102]. However, ALIX is required for the scission in the INM during nuclear egress [80] (Figure 2). In agreement with this observation, ALIX is required for recruitment of ESCRT-III to the NM during infection of cells with Epstein–Barr virus (EBV, a human gammaherpesvirus) [103,104]. In addition, a compound that binds to ESCRT-I protein TSG101 inhibits primary envelopment of HSV-1 in Vero cells [105], supporting the idea that ESCRT-III machinery is required for nuclear egress of HSV-1 capsids. The other ESCRT-III adaptor CHMP7 has a predominant role in reformation of the nuclear membrane in various situations [101,102]. Thus, it is unlikely that ESCRT-III is recruited to repair HSV-1-induced defects in the nuclear envelope. Taken together, these data suggest that, while NEC has ability to promote vesiculation in vitro, it recruits ESCRT-III machinery to complete scission in infected cells. Similarly, Ebola virus VP40 alone can carry out membrane budding in an in vitro system [106], although Ebola virus budding is highly dependent on ESCRT-III machinery [95].

5. De-Envelopment

5.1. Overview of De-Envelopment

De-envelopment is a process in which perinuclear virions fuse with the ONM to release naked capsids into the cytosol (Figure 1B). Membrane fusion and disassembly of the NEC are required in this process. The molecular mechanism of de-envelopment is not well understood. Nonetheless, several viral and cellular proteins have been reported as regulators of HSV-1 de-envelopment (Tables 1 and 2).

In particular, the Us3 protein kinase, which is conserved among *Alphaherpesvirinae*, has a prominent role in the de-envelopment step. Inactivation of Us3 enzymatic activity leads to an accumulation of primary enveloped HSV-1 virions in large invaginations of the INM [83,107,108], indicating that phosphorylation may regulate the de-envelopment process, even though it is not essential. The NEC lattices within the perinuclear virions are stable structures that must be disassembled during de-envelopment. Phosphorylation of the NEC by Us3 may loosen NEC lattices in perinuclear virions to promote de-envelopment [81,87] (Figure 3). In agreement with this hypothesis, Us3 is present in perinuclear virions [83]. As is the case with Us3 kinase inactive mutants, mutations in other viral tegument protein UL51 of HSV-1 also lead to accumulation of primary envelope virions in the perinuclear space [92,109,110], suggesting that the de-envelopment process is modulated by various viral proteins.

Similar to membrane fusion during virus entry, virion glycoproteins may play roles in mediating fusion between the envelope of the primary virions and the ONM. Primary virions accumulate in the perinuclear space during infection with recombinant HSV-1s harboring null or point mutations in both gB and gH [85,90,111]. To date, it is unclear how either gB or gH mediate fusion during de-envelopment, although both are required for fusion during viral entry [112] and Us3 phosphorylates gB [85,86]. In the case of the closely related PRV protein, gB or gH are not present in the perinuclear vesicles or the NMs and they do not have any role on nuclear egress [91]. Thus, gB or gH/gL do not appear to function as conserved components of the fusion machinery during de-envelopment step.

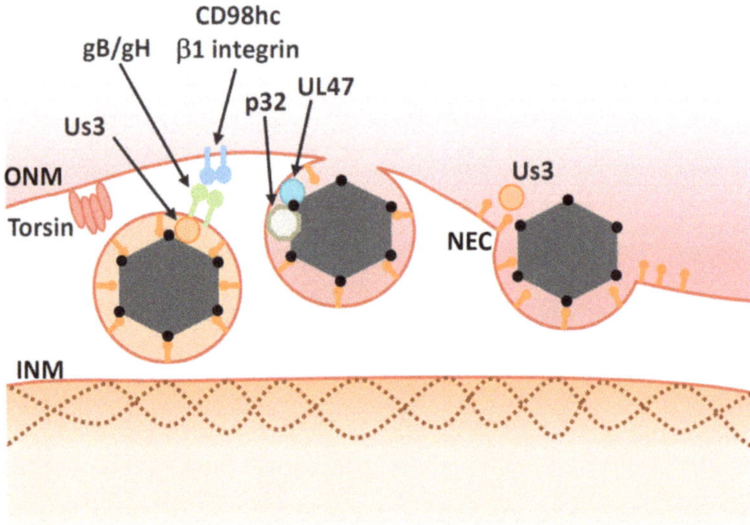

Figure 3. Multiple proteins are involved in the de-envelopment step. Some viral proteins, such as gB/gH and Us3 may be involved in the de-envelopment step, but their roles are unclear. Us3 may induce disassembly of the NEC lattice. Torsin and p32 promote de-envelopment through unidentified mechanisms. Viral protein UL47 interacts with cellular protein p32. The CD98hc/β1 integrin complex may influence fusion activity of the primary virions.

Deletion of either UL31 or UL34 completely halts perinuclear virion formation, which occurs before de-envelopment. Thus, it is conceivable that UL31 and/or UL34 also play an essential role in the de-envelopment step as well as in primary envelopment. UL34 is a transmembrane protein, the C-terminus of which is exposed to the perinuclear space. However, the C-terminal domain (including the transmembrane region) can be replaced by corresponding regions of other viral or cellular proteins without loss of function [113,114]. These observations indicate that the NEC does not directly play a role in the fusion process and that other components of the cellular fusion machinery are involved in de-envelopment. Thus, viral proteins that are involved in the de-envelopment process described above (Table 1) might also modulate cellular fusion machinery at the ONM.

5.2. Cellular Factors Involved in De-Envelopment

In addition to viral factors, cellular factors also regulate the de-envelopment process (Figure 3 and Table 2). Cellular factors including p32, CD98 heavy chain (CD98hc) and β1 integrin are recruited to the NM in HSV-1-infected cells [115,116]. Knockdown of these proteins leads to aberrant accumulation of enveloped virions in the invagination structures derived from the INM; this is phenocopied by inactivation of Us3 protein kinase activity. These observations suggest that the cellular factors mentioned above regulate the de-envelopment step. Accumulation of the multifunctional scaffold protein, p32, at the nuclear rim in infected cells depends on UL47 [116], suggesting that UL47 has an optimal role in the de-envelopment step in addition to its role in primary envelopment [78]. Plasma membrane protein CD98hc may regulate membrane fusion, since anti-CD98hc monoclonal antibodies enhance or inhibit cell–cell fusion, mediated by fusion proteins [117–120]. This CD98hc function in membrane fusion involves its binding partner, β1 integrin [119], suggesting that the CD98hc/β1 integrin interaction regulates fusion between perinuclear virions and the ONM. In the absence of UL34, CD98hc is not redistributed to the NM, but is dispersed throughout the cytoplasm [121]. At the ultrastructural level, HSV-1 infection causes ER compression around the nuclear envelope, whereas the UL34-null mutation causes cytoplasmic dispersion of the ER [121]. These observations suggest that HSV-1 infection redistributes ER around the nuclear envelope through unidentified mechanisms, thereby enabling accumulation of ER-associated de-envelopment factors, such as CD98hc, gB, and gH in the region where de-envelopment occurs.

Table 2. Cellular factors involved in nuclear egress of HSV capsids.

Protein	Interactor	Function	*Drosophila* RNP
Lamins	NEC, Us3, UL13	Prevent nuclear egress [29].	Yes [6,122]
PKC family	NEC	Dissociate lamins [32].	Yes [6]
p32	ICP34.5, UL47	Recruit PKC [42]. Promote de-envelopment [116].	-
PKD	-	Promote nuclear egress indirectly [67].	-
VAPB	-	Promote nuclear egress [94].	-
ESCRT-III	NEC	Mediate scission of INM [80].	Yes [80]
ALIX	NEC	Recruit ESCRT-III to INM [80].	-
TSG101	-	Nuclear egress [105].	
CD98hc β1 integrin	NEC, gB, gH	Promote de-envelopment [115]. (Modulate fusion activity?).	-
Torsins	-	Promote de-envelopment indirectly [123].	Yes [124]

Several lines of evidence indicate that torsins might be involved in HSV-1 nuclear egress. Torsins are members of the AAA+ ATPase superfamily, which is comprised of enzymes that mediate ATP-dependent conformational remodeling of target proteins or protein complexes [125]. Among torsin orthologues in humans, Torsin A (TorA) is the best studied due to its association with a disease designated as early onset dystonia; this association underlies the alternate name for this protein, which is dystonia 1 protein (DYT1) [125]. Torsins are located within the lumen of the ER and NM [126]. In contrast to other AAA+ ATPases, torsins lack a key catalytic residue and need the INM protein lamina-associated polypeptide 1 (LAP1) or the ER protein lumenal domain like LAP1 (LULL1) as a cofactor for full activation [127,128]. Deletion or mutation of TorA resulted in the accumulation of vesicles in the perinuclear space [126]; these vesicles resembled those produced in cells expressing NEC [60]. Ectopic expression of wild-type TorA reduced HSV-1 production and led to accumulation of virus-like structures in the perinuclear space and the lumen of the ER [123], indicating that torsins play a role in de-envelopment. Similarly, perinuclear virions accumulate in TorA-mutant murine cells infected with HSV-1 [129]. Although torsins are involved in nuclear egress, the precise molecular mechanisms by which they regulate this process remain unclear. Another study showed that knockout of torsin itself or the INM-specific torsin cofactor, LAP1, produced only subtle defects in viral replication [130]. In this study, the authors also showed that knockout of LULL1, a torsin cofactor in the ER, reduces HSV-1 growth by one order of magnitude in the absence of nuclear egress defects. Taken together, torsins may be involved in HSV-1 nuclear egress indirectly, as they appear to regulate the nuclear envelope [131,132] and lipid metabolism [133], and both these biological processes have an impact on nuclear egress. Although torsins have been extensively investigated in biochemical studies and animal models, their precise biological functions remain elusive. Further study of their roles in the nuclear egress of herpesvirus might help elucidate these functions.

Neither the de-envelopment step nor viral replication are completely dependent on any of the individual cellular factors described above (Table 2). This suggests a degree of redundancy among these factors, which presents a challenge when attempting to character-ize the precise contribution of each. Future biochemical analyses or fusion assays designed to reconstitute the de-envelopment step are desirable in this regard, as they will help to elucidate molecular mechanisms. Alternatively, it is possible that the major cellular players in vesicle-mediated nuclear egress have not been identified yet. Notably, there have been no reports to date of an a ONM-resident factor that is involved in fusion.

In general, an entry receptor is required for viral glycoprotein-driven fusion. The entry receptor restricts susceptible cells and determines the tropism of a particular virus. A receptor-like molecule that increases affinity for the perinuclear virion would therefore restrict the 'perinuclear tropism', thereby preventing back-fusion to the INM. If viral fusion protein gB mediates fusion between the perinuclear virion and ONM [85,90,111], it might also associate with membrane proteins in the ONM, as gB receptors are important during viral entry [134–136]. As described above, however, gB is not essential for the de-envelopment step. Thus, both a 'ligand' in the perinuclear virion and a 'receptor' in the ONM must be identified to prove this hypothesis.

6. Vesicle-Mediated Nucleocytoplasmic Transport in Uninfected Cells

For a long time, vesicle-mediated nucleocytoplasmic transport was considered to be unique to the infection process used by herpesviruses. However, *Drosophila* RNPs also utilize this nuclear export mechanism [6] (Figure 4A). Interestingly, formation of perinuclear RNPs requires PKC-dependent lamin phosphorylation; parallels can be drawn with the phosphorylation of lamins that occurs during nuclear egress of herpesvirus capsids [6]. Moreover, torsin, lamin and ESCRT-III are important for nucleocytoplasmic transport of *Drosophila* RNPs, which is again similar to the requirements of viral nuclear egress [80,122,124]. These observations indicate that herpesviruses may have expropriated this transport mechanism.

Although perinuclear large vesicles are rarely seen in normal mammalian cells, vesicle-mediated cytoplasmic transport might occur in them. Mutation in torsin produces omega-shaped herniations and vesicles containing NPC proteins within the perinuclear space in murine and human cells [131,132,137]. Thus, torsins may regulate the vesicle-mediated cytoplasmic transport system in normal cells, and impairment of torsins leads to the accumulation of aberrant intermediate structures in the perinuclear space [125,126] (Figure 4B,C). Similarly, aberrant perinuclear blebs and vesicles are observed in ESCRT-III-deficient yeast cells, as ESCRT-III has the ability to clear NPC assembly intermediates [138,139]. In mammalian cells, ESCRT-III is involved in resealing the NM during late anaphase [101,140] and in repairing NM ruptures caused by cell migration through tight interstitial spaces [102,141]. In addition to these effects, impairment of ESCRT-III induces proliferation of the INM in uninfected cells in a cell cycle-independent manner through ALIX, which is an important adaptor for HSV-1 nuclear egress [80]. Thus, ESCRT-III may control INM integrity by regulating vesicle-mediated cytoplasmic transport in normal cells, although this rate is low and the precise mechanisms that govern this activity are unclear (Figure 4B,D).

The role of ESCRT-III in maintenance of INM integrity may be critical in pathological conditions where INM proteins accumulate abnormally. Hutchinson–Gilford progeria syndrome (HGPS) is s premature aging disorder caused by a point mutation that truncates the LMNA gene [142,143]. In fibroblasts from HGPS patients, truncated LMNA (also referred to as 'progerin') accumulates in the nucleus and engenders NM deformations that affect nuclear blebbing and perinuclear vesicle formation [45]. The ALIX-mediated ESCRT-III pathway plays a suppressive role in progerin-induced NM deformation, perhaps through vesicle-mediated cytoplasmic transport [144] (Figure 4B,E). If so, herpesviruses may highjack this system to facilitate nuclear egress of capsids.

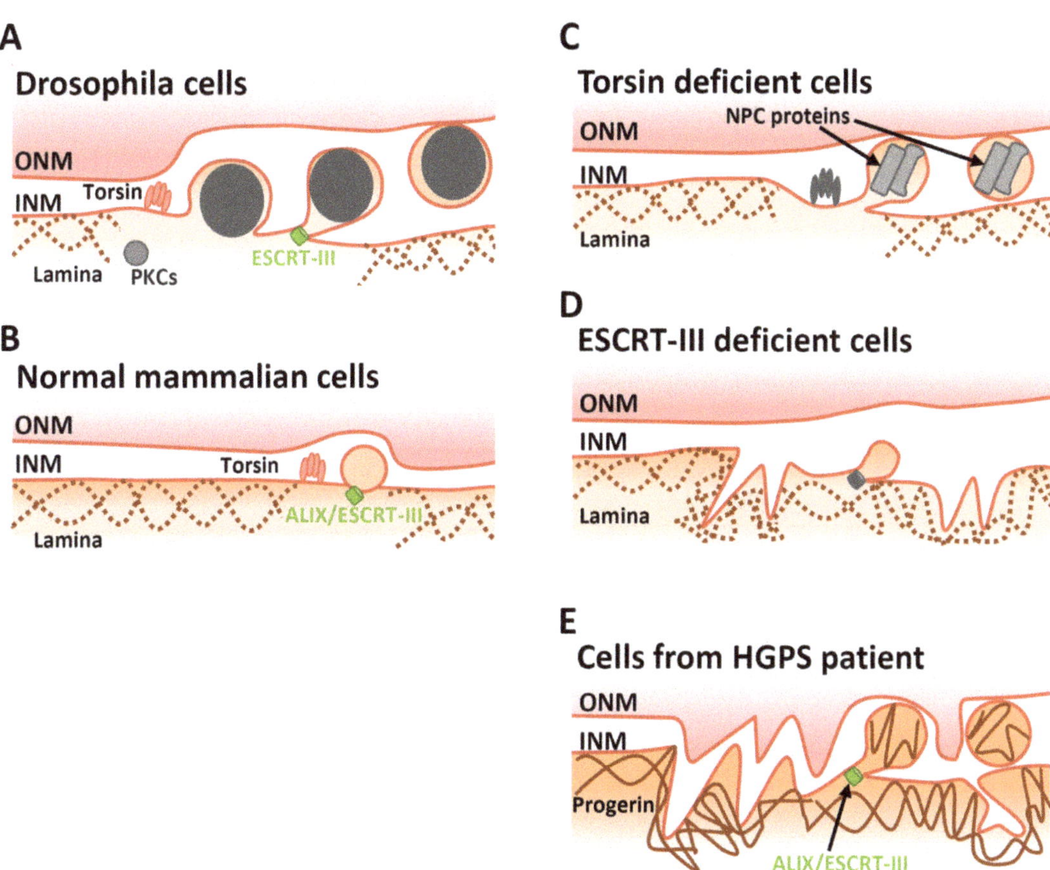

Figure 4. Proposed models of vesicle-mediated cytoplasmic transport in various situations. (**A**) A large RNP complex exits from the nucleus through vesicle-mediated nucleocytoplasmic transport system in *Drosophila* cells. (**B**) In 'normal' mammalian cells, torsins and ESCRT-III might contribute to eliminate excess INM and/or INM proteins. This process might be similar to the vesicle-mediated nucleocytoplasmic transport of HSV-1 nucleocapsids. (**C**) In torsin-deficient cells, herniations and vesicles containing NPC proteins within the perinuclear space are produced. (**D**) In ESCRT-III-deficient cells, INM proliferation is induced. (**E**) In cells from HGPS patients, progerin (mutant lamin A) accumulates in the nucleus, resulting in nuclear envelope deformation and defects including vesicles in the perinuclear space. ESCRT-III represses membrane deformation in these cells.

7. Concluding Remarks

Nuclear egress of viral capsids is essential during the life cycle of herpesviruses. Many groups have extensively studied this mysterious system and uncovered the role of some cellular and viral proteins that are involved in nuclear egress. More recently, the molecular mechanisms that underlie vesicle formation have been revealed by biochemical analysis of the NEC and studies of the cellular scission machinery component, ESCRT-III. In contrast, the precise mechanism of de-envelopment is completely unknown. Moreover, it is not known how and when vesicle formation occurs at the INM in normal cells in the absence of NEC.

Vesicle-mediated nucleocytoplasmic transport might occur rarely and/or be accomplished rapidly in uninfected mammalian cells. If so, it might be difficult to characterize this system without employing exogenous activation or inhibition methods. Studies on the nuclear egress of herpesviruses may help to further elucidate the molecular mechanisms and physiological significance of this transport system in normal situations.

Funding: This research was funded by grants for Scientific Research from the Japan Society for the Promotion of Science (JSPS), grants for Scientific Research on Innovative Areas from the Ministry of Education, Culture, Science, Sports and Technology of Japan (19H05286 and 19H05417), MEXT Leading Initiative for Excellent Young Researchers Grant, contract research funds from Japan Program for Infectious Diseases Research and Infrastructure (20wm0325005h) from the Japan Agency for Medical Research and Development (AMED), and grants from MSD Life Science Foundation, Public Interest Incorporated Foundation.

Conflicts of Interest: The author declares no conflict of interest.

References

1. Hetzer, M.W. The nuclear envelope. *Cold Spring Harb. Perspect. Biol.* **2010**, *2*, a000539. [CrossRef]
2. Crisp, M.; Burke, B. The nuclear envelope as an integrator of nuclear and cytoplasmic architecture. *FEBS Lett.* **2008**, *582*, 2023–2032. [CrossRef] [PubMed]
3. Paci, G.; Caria, J.; Lemke, E.A. Cargo transport through the nuclear pore complex at a glance. *J. Cell Sci.* **2021**, *134*. [CrossRef] [PubMed]
4. Hagen, C.; Dent, K.C.; Zeev-Ben-Mordehai, T.; Grange, M.; Bosse, J.B.; Whittle, C.; Klupp, B.G.; Siebert, C.A.; Vasishtan, D.; Bauerlein, F.J.; et al. Structural Basis of Vesicle Formation at the Inner Nuclear Membrane. *Cell* **2015**, *163*, 1692–1701. [CrossRef] [PubMed]
5. Johnson, D.C.; Baines, J.D. Herpesviruses remodel host membranes for virus egress. *Nat. Rev. Microbiol.* **2011**, *9*, 382–394. [CrossRef] [PubMed]
6. Speese, S.D.; Ashley, J.; Jokhi, V.; Nunnari, J.; Barria, R.; Li, Y.; Ataman, B.; Koon, A.; Chang, Y.T.; Li, Q.; et al. Nuclear envelope budding enables large ribonucleoprotein particle export during synaptic Wnt signaling. *Cell* **2012**, *149*, 832–846. [CrossRef] [PubMed]
7. Pellet, P.E.; Roizman, B. Herpesviridae. In *Fields Virology*, 6th ed.; Knipe, D.M., Howley, P.M., Cohen, J.I., Griffin, D.E., Lamb, R.A., Martin, M.A., Racaniello, V.R., Roizman, B., Eds.; Lippincott-Williams &Wilkins: Philadelphia, PA, USA, 2013; pp. 1802–1822.
8. Roizman, B.; Knipe, D.M.; Whitley, R.J. Herpes simplex viruses. In *Fields Virology*, 6th ed.; Knipe, D.M., Howley, P.M., Cohen, J.I., Griffin, D.E., Lamb, R.A., Martin, M.A., Racaniello, V.R., Roizman, B., Eds.; Lippincott-Williams &Wilkins: Philadelphia, PA, USA, 2013; pp. 1823–1897.
9. Schulz, K.S.; Klupp, B.G.; Granzow, H.; Passvogel, L.; Mettenleiter, T.C. Herpesvirus nuclear egress: Pseudorabies Virus can simultaneously induce nuclear envelope breakdown and exit the nucleus via the envelopment-deenvelopment-pathway. *Virus Res.* **2015**, *209*, 76–86. [CrossRef]
10. Leuzinger, H.; Ziegler, U.; Schraner, E.M.; Fraefel, C.; Glauser, D.L.; Heid, I.; Ackermann, M.; Mueller, M.; Wild, P. Herpes simplex virus 1 envelopment follows two diverse pathways. *J. Virol.* **2005**, *79*, 13047–13059. [CrossRef]
11. Wild, P.; Senn, C.; Manera, C.L.; Sutter, E.; Schraner, E.M.; Tobler, K.; Ackermann, M.; Ziegler, U.; Lucas, M.S.; Kaech, A. Exploring the nuclear envelope of herpes simplex virus 1-infected cells by high-resolution microscopy. *J. Virol.* **2009**, *83*, 408–419. [CrossRef] [PubMed]
12. Mettenleiter, T.C. Vesicular Nucleo-Cytoplasmic Transport-Herpesviruses as Pioneers in Cell Biology. *Viruses* **2016**, *8*, 266. [CrossRef]
13. Chang, Y.E.; Van Sant, C.; Krug, P.W.; Sears, A.E.; Roizman, B. The null mutant of the U(L)31 gene of herpes simplex virus 1: Construction and phenotype in infected cells. *J. Virol.* **1997**, *71*, 8307–8315. [CrossRef]
14. Roller, R.J.; Zhou, Y.; Schnetzer, R.; Ferguson, J.; DeSalvo, D. Herpes simplex virus type 1 U(L)34 gene product is required for viral envelopment. *J. Virol.* **2000**, *74*, 117–129. [CrossRef] [PubMed]
15. Klupp, B.G.; Granzow, H.; Mettenleiter, T.C. Primary envelopment of pseudorabies virus at the nuclear membrane requires the UL34 gene product. *J. Virol.* **2000**, *74*, 10063–10073. [CrossRef] [PubMed]
16. Fuchs, W.; Klupp, B.G.; Granzow, H.; Osterrieder, N.; Mettenleiter, T.C. The interacting UL31 and UL34 gene products of pseudorabies virus are involved in egress from the host-cell nucleus and represent components of primary enveloped but not mature virions. *J. Virol.* **2002**, *76*, 364–378. [CrossRef] [PubMed]
17. Farina, A.; Feederle, R.; Raffa, S.; Gonnella, R.; Santarelli, R.; Frati, L.; Angeloni, A.; Torrisi, M.R.; Faggioni, A.; Delecluse, H.J. BFRF1 of Epstein-Barr virus is essential for efficient primary viral envelopment and egress. *J. Virol.* **2005**, *79*, 3703–3712. [CrossRef] [PubMed]
18. Popa, M.; Ruzsics, Z.; Lotzerich, M.; Dolken, L.; Buser, C.; Walther, P.; Koszinowski, U.H. Dominant negative mutants of the murine cytomegalovirus M53 gene block nuclear egress and inhibit capsid maturation. *J. Virol.* **2010**, *84*, 9035–9046. [CrossRef] [PubMed]

19. Shiba, C.; Daikoku, T.; Goshima, F.; Takakuwa, H.; Yamauchi, Y.; Koiwai, O.; Nishiyama, Y. The UL34 gene product of herpes simplex virus type 2 is a tail-anchored type II membrane protein that is significant for virus envelopment. *J. Gen. Virol.* **2000**, *81*, 2397–2405. [CrossRef]

20. Reynolds, A.E.; Ryckman, B.J.; Baines, J.D.; Zhou, Y.; Liang, L.; Roller, R.J. U(L)31 and U(L)34 proteins of herpes simplex virus type 1 form a complex that accumulates at the nuclear rim and is required for envelopment of nucleocapsids. *J. Virol.* **2001**, *75*, 8803–8817. [CrossRef]

21. Bigalke, J.M.; Heldwein, E.E. Structural basis of membrane budding by the nuclear egress complex of herpesviruses. *EMBO J.* **2015**, *34*, 2921–2936. [CrossRef]

22. Zeev-Ben-Mordehai, T.; Weberruss, M.; Lorenz, M.; Cheleski, J.; Hellberg, T.; Whittle, C.; El Omari, K.; Vasishtan, D.; Dent, K.C.; Harlos, K.; et al. Crystal Structure of the Herpesvirus Nuclear Egress Complex Provides Insights into Inner Nuclear Membrane Remodeling. *Cell Rep.* **2015**, *13*, 2645–2652. [CrossRef]

23. Lye, M.F.; Sharma, M.; El Omari, K.; Filman, D.J.; Schuermann, J.P.; Hogle, J.M.; Coen, D.M. Unexpected features and mechanism of heterodimer formation of a herpesvirus nuclear egress complex. *EMBO J.* **2015**, *34*, 2937–2952. [CrossRef]

24. Leigh, K.E.; Sharma, M.; Mansueto, M.S.; Boeszoermenyi, A.; Filman, D.J.; Hogle, J.M.; Wagner, G.; Coen, D.M.; Arthanari, H. Structure of a herpesvirus nuclear egress complex subunit reveals an interaction groove that is essential for viral replication. *Proc. Natl. Acad. Sci. USA* **2015**, *112*, 9010–9015. [CrossRef]

25. Walzer, S.A.; Egerer-Sieber, C.; Sticht, H.; Sevvana, M.; Hohl, K.; Milbradt, J.; Muller, Y.A.; Marschall, M. Crystal Structure of the Human Cytomegalovirus pUL50-pUL53 Core Nuclear Egress Complex Provides Insight into a Unique Assembly Scaffold for Virus-Host Protein Interactions. *J. Biol. Chem.* **2015**, *290*, 27452–27458. [CrossRef]

26. Gruenbaum, Y.; Margalit, A.; Goldman, R.D.; Shumaker, D.K.; Wilson, K.L. The nuclear lamina comes of age. *Nat. Rev. Mol. Cell Biol.* **2005**, *6*, 21–31. [CrossRef]

27. Muranyi, W.; Haas, J.; Wagner, M.; Krohne, G.; Koszinowski, U.H. Cytomegalovirus recruitment of cellular kinases to dissolve the nuclear lamina. *Science* **2002**, *297*, 854–857. [CrossRef]

28. Scott, E.S.; O'Hare, P. Fate of the inner nuclear membrane protein lamin B receptor and nuclear lamins in herpes simplex virus type 1 infection. *J. Virol.* **2001**, *75*, 8818–8830. [CrossRef]

29. Reynolds, A.E.; Liang, L.; Baines, J.D. Conformational changes in the nuclear lamina induced by herpes simplex virus type 1 require genes U(L)31 and U(L)34. *J. Virol.* **2004**, *78*, 5564–5575. [CrossRef] [PubMed]

30. Simpson-Holley, M.; Baines, J.; Roller, R.; Knipe, D.M. Herpes simplex virus 1 U(L)31 and U(L)34 gene products promote the late maturation of viral replication compartments to the nuclear periphery. *J. Virol.* **2004**, *78*, 5591–5600. [CrossRef] [PubMed]

31. Simpson-Holley, M.; Colgrove, R.C.; Nalepa, G.; Harper, J.W.; Knipe, D.M. Identification and functional evaluation of cellular and viral factors involved in the alteration of nuclear architecture during herpes simplex virus 1 infection. *J. Virol.* **2005**, *79*, 12840–12851. [CrossRef] [PubMed]

32. Park, R.; Baines, J.D. Herpes simplex virus type 1 infection induces activation and recruitment of protein kinase C to the nuclear membrane and increased phosphorylation of lamin B. *J. Virol.* **2006**, *80*, 494–504. [CrossRef]

33. Bjerke, S.L.; Roller, R.J. Roles for herpes simplex virus type 1 UL34 and US3 proteins in disrupting the nuclear lamina during herpes simplex virus type 1 egress. *Virology* **2006**, *347*, 261–276. [CrossRef] [PubMed]

34. Mou, F.; Forest, T.; Baines, J.D. US3 of herpes simplex virus type 1 encodes a promiscuous protein kinase that phosphorylates and alters localization of lamin A/C in infected cells. *J. Virol.* **2007**, *81*, 6459–6470. [CrossRef] [PubMed]

35. Mou, F.; Wills, E.G.; Park, R.; Baines, J.D. Effects of lamin A/C, lamin B1, and viral US3 kinase activity on viral infectivity, virion egress, and the targeting of herpes simplex virus U(L)34-encoded protein to the inner nuclear membrane. *J. Virol.* **2008**, *82*, 8094–8104. [CrossRef] [PubMed]

36. Kuny, C.V.; Chinchilla, K.; Culbertson, M.R.; Kalejta, R.F. Cyclin-dependent kinase-like function is shared by the beta- and gamma-subset of the conserved herpesvirus protein kinases. *PLoS Pathog.* **2010**, *6*, e1001092. [CrossRef] [PubMed]

37. Lee, C.P.; Huang, Y.H.; Lin, S.F.; Chang, Y.; Chang, Y.H.; Takada, K.; Chen, M.R. Epstein-Barr virus BGLF4 kinase induces disassembly of the nuclear lamina to facilitate virion production. *J. Virol.* **2008**, *82*, 11913–11926. [CrossRef]

38. Hamirally, S.; Kamil, J.P.; Ndassa-Colday, Y.M.; Lin, A.J.; Jahng, W.J.; Baek, M.C.; Noton, S.; Silva, L.A.; Simpson-Holley, M.; Knipe, D.M.; et al. Viral mimicry of Cdc2/cyclin-dependent kinase 1 mediates disruption of nuclear lamina during human cytomegalovirus nuclear egress. *PLoS Pathog.* **2009**, *5*, e1000275. [CrossRef]

39. Leach, N.; Bjerke, S.L.; Christensen, D.K.; Bouchard, J.M.; Mou, F.; Park, R.; Baines, J.; Haraguchi, T.; Roller, R.J. Emerin is hyperphosphorylated and redistributed in herpes simplex virus type 1-infected cells in a manner dependent on both UL34 and US3. *J. Virol.* **2007**, *81*, 10792–10803. [CrossRef]

40. Morris, J.B.; Hofemeister, H.; O'Hare, P. Herpes simplex virus infection induces phosphorylation and delocalization of emerin, a key inner nuclear membrane protein. *J. Virol.* **2007**, *81*, 4429–4437. [CrossRef]

41. Wu, S.; Pan, S.; Zhang, L.; Baines, J.; Roller, R.; Ames, J.; Yang, M.; Wang, J.; Chen, D.; Liu, Y.; et al. Herpes Simplex Virus 1 Induces Phosphorylation and Reorganization of Lamin A/C through the gamma134.5 Protein That Facilitates Nuclear Egress. *J. Virol.* **2016**, *90*, 10414–10422. [CrossRef]

42. Wang, Y.; Yang, Y.; Wu, S.; Pan, S.; Zhou, C.; Ma, Y.; Ru, Y.; Dong, S.; He, B.; Zhang, C.; et al. p32 is a novel target for viral protein ICP34.5 of herpes simplex virus type 1 and facilitates viral nuclear egress. *J. Biol. Chem.* **2014**, *289*, 35795–35805. [CrossRef]

43. Turan, A.; Grosche, L.; Krawczyk, A.; Muhl-Zurbes, P.; Drassner, C.; Duthorn, A.; Kummer, M.; Hasenberg, M.; Voortmann, S.; Jastrow, H.; et al. Autophagic degradation of lamins facilitates the nuclear egress of herpes simplex virus type 1. *J. Cell Biol.* **2019**, *218*, 508–523. [CrossRef]

44. Full, F.; van Gent, M.; Sparrer, K.M.J.; Chiang, C.; Zurenski, M.A.; Scherer, M.; Brockmeyer, N.H.; Heinzerling, L.; Sturzl, M.; Korn, K.; et al. Centrosomal protein TRIM43 restricts herpesvirus infection by regulating nuclear lamina integrity. *Nat. Microbiol.* **2019**, *4*, 164–176. [CrossRef] [PubMed]

45. Schreiber, K.H.; Kennedy, B.K. When lamins go bad: Nuclear structure and disease. *Cell* **2013**, *152*, 1365–1375. [CrossRef]

46. Silva, L.; Cliffe, A.; Chang, L.; Knipe, D.M. Role for A-type lamins in herpesviral DNA targeting and heterochromatin modulation. *PLoS Pathog.* **2008**, *4*, e1000071. [CrossRef] [PubMed]

47. Silva, L.; Oh, H.S.; Chang, L.; Yan, Z.; Triezenberg, S.J.; Knipe, D.M. Roles of the nuclear lamina in stable nuclear association and assembly of a herpesviral transactivator complex on viral immediate-early genes. *mBio* **2012**, *3*. [CrossRef]

48. Oh, H.S.; Traktman, P.; Knipe, D.M. Barrier-to-Autointegration Factor 1 (BAF/BANF1) Promotes Association of the SETD1A Histone Methyltransferase with Herpes Simplex Virus Immediate-Early Gene Promoters. *mBio* **2015**, *6*, e00345-15. [CrossRef]

49. Roller, R.J.; Bjerke, S.L.; Haugo, A.C.; Hanson, S. Analysis of a charge cluster mutation of herpes simplex virus type 1 UL34 and its extragenic suppressor suggests a novel interaction between pUL34 and pUL31 that is necessary for membrane curvature around capsids. *J. Virol.* **2010**, *84*, 3921–3934. [CrossRef] [PubMed]

50. Funk, C.; Ott, M.; Raschbichler, V.; Nagel, C.H.; Binz, A.; Sodeik, B.; Bauerfeind, R.; Bailer, S.M. The Herpes Simplex Virus Protein pUL31 Escorts Nucleocapsids to Sites of Nuclear Egress, a Process Coordinated by Its N-Terminal Domain. *PLoS Pathog.* **2015**, *11*, e1004957. [CrossRef]

51. Trus, B.L.; Newcomb, W.W.; Cheng, N.; Cardone, G.; Marekov, L.; Homa, F.L.; Brown, J.C.; Steven, A.C. Allosteric signaling and a nuclear exit strategy: Binding of UL25/UL17 heterodimers to DNA-Filled HSV-1 capsids. *Mol. Cell* **2007**, *26*, 479–489. [CrossRef]

52. Sheaffer, A.K.; Newcomb, W.W.; Gao, M.; Yu, D.; Weller, S.K.; Brown, J.C.; Tenney, D.J. Herpes simplex virus DNA cleavage and packaging proteins associate with the procapsid prior to its maturation. *J. Virol.* **2001**, *75*, 687–698. [CrossRef]

53. Thurlow, J.K.; Rixon, F.J.; Murphy, M.; Targett-Adams, P.; Hughes, M.; Preston, V.G. The herpes simplex virus type 1 DNA packaging protein UL17 is a virion protein that is present in both the capsid and the tegument compartments. *J. Virol.* **2005**, *79*, 150–158. [CrossRef] [PubMed]

54. Newcomb, W.W.; Homa, F.L.; Brown, J.C. Herpes simplex virus capsid structure: DNA packaging protein UL25 is located on the external surface of the capsid near the vertices. *J. Virol.* **2006**, *80*, 6286–6294. [CrossRef]

55. Yang, K.; Baines, J.D. Selection of HSV capsids for envelopment involves interaction between capsid surface components pUL31, pUL17, and pUL25. *Proc. Natl. Acad. Sci. USA* **2011**, *108*, 14276–14281. [CrossRef] [PubMed]

56. Yang, K.; Wills, E.; Lim, H.Y.; Zhou, Z.H.; Baines, J.D. Association of herpes simplex virus pUL31 with capsid vertices and components of the capsid vertex-specific complex. *J. Virol.* **2014**, *88*, 3815–3825. [CrossRef] [PubMed]

57. Takeshima, K.; Arii, J.; Maruzuru, Y.; Koyanagi, N.; Kato, A.; Kawaguchi, Y. Identification of the Capsid Binding Site in the Herpes Simplex Virus 1 Nuclear Egress Complex and Its Role in Viral Primary Envelopment and Replication. *J. Virol.* **2019**, *93*. [CrossRef]

58. Newcomb, W.W.; Fontana, J.; Winkler, D.C.; Cheng, N.; Heymann, J.B.; Steven, A.C. The Primary Enveloped Virion of Herpes Simplex Virus 1: Its Role in Nuclear Egress. *mBio* **2017**, *8*. [CrossRef] [PubMed]

59. Draganova, E.B.; Zhang, J.; Zhou, Z.H.; Heldwein, E.E. Structural basis for capsid recruitment and coat formation during HSV-1 nuclear egress. *Elife* **2020**, *9*. [CrossRef]

60. Klupp, B.G.; Granzow, H.; Fuchs, W.; Keil, G.M.; Finke, S.; Mettenleiter, T.C. Vesicle formation from the nuclear membrane is induced by coexpression of two conserved herpesvirus proteins. *Proc. Natl. Acad. Sci. USA* **2007**, *104*, 7241–7246. [CrossRef] [PubMed]

61. Luitweiler, E.M.; Henson, B.W.; Pryce, E.N.; Patel, V.; Coombs, G.; McCaffery, J.M.; Desai, P.J. Interactions of the Kaposi's Sarcoma-associated herpesvirus nuclear egress complex: ORF69 is a potent factor for remodeling cellular membranes. *J. Virol.* **2013**, *87*, 3915–3929. [CrossRef]

62. Desai, P.J.; Pryce, E.N.; Henson, B.W.; Luitweiler, E.M.; Cothran, J. Reconstitution of the Kaposi's sarcoma-associated herpesvirus nuclear egress complex and formation of nuclear membrane vesicles by coexpression of ORF67 and ORF69 gene products. *J. Virol.* **2012**, *86*, 594–598. [CrossRef]

63. Bigalke, J.M.; Heuser, T.; Nicastro, D.; Heldwein, E.E. Membrane deformation and scission by the HSV-1 nuclear egress complex. *Nat. Commun* **2014**, *5*, 4131. [CrossRef]

64. Lorenz, M.; Vollmer, B.; Unsay, J.D.; Klupp, B.G.; Garcia-Saez, A.J.; Mettenleiter, T.C.; Antonin, W. A single herpesvirus protein can mediate vesicle formation in the nuclear envelope. *J. Biol. Chem.* **2015**, *290*, 6962–6974. [CrossRef]

65. Arii, J.; Takeshima, K.; Maruzuru, Y.; Koyanagi, N.; Kato, A.; Kawaguchi, Y. Roles of the Interhexamer Contact Site for Hexagonal Lattice Formation of the Herpes Simplex Virus 1 Nuclear Egress Complex in Viral Primary Envelopment and Replication. *J. Virol.* **2019**, *93*. [CrossRef]

66. Thaller, D.J.; Tong, D.; Marklew, C.J.; Ader, N.R.; Mannino, P.J.; Borah, S.; King, M.C.; Ciani, B.; Lusk, C.P. Direct binding of ESCRT protein Chm7 to phosphatidic acid-rich membranes at nuclear envelope herniations. *J. Cell Biol.* **2021**, *220*. [CrossRef]

67. Roussel, E.; Lippe, R. Cellular Protein Kinase D Modulators Play a Role during Multiple Steps of Herpes Simplex Virus 1 Egress. *J. Virol.* **2018**, *92*. [CrossRef]

68. Arii, J.; Fukui, A.; Shimanaka, Y.; Kono, N.; Arai, H.; Maruzuru, Y.; Koyanagi, N.; Kato, A.; Mori, Y.; Kawaguchi, Y. Role of Phosphatidylethanolamine Biosynthesis in Herpes Simplex Virus 1-Infected Cells in Progeny Virus Morphogenesis in the Cytoplasm and in Viral Pathogenicity In Vivo. *J. Virol.* **2020**, *94*. [CrossRef] [PubMed]
69. Naldinho-Souto, R.; Browne, H.; Minson, T. Herpes simplex virus tegument protein VP16 is a component of primary enveloped virions. *J. Virol.* **2006**, *80*, 2582–2584. [CrossRef] [PubMed]
70. Bucks, M.A.; O'Regan, K.J.; Murphy, M.A.; Wills, J.W.; Courtney, R.J. Herpes simplex virus type 1 tegument proteins VP1/2 and UL37 are associated with intranuclear capsids. *Virology* **2007**, *361*, 316–324. [CrossRef] [PubMed]
71. Donnelly, M.; Elliott, G. Fluorescent tagging of herpes simplex virus tegument protein VP13/14 in virus infection. *J. Virol.* **2001**, *75*, 2575–2583. [CrossRef]
72. Pomeranz, L.E.; Blaho, J.A. Modified VP22 localizes to the cell nucleus during synchronized herpes simplex virus type 1 infection. *J. Virol.* **1999**, *73*, 6769–6781. [CrossRef]
73. Henaff, D.; Remillard-Labrosse, G.; Loret, S.; Lippe, R. Analysis of the early steps of herpes simplex virus 1 capsid tegumentation. *J. Virol.* **2013**, *87*, 4895–4906. [CrossRef] [PubMed]
74. Luxton, G.W.; Lee, J.I.; Haverlock-Moyns, S.; Schober, J.M.; Smith, G.A. The pseudorabies virus VP1/2 tegument protein is required for intracellular capsid transport. *J. Virol.* **2006**, *80*, 201–209. [CrossRef]
75. Leelawong, M.; Lee, J.I.; Smith, G.A. Nuclear egress of pseudorabies virus capsids is enhanced by a subspecies of the large tegument protein that is lost upon cytoplasmic maturation. *J. Virol.* **2012**, *86*, 6303–6314. [CrossRef]
76. Huet, A.; Makhov, A.M.; Huffman, J.B.; Vos, M.; Homa, F.L.; Conway, J.F. Extensive subunit contacts underpin herpesvirus capsid stability and interior-to-exterior allostery. *Nat. Struct. Mol. Biol.* **2016**, *23*, 531–539. [CrossRef]
77. Cardone, G.; Newcomb, W.W.; Cheng, N.; Wingfield, P.T.; Trus, B.L.; Brown, J.C.; Steven, A.C. The UL36 tegument protein of herpes simplex virus 1 has a composite binding site at the capsid vertices. *J. Virol.* **2012**, *86*, 4058–4064. [CrossRef]
78. Liu, Z.; Kato, A.; Shindo, K.; Noda, T.; Sagara, H.; Kawaoka, Y.; Arii, J.; Kawaguchi, Y. Herpes simplex virus 1 UL47 interacts with viral nuclear egress factors UL31, UL34, and Us3 and regulates viral nuclear egress. *J. Virol.* **2014**, *88*, 4657–4667. [CrossRef]
79. Maruzuru, Y.; Shindo, K.; Liu, Z.; Oyama, M.; Kozuka-Hata, H.; Arii, J.; Kato, A.; Kawaguchi, Y. Role of herpes simplex virus 1 immediate early protein ICP22 in viral nuclear egress. *J. Virol.* **2014**, *88*, 7445–7454. [CrossRef]
80. Arii, J.; Watanabe, M.; Maeda, F.; Tokai-Nishizumi, N.; Chihara, T.; Miura, M.; Maruzuru, Y.; Koyanagi, N.; Kato, A.; Kawaguchi, Y. ESCRT-III mediates budding across the inner nuclear membrane and regulates its integrity. *Nat. Commun.* **2018**, *9*, 3379. [CrossRef] [PubMed]
81. Mou, F.; Wills, E.; Baines, J.D. Phosphorylation of the U(L)31 protein of herpes simplex virus 1 by the U(S)3-encoded kinase regulates localization of the nuclear envelopment complex and egress of nucleocapsids. *J. Virol.* **2009**, *83*, 5181–5191. [CrossRef] [PubMed]
82. Cano-Monreal, G.L.; Wylie, K.M.; Cao, F.; Tavis, J.E.; Morrison, L.A. Herpes simplex virus 2 UL13 protein kinase disrupts nuclear lamins. *Virology* **2009**, *392*, 137–147. [CrossRef] [PubMed]
83. Reynolds, A.E.; Wills, E.G.; Roller, R.J.; Ryckman, B.J.; Baines, J.D. Ultrastructural localization of the herpes simplex virus type 1 UL31, UL34, and US3 proteins suggests specific roles in primary envelopment and egress of nucleocapsids. *J. Virol.* **2002**, *76*, 8939–8952. [CrossRef]
84. Purves, F.C.; Spector, D.; Roizman, B. The herpes simplex virus 1 protein kinase encoded by the US3 gene mediates posttranslational modification of the phosphoprotein encoded by the UL34 gene. *J. Virol.* **1991**, *65*, 5757–5764. [CrossRef]
85. Wisner, T.W.; Wright, C.C.; Kato, A.; Kawaguchi, Y.; Mou, F.; Baines, J.D.; Roller, R.J.; Johnson, D.C. Herpesvirus gB-induced fusion between the virion envelope and outer nuclear membrane during virus egress is regulated by the viral US3 kinase. *J. Virol.* **2009**, *83*, 3115–3126. [CrossRef]
86. Kato, A.; Arii, J.; Shiratori, I.; Akashi, H.; Arase, H.; Kawaguchi, Y. Herpes simplex virus 1 protein kinase Us3 phosphorylates viral envelope glycoprotein B and regulates its expression on the cell surface. *J. Virol.* **2009**, *83*, 250–261. [CrossRef] [PubMed]
87. Kato, A.; Yamamoto, M.; Ohno, T.; Kodaira, H.; Nishiyama, Y.; Kawaguchi, Y. Identification of proteins phosphorylated directly by the Us3 protein kinase encoded by herpes simplex virus 1. *J. Virol.* **2005**, *79*, 9325–9331. [CrossRef]
88. Kopp, M.; Klupp, B.G.; Granzow, H.; Fuchs, W.; Mettenleiter, T.C. Identification and characterization of the pseudorabies virus tegument proteins UL46 and UL47: Role for UL47 in virion morphogenesis in the cytoplasm. *J. Virol.* **2002**, *76*, 8820–8833. [CrossRef]
89. Fuchs, W.; Granzow, H.; Mettenleiter, T.C. A pseudorabies virus recombinant simultaneously lacking the major tegument proteins encoded by the UL46, UL47, UL48, and UL49 genes is viable in cultured cells. *J. Virol.* **2003**, *77*, 12891–12900. [CrossRef] [PubMed]
90. Farnsworth, A.; Wisner, T.W.; Webb, M.; Roller, R.; Cohen, G.; Eisenberg, R.; Johnson, D.C. Herpes simplex virus glycoproteins gB and gH function in fusion between the virion envelope and the outer nuclear membrane. *Proc. Natl. Acad. Sci. USA* **2007**, *104*, 10187–10192. [CrossRef] [PubMed]
91. Klupp, B.; Altenschmidt, J.; Granzow, H.; Fuchs, W.; Mettenleiter, T.C. Glycoproteins required for entry are not necessary for egress of pseudorabies virus. *J. Virol.* **2008**, *82*, 6299–6309. [CrossRef]
92. Nozawa, N.; Daikoku, T.; Koshizuka, T.; Yamauchi, Y.; Yoshikawa, T.; Nishiyama, Y. Subcellular localization of herpes simplex virus type 1 UL51 protein and role of palmitoylation in Golgi apparatus targeting. *J. Virol.* **2003**, *77*, 3204–3216. [CrossRef] [PubMed]

93. Klupp, B.G.; Granzow, H.; Klopfleisch, R.; Fuchs, W.; Kopp, M.; Lenk, M.; Mettenleiter, T.C. Functional analysis of the pseudorabies virus UL51 protein. *J. Virol.* **2005**, *79*, 3831–3840. [CrossRef]

94. Saiz-Ros, N.; Czapiewski, R.; Epifano, I.; Stevenson, A.; Swanson, S.K.; Dixon, C.R.; Zamora, D.B.; McElwee, M.; Vijayakrishnan, S.; Richardson, C.A.; et al. Host Vesicle Fusion Protein VAPB Contributes to the Nuclear Egress Stage of Herpes Simplex Virus Type-1 (HSV-1) Replication. *Cells* **2019**, *8*, 120. [CrossRef]

95. Votteler, J.; Sundquist, W.I. Virus budding and the ESCRT pathway. *Cell Host Microbe* **2013**, *14*, 232–241. [CrossRef]

96. Vietri, M.; Radulovic, M.; Stenmark, H. The many functions of ESCRTs. *Nat. Rev. Mol. Cell Biol.* **2020**, *21*, 25–42. [CrossRef]

97. Crump, C.M.; Yates, C.; Minson, T. Herpes simplex virus type 1 cytoplasmic envelopment requires functional Vps4. *J. Virol.* **2007**, *81*, 7380–7387. [CrossRef] [PubMed]

98. Tandon, R.; AuCoin, D.P.; Mocarski, E.S. Human cytomegalovirus exploits ESCRT machinery in the process of virion maturation. *J. Virol.* **2009**, *83*, 10797–10807. [CrossRef] [PubMed]

99. Kharkwal, H.; Smith, C.G.; Wilson, D.W. Blocking ESCRT-mediated envelopment inhibits microtubule-dependent trafficking of alphaherpesviruses in vitro. *J. Virol.* **2014**, *88*, 14467–14478. [CrossRef] [PubMed]

100. Pawliczek, T.; Crump, C.M. Herpes simplex virus type 1 production requires a functional ESCRT-III complex but is independent of TSG101 and ALIX expression. *J. Virol.* **2009**, *83*, 11254–11264. [CrossRef] [PubMed]

101. Vietri, M.; Schink, K.O.; Campsteijn, C.; Wegner, C.S.; Schultz, S.W.; Christ, L.; Thoresen, S.B.; Brech, A.; Raiborg, C.; Stenmark, H. Spastin and ESCRT-III coordinate mitotic spindle disassembly and nuclear envelope sealing. *Nature* **2015**, *522*, 231–235. [CrossRef] [PubMed]

102. Denais, C.M.; Gilbert, R.M.; Isermann, P.; McGregor, A.L.; te Lindert, M.; Weigelin, B.; Davidson, P.M.; Friedl, P.; Wolf, K.; Lammerding, J. Nuclear envelope rupture and repair during cancer cell migration. *Science* **2016**, *352*, 353–358. [CrossRef]

103. Lee, C.P.; Liu, P.T.; Kung, H.N.; Su, M.T.; Chua, H.H.; Chang, Y.H.; Chang, C.W.; Tsai, C.H.; Liu, F.T.; Chen, M.R. The ESCRT machinery is recruited by the viral BFRF1 protein to the nucleus-associated membrane for the maturation of Epstein-Barr Virus. *PLoS Pathog.* **2012**, *8*, e1002904. [CrossRef] [PubMed]

104. Lee, C.P.; Liu, G.T.; Kung, H.N.; Liu, P.T.; Liao, Y.T.; Chow, L.P.; Chang, L.S.; Chang, Y.H.; Chang, C.W.; Shu, W.C.; et al. The Ubiquitin Ligase Itch and Ubiquitination Regulate BFRF1-Mediated Nuclear Envelope Modification for Epstein-Barr Virus Maturation. *J. Virol.* **2016**, *90*, 8994–9007. [CrossRef] [PubMed]

105. Leis, J.; Luan, C.H.; Audia, J.E.; Dunne, S.F.; Heath, C.M. Ilaprazole and other novel prazole-based compounds that bind Tsg101 inhibit viral budding of HSV-1/2 and HIV from cells. *J. Virol.* **2021**. [CrossRef] [PubMed]

106. Soni, S.P.; Stahelin, R.V. The Ebola virus matrix protein VP40 selectively induces vesiculation from phosphatidylserine-enriched membranes. *J. Biol. Chem.* **2014**, *289*, 33590–33597. [CrossRef] [PubMed]

107. Klupp, B.G.; Granzow, H.; Mettenleiter, T.C. Effect of the pseudorabies virus US3 protein on nuclear membrane localization of the UL34 protein and virus egress from the nucleus. *J. Gen. Virol* **2001**, *82*, 2363–2371. [CrossRef] [PubMed]

108. Ryckman, B.J.; Roller, R.J. Herpes simplex virus type 1 primary envelopment: UL34 protein modification and the US3-UL34 catalytic relationship. *J. Virol.* **2004**, *78*, 399–412. [CrossRef] [PubMed]

109. Kato, A.; Oda, S.; Watanabe, M.; Oyama, M.; Kozuka-Hata, H.; Koyanagi, N.; Maruzuru, Y.; Arii, J.; Kawaguchi, Y. Roles of the Phosphorylation of Herpes Simplex Virus 1 UL51 at a Specific Site in Viral Replication and Pathogenicity. *J. Virol.* **2018**, *92*. [CrossRef] [PubMed]

110. Riva, L.; Thiry, M.; Lebrun, M.; L'Homme, L.; Piette, J.; Sadzot-Delvaux, C. Deletion of the ORF9p acidic cluster impairs the nuclear egress of varicella-zoster virus capsids. *J. Virol.* **2015**, *89*, 2436–2441. [CrossRef] [PubMed]

111. Wright, C.C.; Wisner, T.W.; Hannah, B.P.; Eisenberg, R.J.; Cohen, G.H.; Johnson, D.C. Fusion between perinuclear virions and the outer nuclear membrane requires the fusogenic activity of herpes simplex virus gB. *J. Virol.* **2009**, *83*, 11847–11856. [CrossRef] [PubMed]

112. Arii, J.; Kawaguchi, Y. The Role of HSV Glycoproteins in Mediating Cell Entry. *Adv. Exp. Med. Biol.* **2018**, *1045*, 3–21. [CrossRef]

113. Ott, M.; Tascher, G.; Hassdenteufel, S.; Zimmermann, R.; Haas, J.; Bailer, S.M. Functional characterization of the essential tail anchor of the herpes simplex virus type 1 nuclear egress protein pUL34. *J. Gen. Virol.* **2011**, *92*, 2734–2745. [CrossRef] [PubMed]

114. Schuster, F.; Klupp, B.G.; Granzow, H.; Mettenleiter, T.C. Structural determinants for nuclear envelope localization and function of pseudorabies virus pUL34. *J. Virol.* **2012**, *86*, 2079–2088. [CrossRef] [PubMed]

115. Hirohata, Y.; Arii, J.; Liu, Z.; Shindo, K.; Oyama, M.; Kozuka-Hata, H.; Sagara, H.; Kato, A.; Kawaguchi, Y. Herpes Simplex Virus 1 Recruits CD98 Heavy Chain and beta1 Integrin to the Nuclear Membrane for Viral De-Envelopment. *J. Virol.* **2015**, *89*, 7799–7812. [CrossRef] [PubMed]

116. Liu, Z.; Kato, A.; Oyama, M.; Kozuka-Hata, H.; Arii, J.; Kawaguchi, Y. Role of Host Cell p32 in Herpes Simplex Virus 1 De-Envelopment during Viral Nuclear Egress. *J. Virol.* **2015**, *89*, 8982–8998. [CrossRef] [PubMed]

117. Ohgimoto, S.; Tabata, N.; Suga, S.; Nishio, M.; Ohta, H.; Tsurudome, M.; Komada, H.; Kawano, M.; Watanabe, N.; Ito, Y. Molecular characterization of fusion regulatory protein-1 (FRP-1) that induces multinucleated giant cell formation of monocytes and HIV gp160-mediated cell fusion. FRP-1 and 4F2/CD98 are identical molecules. *J. Immunol.* **1995**, *155*, 3585–3592. [PubMed]

118. Okamoto, K.; Tsurudome, M.; Ohgimoto, S.; Kawano, M.; Nishio, M.; Komada, H.; Ito, M.; Sakakura, Y.; Ito, Y. An anti-fusion regulatory protein-1 monoclonal antibody suppresses human parainfluenza virus type 2-induced cell fusion. *J. Gen. Virol.* **1997**, *78*, 83–89. [CrossRef]

119. Ohta, H.; Tsurudome, M.; Matsumura, H.; Koga, Y.; Morikawa, S.; Kawano, M.; Kusugawa, S.; Komada, H.; Nishio, M.; Ito, Y. Molecular and biological characterization of fusion regulatory proteins (FRPs): Anti-FRP mAbs induced HIV-mediated cell fusion via an integrin system. *EMBO J.* **1994**, *13*, 2044–2055. [CrossRef]
120. Ito, Y.; Komada, H.; Kusagawa, S.; Tsurudome, M.; Matsumura, H.; Kawano, M.; Ohta, H.; Nishio, M. Fusion regulation proteins on the cell surface: Isolation and characterization of monoclonal antibodies which enhance giant polykaryocyte formation in Newcastle disease virus-infected cell lines of human origin. *J. Virol.* **1992**, *66*, 5999–6007. [CrossRef]
121. Maeda, F.; Arii, J.; Hirohata, Y.; Maruzuru, Y.; Koyanagi, N.; Kato, A.; Kawaguchi, Y. Herpes Simplex Virus 1 UL34 Protein Regulates the Global Architecture of the Endoplasmic Reticulum in Infected Cells. *J. Virol.* **2017**, *91*. [CrossRef]
122. Li, Y.; Hassinger, L.; Thomson, T.; Ding, B.; Ashley, J.; Hassinger, W.; Budnik, V. Lamin Mutations Accelerate Aging via Defective Export of Mitochondrial mRNAs through Nuclear Envelope Budding. *Curr. Biol.* **2016**, *26*, 2052–2059. [CrossRef]
123. Maric, M.; Shao, J.; Ryan, R.J.; Wong, C.S.; Gonzalez-Alegre, P.; Roller, R.J. A functional role for TorsinA in herpes simplex virus 1 nuclear egress. *J. Virol.* **2011**, *85*, 9667–9679. [CrossRef]
124. Jokhi, V.; Ashley, J.; Nunnari, J.; Noma, A.; Ito, N.; Wakabayashi-Ito, N.; Moore, M.J.; Budnik, V. Torsin mediates primary envelopment of large ribonucleoprotein granules at the nuclear envelope. *Cell Rep.* **2013**, *3*, 988–995. [CrossRef]
125. Rampello, A.J.; Prophet, S.M.; Schlieker, C. The Role of Torsin AAA+ Proteins in Preserving Nuclear Envelope Integrity and Safeguarding Against Disease. *Biomolecules* **2020**, *10*, 468. [CrossRef]
126. Laudermilch, E.; Schlieker, C. Torsin ATPases: Structural insights and functional perspectives. *Curr. Opin. Cell Biol.* **2016**, *40*, 1–7. [CrossRef]
127. Brown, R.S.; Zhao, C.; Chase, A.R.; Wang, J.; Schlieker, C. The mechanism of Torsin ATPase activation. *Proc. Natl. Acad. Sci. USA* **2014**, *111*, E4822–E4831. [CrossRef] [PubMed]
128. Sosa, B.A.; Demircioglu, F.E.; Chen, J.Z.; Ingram, J.; Ploegh, H.L.; Schwartz, T.U. How lamina-associated polypeptide 1 (LAP1) activates Torsin. *Elife* **2014**, *3*, e03239. [CrossRef] [PubMed]
129. Gyorgy, B.; Cruz, L.; Yellen, D.; Aufiero, M.; Alland, I.; Zhang, X.; Ericsson, M.; Fraefel, C.; Li, Y.C.; Takeda, S.; et al. Mutant torsinA in the heterozygous DYT1 state compromises HSV propagation in infected neurons and fibroblasts. *Sci. Rep.* **2018**, *8*, 2324. [CrossRef] [PubMed]
130. Turner, E.M.; Brown, R.S.; Laudermilch, E.; Tsai, P.L.; Schlieker, C. The Torsin Activator LULL1 Is Required for Efficient Growth of Herpes Simplex Virus 1. *J. Virol.* **2015**, *89*, 8444–8452. [CrossRef]
131. Goodchild, R.E.; Kim, C.E.; Dauer, W.T. Loss of the dystonia-associated protein torsinA selectively disrupts the neuronal nuclear envelope. *Neuron* **2005**, *48*, 923–932. [CrossRef]
132. Liang, C.C.; Tanabe, L.M.; Jou, S.; Chi, F.; Dauer, W.T. TorsinA hypofunction causes abnormal twisting movements and sensorimotor circuit neurodegeneration. *J. Clin. Investig.* **2014**, *124*, 3080–3092. [CrossRef] [PubMed]
133. Grillet, M.; Dominguez Gonzalez, B.; Sicart, A.; Pottler, M.; Cascalho, A.; Billion, K.; Hernandez Diaz, S.; Swerts, J.; Naismith, T.V.; Gounko, N.V.; et al. Torsins Are Essential Regulators of Cellular Lipid Metabolism. *Dev. Cell* **2016**, *38*, 235–247. [CrossRef] [PubMed]
134. Arii, J.; Goto, H.; Suenaga, T.; Oyama, M.; Kozuka-Hata, H.; Imai, T.; Minowa, A.; Akashi, H.; Arase, H.; Kawaoka, Y.; et al. Non-muscle myosin IIA is a functional entry receptor for herpes simplex virus-1. *Nature* **2010**, *467*, 859–862. [CrossRef] [PubMed]
135. Arii, J.; Hirohata, Y.; Kato, A.; Kawaguchi, Y. Nonmuscle myosin heavy chain IIb mediates herpes simplex virus 1 entry. *J. Virol.* **2015**, *89*, 1879–1888. [CrossRef] [PubMed]
136. Satoh, T.; Arii, J.; Suenaga, T.; Wang, J.; Kogure, A.; Uehori, J.; Arase, N.; Shiratori, I.; Tanaka, S.; Kawaguchi, Y.; et al. PILRalpha is a herpes simplex virus-1 entry coreceptor that associates with glycoprotein B. *Cell* **2008**, *132*, 935–944. [CrossRef] [PubMed]
137. Laudermilch, E.; Tsai, P.L.; Graham, M.; Turner, E.; Zhao, C.; Schlieker, C. Dissecting Torsin/cofactor function at the nuclear envelope: A genetic study. *Mol. Biol. Cell* **2016**, *27*, 3964–3971. [CrossRef] [PubMed]
138. Webster, B.M.; Colombi, P.; Jager, J.; Lusk, C.P. Surveillance of nuclear pore complex assembly by ESCRT-III/Vps4. *Cell* **2014**, *159*, 388–401. [CrossRef] [PubMed]
139. Webster, B.M.; Thaller, D.J.; Jager, J.; Ochmann, S.E.; Borah, S.; Lusk, C.P. Chm7 and Heh1 collaborate to link nuclear pore complex quality control with nuclear envelope sealing. *EMBO J.* **2016**, *35*, 2447–2467. [CrossRef]
140. Olmos, Y.; Hodgson, L.; Mantell, J.; Verkade, P.; Carlton, J.G. ESCRT-III controls nuclear envelope reformation. *Nature* **2015**, *522*, 236–239. [CrossRef]
141. Raab, M.; Gentili, M.; de Belly, H.; Thiam, H.R.; Vargas, P.; Jimenez, A.J.; Lautenschlaeger, F.; Voituriez, R.; Lennon-Dumenil, A.M.; Manel, N.; et al. ESCRT III repairs nuclear envelope ruptures during cell migration to limit DNA damage and cell death. *Science* **2016**, *352*, 359–362. [CrossRef]
142. Eriksson, M.; Brown, W.T.; Gordon, L.B.; Glynn, M.W.; Singer, J.; Scott, L.; Erdos, M.R.; Robbins, C.M.; Moses, T.Y.; Berglund, P.; et al. Recurrent de novo point mutations in lamin A cause Hutchinson-Gilford progeria syndrome. *Nature* **2003**, *423*, 293–298. [CrossRef]
143. De Sandre-Giovannoli, A.; Bernard, R.; Cau, P.; Navarro, C.; Amiel, J.; Boccaccio, I.; Lyonnet, S.; Stewart, C.L.; Munnich, A.; Le Merrer, M.; et al. Lamin a truncation in Hutchinson-Gilford progeria. *Science* **2003**, *300*, 2055. [CrossRef] [PubMed]
144. Arii, J.; Maeda, F.; Maruzuru, Y.; Koyanagi, N.; Kato, A.; Mori, Y.; Kawaguchi, Y. ESCRT-III controls nuclear envelope deformation induced by progerin. *Sci. Rep.* **2020**, *10*, 18877. [CrossRef]

 *viruses*

Review

When in Need of an ESCRT: The Nature of Virus Assembly Sites Suggests Mechanistic Parallels between Nuclear Virus Egress and Retroviral Budding

Kevin M. Rose [1],* , Stephanie J. Spada [2], Vanessa M. Hirsch [2] and Fadila Bouamr [2]

[1] Department of Molecular and Cell Biology, California Institute for Quantitative Biosciences, University of California—Berkeley, Berkeley, CA 94720, USA

[2] Laboratory of Molecular Microbiology, National Institute of Allergy and Infectious Diseases, National Institutes of Health, Bethesda, Rockville, MD 20894, USA; stephanie.spada@nih.gov (S.J.S.); vhirsch@niaid.nih.gov (V.M.H.); bouamrf@mail.nih.gov (F.B.)

* Correspondence: kevin_rose@berkeley.edu

Abstract: The proper assembly and dissemination of progeny virions is a fundamental step in virus replication. As a whole, viruses have evolved a myriad of strategies to exploit cellular compartments and mechanisms to ensure a successful round of infection. For enveloped viruses such as retroviruses and herpesviruses, acquisition and incorporation of cellular membrane is an essential process during the formation of infectious viral particles. To do this, these viruses have evolved to hijack the host Endosomal Sorting Complexes Required for Transport (ESCRT-I, -II, and -III) to coordinate the sculpting of cellular membrane at virus assembly and dissemination sites, in seemingly different, yet fundamentally similar ways. For instance, at the plasma membrane, ESCRT-I recruitment is essential for HIV-1 assembly and budding, while it is dispensable for the release of HSV-1. Further, HSV-1 was shown to recruit ESCRT-III for nuclear particle assembly and egress, a process not used by retroviruses during replication. Although the cooption of ESCRTs occurs in two separate subcellular compartments and at two distinct steps for these viral lifecycles, the role fulfilled by ESCRTs at these sites appears to be conserved. This review discusses recent findings that shed some light on the potential parallels between retroviral budding and nuclear egress and proposes a model where HSV-1 nuclear egress may occur through an ESCRT-dependent mechanism.

Keywords: enveloped virus budding; nuclear egress; ESCRT; membrane scission

Citation: Rose, K.M.; Spada, S.J.; Hirsch, V.M.; Bouamr, F. When in Need of an ESCRT: The Nature of Virus Assembly Sites Suggests Mechanistic Parallels between Nuclear Virus Egress and Retroviral Budding. *Viruses* **2021**, *13*, 1138. https://doi.org/10.3390/v13061138

Academic Editor: Donald M. Coen

Received: 22 April 2021
Accepted: 2 June 2021
Published: 13 June 2021

Publisher's Note: MDPI stays neutral with regard to jurisdictional claims in published maps and institutional affiliations.

1. Introduction

ESCRTs Are Recruited to Sites of Viral Replication on Membranes

One of the most important trafficking pathways in the cell is the sorting of cargo once it has been internalized at the plasma membrane [1]. Cargo destined for degradation is typically modified with a small protein tag known as ubiquitin, a process referred to as ubiquitylation, and is compartmentalized within an internal vesicle where it will become part of the multivesicular body (MVB) on its way to the late endosome [2,3]. To ensure cargo has been compartmentalized during MVB biogenesis, the endosomal sorting complex required for transport (ESCRT) is recruited to the cargo by virtue of the ubiquitylation signal and directs cargo sorting [2,4]. The ubiquitin moiety from the cargo is detected by the ESCRT-I component TSG101 as well as the ESCRT-associated protein ALIX [5,6]. Once the cargo has been recognized, ESCRT-I can direct the sealing of the membrane compartment through sequential interactions with ESCRT-II and ESCRT-III, while ALIX can recruit ESCRT-III directly (Figure 1) [7]. ESCRT-III recruitment is temporally and spatially dynamic within the cell where it forms filaments that mediate the scission and sealing of a multitude of cellular membranes ranging from the nuclear envelope to the plasma membrane and many membranes in between [8–12]. Typically responsible for orchestrating the early events of membrane trafficking and scission in the cell, TSG101 and ALIX are commonly hijacked by viruses to facilitate viral replication and dissemination, one of the most well studied of cases coming from retroviruses like the Human Immunodeficiency Virus type 1 (HIV-1) [5,13]. In the case of HIV-1, the retroviral Gag protein encodes two late budding domains (L-domains). These short peptide motifs independently recruit ALIX (via a LYPXnL motif) and ESCRT-I (via a PTAP motif) to sites of viral budding and assembly by mimicking the peptide signal found within the cognate cellular partners of ALIX and TSG101 [5,13–15] (For review see [16]). Additionally, retroviral Gag is ubiquitylated by the ubiquitin ligase NEDD4, and ubiquitin modification is essential for ESCRT-mediated budding [17]. Recruitment of ESCRTs therein ensures genome packaging and proper maturation of the newly assembling retrovirus particle [18] (Figure 2a).

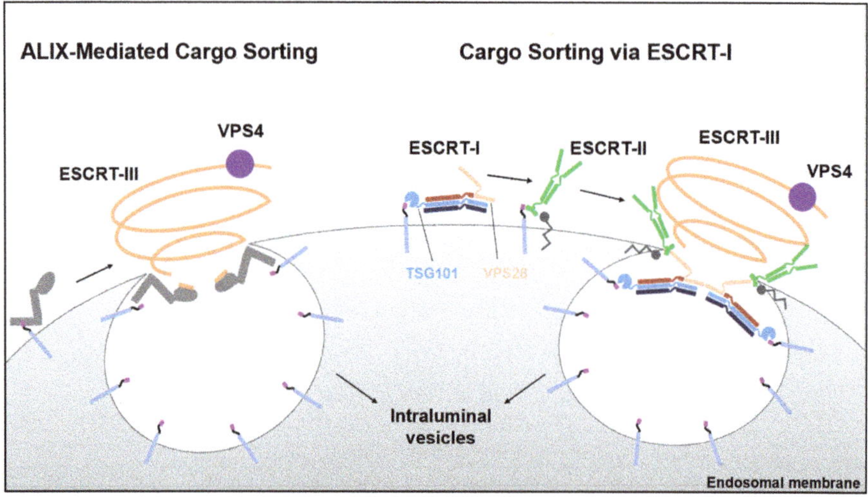

Figure 1. The respective roles of ALIX and ESCRT-I in the sorting of membranous cargo. Upon internalization, ubiquitylated cargo is detected by ALIX (**left**) and ESCRT-I (**right**) for compartmentalization into intraluminal vesicles that are destined for degradation via the late endosome. Both ALIX and ESCRT-I contain ubiquitin binding domains that facilitate this first step. Unlike ESCRT-I, ALIX possesses an ESCRT-III binding domain that allows for the direct recruitment of ESCRT-III and VPS4, the machinery required for sealing of cargo within intraluminal vesicles and abscising these vesicles from the endosomal membrane. In a similar fashion, the ESCRT-I component TSG101 binds ubiquitylated cargo, while the VPS28 component can recruit ESCRT-III through ESCRT-II which also binds ubiquitylated cargo as well as phospho-inositol lipids.

Despite the conservation of L-Domains seen in HIV-1 and host factors, there are notable exceptions to the consensus motifs commonly associated with ESCRT recruitment, namely in Simian Immunodeficiency Viruses (SIVs), or HIV-1-progenitor viruses, that encode divergent ALIX recruitment domains as determined by X-ray crystallography [19]. SIVs encode surprising species-dependent sequence variations in their ALIX recruitment domains with virus isolated from rhesus macaque (SIVmac) requiring a PYKEVTEDL motif, isolated from African green monkey (SIVagm) relying on a AYDPARKLL motif, and virus from sooty mangabey (SIVsmm) harboring a PYKEVTEDLLHLNSLF sequence for ALIX recruitment, each uniquely different from the consensus motif [20,21]. ALIX interactions with LYPXnL late domains are mediated by the central V domain of ALIX that has a hydrophobic cavity capable of accepting and stabilizing hydrophobic residues from an LYPXnL-containing binding partner [21]. ALIX recruitment by HIV-1 and SIVs with atypical motifs can be accomplished so long as the viruses encode a phenylalanine or tyrosine residue, with the proper flexibility to embed within the V domain of ALIX, independently of the immediate amino acid residues flanking the canonical LYPXnL motif [21,22].

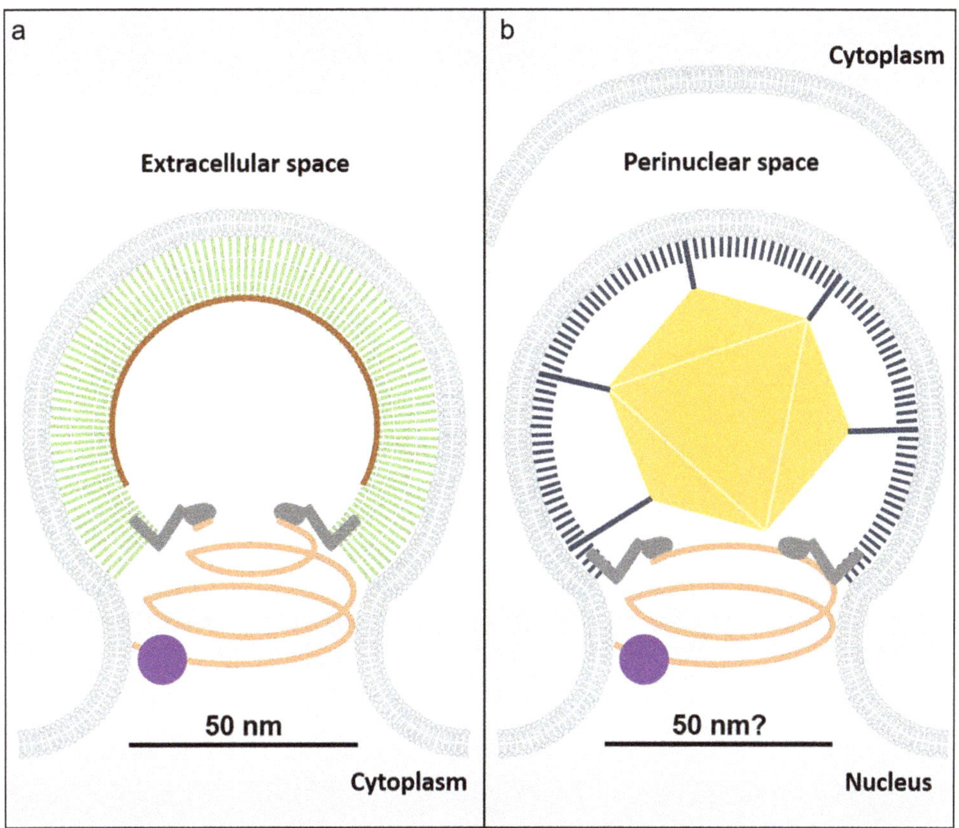

Figure 2. The process of HSV-1 nuclear egress and scission resembles the retroviral egress pathway. (**a**) Diagram of the well characterized HIV-1 budding site. HIV-1 Gag (green sticks) binds RNA (dark orange) and assembles on the membrane and recruits ALIX (dark grey) via the L-Domain located in its C-terminal p6 domain. HIV-1 budding necks have been shown typically to adopt a diameter of approximately 50 nm prior to membrane scission and this topology is likely to be shared by the NEC given the mechanistic constraints of the ESCRT-III membrane scission machinery. The downstream ESCRT-III is shown in salmon and the VPS4 membrane scission machinery in purple. (**b**) HSV-1 particle assembly begins in the nucleus where a newly formed capsid must transit the nuclear envelope by assembly and budding. Schematic of the HSV-1 nuclear egress complex prior to membrane scission assembled with its capsid as it buds through the nuclear envelope. A single virus capsid is shown as a gold hexagon and the NEC UL31-UL34 heterodimer is shown as membrane-associated sticks (black) that occasionally interact with the vertices of the capsid. ALIX is recruited to sites of nuclear egress through interactions with NEC and mediates the recruitment of the downstream ESCRT components to facilitate membrane scission.

Flaviviruses, another class of RNA virus that recruit ESCRT proteins, are classified on the basis of their tripartite use of a single stranded RNA genome as an mRNA, a genomic copy, and as a parental strand for transcription [23]. This is achieved through the encoding of a viral RNA-dependent RNA Polymerase (RdRp) that replicates and transcribes the RNA genome through a double stranded RNA (dsRNA) intermediate [24]. As part of the evolutionary arms race between viruses and host, mammalian cells are equipped with cytosolic sensors for the detection and degradation of large viral dsRNA [25]. To avoid detection, it is thought that members of this virus group have evolved to distort cellular membranes for the construction of immune system-naive membrane invaginations, called vesicle packets, that facilitate viral replication despite surveillance by the innate immune sensors [26]. Within these viral compartments, viral RNA is synthesized and exported through the lumen of the vesicle packet and into the cytoplasm for the translation of viral proteins [27]. Although the precise molecular details underlying vesicle packet formation are not fully understood, the recruitment of ESCRTs to sites of replication has been established for several emerging mosquito-borne flaviviruses including Dengue virus, all of which are thought to be facilitated through the viral helicase protein NS3 [28,29]. However, only in the case of Yellow Fever Virus replication was a canonical L-Domain sequence identified. In this scenario, the upstream ESCRTs may act in a membrane scaffolding capacity where they serve to keep the vesicle packet lumen open to facilitate the export of viral RNA transcripts and genomes into the cytoplasm. Lack of obvious L-Domain sequences despite clear biochemical association raises intriguing questions, about how dynamic ALIX recruitment is to sites of viral assembly on membranes, and whether or not other enveloped viruses of clinical importance, like Herpes Simplex Virus 1 (HSV-1), harbor biologically relevant ALIX recruitment L-domains within their structural proteins to facilitate essential membrane-associated steps in their life cycles [30].

Unlike RNA viruses, particle formation of the DNA virus HSV-1 begins in the nucleus (Figure 2b) [31]. Therein, after having undergone lytic reactivation, the newly synthesized viral DNA is packaged within the viral capsid [32]. In order to produce infectious progeny virions, it is theorized that the nuclear viral capsid must pass through the nuclear envelope through a membrane budding pathway, since HSV-1 capsids are too large to traverse the nuclear pore complex [33]. To do this, herpesviruses assemble a nuclear egress complex (NEC) from viral UL31 and UL34 proteins (HSV-1 nomenclature) [30]. UL34 localizes to the inner nuclear envelope via its C-terminal type II single-pass membrane anchor [34]. UL31 binds to HSV-1 capsid vertices but also associates with membrane bound UL34 and effectively acts as a dynamic scaffold between the inner nuclear membrane and the viral capsid [31]. Additionally, the UL31-UL34 heterodimer forms a lattice about the inner nuclear membrane in a manner structurally reminiscent of HIV-1 Gag assembly sites (Figure 2a,b) [31,35,36]. In order to efficiently bud across the nuclear envelope, HSV-1 and Epstein-Barr Virus (EBV) were recently shown to recruit the early acting ESCRT protein ALIX and ESCRT-III associated protein VPS4 directly to the NEC via interactions with UL34 [37,38]. Remarkably, EBV BFRF1 (UL34 in HSV-1), was also shown to recruit the NEDD4-like ubiquitin ligase ITCH, and that ubiquitination of BFRF1 regulates the formation of nuclear envelope-derived vesicles during virus maturation, drawing yet another parallel with retroviral budding [39]. Depletion of ALIX resulted in the loss of the downstream ESCRT-III component required for membrane scission, CHMP4B, and reduced HSV-1 and EBV replication. Additionally, in the case of HSV-1, impairment of the membrane scission machinery VPS4 resulted in a significant accumulation of enveloped virions in the nucleus, some of which remained tethered to the inner nuclear membrane [37], suggesting a role for ESCRTs in herpesvirus nuclear egress. However, despite clear biochemical data describing physical interactions between ALIX and UL34, no L-Domain sequence was identified. The remainder of this review focuses on comparing the nuclear egress of HSV-1 capsids to that of retroviral budding sites and proposes the existence of an ALIX binding site within UL34 that could support the model of ALIX recruitment to herpes-viral NECs.

2. A Hexagonal Assembly of HIV-1 Gag Drives Particle Maturation and ESCRT Recruitment

HIV-1 particle assembly at the membrane has been the focus of intensive study for more than two decades [5]. The retroviral Gag protein traffics to the membrane and forms higher-order oligomers via its capsid domain and assembles into an immature lattice that exhibits a hexagonal geometry (Figure 3a) [40]. The Gag lattice is intrinsically curved which is thought to assist in the shaping of the plasma membrane into a spherical membrane bud [41]. Maturation of the lattice is an essential step in the production of infectious virus particles and is still to this day not fully understood [42]. What is known is that the viral genome along with the viral proteins required for maturation are packaged as a ribonucleoprotein (RNP) within the center of the immature lattice, and packaging of the RNP is what triggers Gag maturation [18]. During or after this step, the upstream ESCRT proteins are recruited to the lumen of the retroviral budding neck via interactions with the Gag p6 domains located at the cytoplasmic face and luminal opening to the immature lattice [43]. ESCRT-I then assembles into a ring that templates the oligomerization of downstream ESCRT-III and the membrane scission machinery which abscises membranes by forming filaments within membrane necks [18,44–46]. ALIX is also an essential component of this process and exhibits the same localization pattern, as does ESCRT-I [47]. By binding to a spatially separated L-domain from that of ESCRT-I, ALIX can localize to the budding neck lumen and recruit ESCRT-III directly by virtue of its p6-binding V domain and CHMP4B-binding Bro1 domain, respectively [47–49]. Evidence of these structures have been captured in relevant cell types as well as in vitro [44,50,51]. Although separate in vitro studies have revealed that Gag assembly alone may be sufficient to drive membrane curvature of reconstituted vesicles, [52] the presence of L domain sequences is a requirement for proper retroviral assembly and release in cells [5].

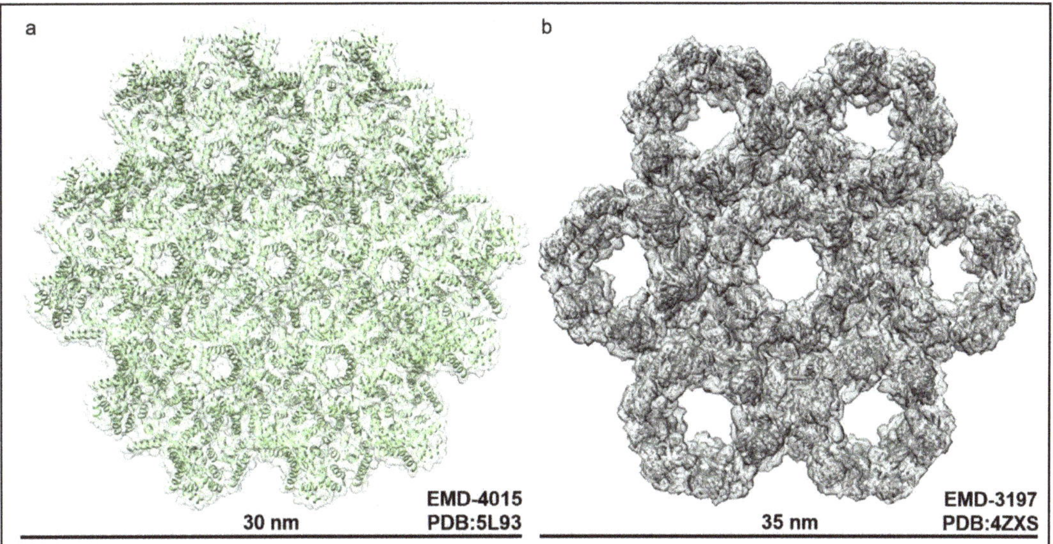

Figure 3. Both HSV-1 and HIV-1 form membrane bound hexagonal lattices that are involved in envelopment. HIV-1 Gag forms higher-order assemblies on the plasma membrane during viral budding. (**a**) The immature Gag lattice in green adopts a hexagonal geometry when bound to the plasma membrane and is found within immature virus particles. This viral complex must assemble on membranes in order to trigger the recruitment and activity of the membrane scission and sealing machinery. (**b**) The HSV-1 NEC complex binds to the inner nuclear membrane and assembles into a hexagonal lattice containing repeating units of a UL31-UL34 heterodimer. Structure of the HSV-1 NEC by fitting of the NEC crystal structure (PDB code: 4ZXS) into the 3D average of the NEC coat that was recently resolved via cryo-electron microscopy for Pseudorabies virus [36]. One repeating unit of the hexagonal lattice is shown, which has a diameter of approximately 35 nm. Note the differences in inter-subunit compaction and diameters between the assemblies.

Retroviral budding is a highly organized process in which the formation of the Gag lattice is critical for the proper recruitment and function of ESCRT in virus release [44,53]. Although ESCRT-I is transiently recruited with Gag to sites of active assembly, its functional significance only comes into play as an aggregate of individual recruitment events [46,50]. Indeed, it is only after recruitment of TSG101 or ALIX that ESCRT-III can be recruited and induce membrane scission [54]. This was shown recently through disruption of the ESCRT-I helical interface that left autophagosomes unsealed and prevented the release of HIV-1, indicative of a lack of downstream recruitment and scission [44]. This suggested that the budding site, and more importantly the hexagonal lattice, of HIV-1 is a well-evolved molecular signature that topologically mimics the cellular recruitment signal behind ESCRT function in cellular cargo trafficking pathways [55].

The immature hexagonal lattice exhibited by HIV-1 Gag during assembly is such an essential component of the particle maturation pathway that it has even become the target of novel antiretroviral drugs [56]. Bevirimat, the first maturation inhibitor, was designed to block the complete maturation of HIV-1 Gag such that the Gag-derived capsid hexamers that ultimately assemble into the core cannot be liberated from the Gag polyprotein [56]. Assembly sites targeted by this drug exhibited thick and mostly unprocessed Gag shells that remained associated with the membrane and packaged the genome eccentrically [56]. Additional inhibitors have been raised against the luminal opening of the Gag lattice [57]. More specifically, it was shown that HIV-1 assembly could be inhibited by targeting the N-terminal domain of TSG101 with the drug N16 that binds p6 domains within the Gag hexagonal lattice [57]. Intriguingly, in determining the mechanism of action of N16, it was found that this compound disrupted interactions outside of the p6 binding site in TSG101 [57]. The N-terminus of TSG101 contains an ancestral and inactivated domain that may have once been used as part of the ubiquitin conjugation pathway [58]. Despite having lost its catalytic ability, the TSG101 N-terminus still retains the ability to bind ubiquitin. N16 impairs HIV-1 maturation by binding to this secondary ubiquitin binding site, as opposed to the p6 binding pocket [57]. Given that ubiquitin conjugation to Gag is essential for ESCRT-I mediated virus release [17], the dramatic effect of N16 on virus release underscores the importance of ubiquitin in the process of virus particle maturation.

3. The NECs of Herpesviruses form Gag-Like Assemblies That Interact with ALIX and Are Required for Efficient Nuclear Egress

Unlike in retroviral budding where infectious particles are released from the plasma membrane, HSV-1 particle assembly and egress occurs first at the inner nuclear envelope and then again at the plasma membrane, for the release of infectious particles [32]. This first process is carried out by two viral proteins UL31 and UL34 which form the inner nuclear membrane-associated NEC [59]. UL34 anchors this complex to the membrane via a C-terminal transmembrane domain where it recruits the capsid-binding UL31 protein and forms a stable heterodimer [59]. Recently, it was determined through high resolution structural studies that multiple copies of the UL31-UL34 heterodimer assemble into a retroviral Gag-like hexagonal lattice on the surface of the inner nuclear membrane (Figure 3b) [35,36,60]. Interestingly, like HIV-1 Gag, in vitro studies using purified NECs on reconstituted membranes showed that the NECs themselves are sufficient to drive membrane deformation and egress [60]. The relevance of these higher order structures in virus replication was validated via mutagenesis in which inter-hexamer contact sites were abrogated [61]. In doing so, it was proposed that the hexagonal lattice plays an important role in regulating primary envelopment, and therefore viral replication [31,61]. Additionally, there are notable cellular binding partners that also promote herpesvirus primary envelopment including the WD repeat-containing protein 5 (WDR5) and the ESCRT-associated protein ALIX [37,62]. ALIX was shown to interact with both recombinant UL34 and the UL34 from HSV-1-infected cell lysate, an interaction that drives nuclear virus egress in cells [37]. In parallel, disruption of the membrane scission machinery VPS4 by mutagenesis also reduced virus nuclear egress [37]. In contrast, one previous study showed that in a different cell line, HSV-1 nuclear egress was unaffected by VPS4 inactivation [63]. This hints at a potential cell-type specific involvement of ESCRT proteins in virus nuclear egress. Intriguingly, both Bro1 and V domains of ALIX were capable of interacting with UL34, hinting at the possible presence of an L-domain sequence, but also that there may be additional electrostatic interactions between the proteins as is seen in ALIX interactions with HIV-1 Gag [64,65]. However, although HSV-1 harbors almost two dozen potential ESCRT-I TSG101 and ALIX binding sites across its viral proteins [66], TSG101 and ALIX do not appear to be essential for infectious virus release despite both being detected within extracellular virions [66]. Further still, no ALIX binding sites were identified in either component of the NEC which complicates the interpretation of ALIX–

NEC interactions [66,67]. One possible explanation for the results of these studies are that their sequencing surveys relied on canonical L-domain sequences that are typically attributed to ESCRT and ALIX interactions based on the most well characterized cellular and viral binding partners [15,21,68]. Regardless, more evidence is needed before any definitive role for ESCRTs in nuclear egress can be established.

In analyzing the NEC for potential non-canonical L-domains, like those found in the RNA viruses mentioned above that may promote its association with ALIX, sequence alignments and structural modelling were used to uncover such potential motifs. The sequence of HSV-1 UL34 was compared to that of several herpesviruses across three herpesvirus families (alpha, beta, and gamma-herpesviruses) (Figure 4a). Sequence alignments revealed little homology across virus families and family members. HSV-1 and HSV-1-2, both alpha-herpesviruses, share almost 80% identity, while the same protein from Varicella zoster virus (VZV), another alpha-herpesvirus, had less than 40% sequence identity with HSV-1. Additionally, less than 15% identity was seen between HSV-1 and members of the beta-herpesviruses family like Human Cytomegalovirus (HCMV), Human Herpesvirus 7 (HHV7), and Murine Cytomegalovirus (MCMV), as well as the gamma-herpesvirus family members Kaposi's Sarcoma-Associated Herpesvirus (KSHV) and EBV. Three putative L-domain motifs were identified in HSV-1 and were highlighted in yellow (Figure 4a). The sequences were also placed in a table along with known viral L-domain motifs (Figure 4b). The first potential L-Domain sequence was the near N-terminal sequence $_8$YPGHPGDAFEGL$_{19}$. However, despite the predicted flexibility of this sequence, it sterically clashed with UL31, suggesting that it would not likely be accessible for binding to ALIX, or could be engaged in interactions with UL31. The second sequence just downstream was $_{41}$YSPSSL$_{46}$. However, this motif was shown to embed its tyrosine residue within secondary structure elements of UL34 where it makes hydrophobic contacts with neighboring phenylalanine and tryptophan residues in the context of the UL31-UL34 heterodimer, making it inaccessible for binding to ALIX [31]. The third and most likely candidate for an ALIX-binding L-domain sequence was $_{202}$YGAEAGL$_{208}$. However, this region of UL34 has not been previously characterized structurally, likely due to its intrinsic flexibility and location proximal to the C-terminal transmembrane domain. Therefore, a molecular model of this region of the HSV-1 NEC was generated using the resolved residues from the HSV-1 NEC crystal structure (PDB:4ZXS), combined with structural predictions from Phyre2 and MODELLER in order to structurally analyze the missing regions of the protein [31,69,70] (Figure 4c). The prediction of this flexible region revealed the presence of a loop that faces away from the rest of the NEC and contains the putative L-Domain sequence $_{202}$YGAEAGL$_{208}$. Curiously, sequence alignments revealed little conservation of putative L-Domains across viruses, suggesting that the model for ALIX recruitment may only occur in select herpesvirus nuclear egress pathways. Perhaps in cases of ESCRT-III dependent nuclear egress, an edge of the NEC containing the L-Domain sequence is exposed to the nucleoplasm to be recognized for ALIX-mediated membrane budding and scission, much like how HIV-1 Gag recruits ALIX to the plasma membrane [71]. Although this model places the L-Domain of the NEC proximal to the nuclear envelope, this proposed mechanism is not unlike the recruitment of ESCRTs seen for beta-retroviruses and spuma-retroviruses where ESCRT binding is predicted to occur close to the plasma membrane [72,73]. This model could serve as a guide for future studies looking to further our understanding of the molecular basis and relevance behind ALIX recruitment and function at sites of viral nuclear egress.

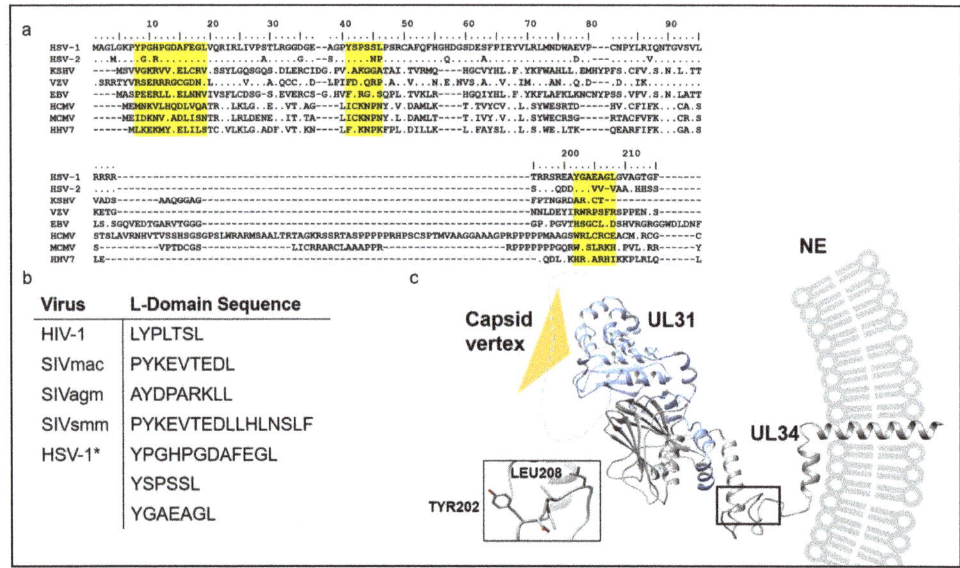

Figure 4. Sequence and structure analysis of HSV-1 UL34 reveals a putative L-Domain-like sequence that may promote its interactions with ALIX. (**a**) Sequence alignment of the HSV-1 NEC complex component UL34 across herpesvirus families. Putative L-Domain residues seen in HSV-1 are highlighted in yellow. (**b**) Sequence comparison between previously identified L-Domains with those identified from the sequence analysis herein (marked with an asterisk). (**c**) Molecular modeling of the membrane-bound NEC complex from HSV-1. The presence of a potential ALIX binding site located on a flexible inter-domain linker is shown boxed and with an inset. This linker region within UL34 contains a tyrosine residue (TYR202) that extends outward and away from the rest of the protein. At the edges of the NEC hexagonal assembly, these flexible loops could be exposed to mediate recruitment of ALIX to sites of nuclear egress.

4. ESCRT Is Recruited to Sites of RNA and DNA-Protein Complex Egress via a Shared Molecular Signature

The parallels between retroviral budding and viral nuclear egress described here point toward a remarkable replication mechanism reached by divergent virus families. These viruses operate via a particle assembly pathway that is topologically equivalent to cellular mechanisms [74–76]. Using the assembly components of either of these viruses as tools to study the role of the ESCRTs in these processes could help address longstanding questions in the field of membrane biology as to why some damaged membranes are reparable and sealable, while others are recycled, both via an ESCRT-dependent mechanism [76,77]. Simultaneously, these studies may also uncover previously overlooked mechanisms of how viruses selectively recruit the ESCRT machinery at very discrete steps to avoid being recycled themselves [12]. More specifically, viral mechanisms of antagonism of regulatory proteins associated with the ESCRTs and their intersection in cellular pathways such as autophagy could be identified [78,79].

One example of a scenario in which the nuclear envelope is recycled rather than repaired is in the membrane damage induced by defects in Lamin A in cases of Progeria [76]. Progeria is a disease caused by the accumulation of a truncated form of Lamin A that severely compromises nuclear architecture [80]. In the analysis of patient samples with clinical manifestations of the disease, it was shown that ALIX was the principal recruitment factor for the autophagy machinery used for the degradation of membrane-damaged areas, rather than for the repair of these areas. This study did not investigate the molecular basis behind ALIX recruitment to these sites, but proposed a mechanism by which excess membrane protein is sensed and marked for degradation via ALIX as part of nuclear envelope quality control surveillance [76]. Lamins readily adopt intermediate filaments as part of the nuclear cytoskeleton [81]. The truncation seen in Progeria occurs in the C-terminal portion of the protein which does not compromise the ability of the pathological Lamin A variant, progerin, from forming filaments, but may alter its ability to associate with DNA [82]. Curiously, lattice-like intranuclear aggregates were observed when Lamin A or a Lamin A D446V mutant from Polish patients with Emery-Dreifuss muscular dystrophy type 2 (EDMD2) were over-expressed [83]. Nuclear blebbing is also a frequently observed phenotype due to progerin expression or when other Lamin proteins are depleted [84]. Further, these nuclear blebs appear to be enriched in euchromatin, despite being transcriptionally inactive [85]. The phenotypes observed in these conditions raises the question of whether or not aberrantly associated genomic nucleic acid is also a component of these defective Lamin A assemblies that are sensed and degraded via the ESCRT pathway. Evidence for this association can be seen in the genomic instability and impaired DNA damage response seen in Progeria patients [86,87]. This quality control process could be another instance in which a nucleic acid-associated protein assembly distorts a membrane compartment which triggers ALIX or ESCRT-III recruitment.

In a phenotypically normal scenario, large RNPs required for neuronal plasticity can be seen budding across the inner nuclear membrane of *Drosophila* cells in a similar manner to HSV-1 capsids [88]. Like HSV-1 particles, these RNA-protein aggregates are too large to passively diffuse through the nuclear pore complex and require the assistance of the ESCRT proteins to bud through the inner nuclear membrane [37]. However, only the downstream ESCRT-III protein Shrub (CHMP4B in humans) has been found to regulate this process [37,89]. Further, the current understanding of these RNPs points to their assembly as granules as opposed to highly organized lattices like that seen by HSV-1 NEC or HIV-1 Gag [90]. Yet these structures represent another example of a membrane bound RNP that triggers ESCRT recruitment. Since ALIX has been shown to recruit Shrub during cell division in *Drosophila* germline cells, it plausible that ALIX could also serve as a major recruitment factor of Shrub for large RNP nuclear egress [91] as well. Like with HSV-1 nuclear egress, more insight is needed before the role of ESCRTs in large RNP nuclear egress is determined with certainty.

5. Conclusions

Despite the tremendous diversity seen across their genomes and morphologies, viruses are bound by the molecular machinery of their cellular hosts. The outcome is the evolution of several independent viral mechanisms that converge on cooption of conserved host mechanisms [92]. One of the most well studied of these pathways hijacked by viruses is the endosomal sorting pathway which is orchestrated by the ESCRT proteins [8,75]. While HIV-1 requires this cellular machinery for the last step in its replication cycle, is proposed to hijack these proteins at a much earlier step in the viral replication cycle stage that is distinct from HIV-1 [37,66]. In parallel, flaviviruses coopt the ESCRTs in yet a third way to sites of replication in the ER [12]. In all three cases, it appears that the role the ESCRTs play is to facilitate the formation and subsequent abscission of a budding neck structure that is essential to the replication of the virus utilizing it [16]. However, the nature by which these assembly sites are established and maintained remains obscure. What has become clear from recent structural studies of HIV-1 and HSV-1 is that the hexagonal lattices formed during particle maturation for both of these viruses bear striking structural and mechanistic similarities [31,40,52,60]. Whether they be in the nucleus as part of the ESCRT-mediated model of nuclear egress for HSV-1 or at the cytoplasmic periphery during the final moments of retroviral budding, these lattices represent fundamental convergence on a mechanism used by viruses for usurpation of a cellular pathway [16,61,77]. These structures operate with no consensus sequence or structure but more likely create a precise molecular signature within membrane compartments that is ideal for ESCRT recruitment. Numerous insights pertinent to health and disease processes are to be gained from studying assemblies such as these so that we may better our understanding of the fundamental cellular pathway of membrane trafficking. Indeed, work is already being done aimed at understanding the damage caused by the aggregation of proteins such as tau in cases of Alzheimer's disease [93]. In an attempt to mitigate the cytotoxicity of the prion-like protein, tau, the ESCRTs, and ALIX, are involved in the recycling of tau via autophagy [93]. However, tau aggregates are capable of breaching this membrane barrier in a manner irreparable by the ESCRT machinery, thereby increasing tau aggregation propensity and worsening disease states [93]. The molecular insights gained from studying the sealing of virus particles could help answer fundamental questions about membrane biology and improve our understanding of why the damage inflicted by tau on the endosome is un-sealable. Furthering our understanding of these mechanisms can potentially lead to the generation of novel therapeutics that promote tau clearance and mitigate disease, or interfere with virus assembly pathways.

Author Contributions: S.J.S. conceptualized nuclear envelope mechanisms. F.B. co-wrote the manuscript. K.M.R., V.M.H.—writing. All authors have read and agreed to the published version of the manuscript.

Funding: This research was funded by the NIH Molecular Biophysics Training Grant GM008295.

Data Availability Statement: All molecular models and data generated during the construction of this review are available upon request.

Conflicts of Interest: The author declares no conflict of interest.

References

1. Hanson, P.I.; Shim, S.; Merrill, S.A. Cell biology of the ESCRT machinery. *Curr. Opin. Cell Biol.* **2009**, *21*, 568–574. [CrossRef] [PubMed]
2. Katzmann, D.J.; Babst, M.; Emr, S.D. Ubiquitin-Dependent Sorting into the Multivesicular Body Pathway Requires the Function of a Conserved Endosomal Protein Sorting Complex, ESCRT-I. *Cell* **2001**, *106*, 145–155. [CrossRef]
3. Hurley, J.H. ESCRT complexes and the biogenesis of multivesicular bodies. *Curr. Opin. Cell Biol.* **2008**, *20*, 4–11. [CrossRef] [PubMed]
4. Raiborg, C.; Stenmark, H. The ESCRT machinery in endosomal sorting of ubiquitylated membrane proteins. *Nature* **2009**, *458*, 445–452. [CrossRef] [PubMed]
5. Garrus, J.E.; von Schwedler, U.K.; Pornillos, O.W.; Morham, S.G.; Zavitz, K.H.; Wang, H.E.; Wettstein, D.A.; Stray, K.M.; Côté, M.; Rich, R.L.; et al. Tsg101 and the vacuolar protein sorting pathway are essential for HIV-1 budding. *Cell* **2001**, *107*, 55–65. [CrossRef]
6. Pashkova, N.; Gakhar, L.; Winistorfer, S.C.; Sunshine, A.B.; Rich, M.; Dunham, M.J.; Yu, L.; Piper, R.C. The yeast Alix homolog Bro1 functions as a ubiquitin receptor for protein sorting into multivesicular endosomes. *Dev. Cell* **2013**, *25*, 520–533. [CrossRef]
7. Teis, D.; Saksena, S.; Emr, S.D. Ordered assembly of the ESCRT-III complex on endosomes is required to sequester cargo during MVB formation. *Dev. Cell* **2008**, *15*, 578–589. [CrossRef] [PubMed]
8. Hurley, J.H. ESCRTs are everywhere. *EMBO J.* **2015**, *34*, 2398–2407. [CrossRef] [PubMed]
9. Vietri, M.; Schink, K.O.; Campsteijn, C.; Wegner, C.S.; Schultz, S.W.; Christ, L.; Thoresen, S.B.; Brech, A.; Raiborg, C.; Stenmark, H. Spastin and ESCRT-III coordinate mitotic spindle disassembly and nuclear envelope sealing. *Nature* **2015**, *522*, 231–235. [CrossRef]
10. Jimenez, A.J.; Maiuri, P.; Lafaurie-Janvore, J.; Divoux, S.; Piel, M.; Perez, F. ESCRT Machinery Is Required for Plasma Membrane Repair. *Science* **2014**, *343*, 1247136. [CrossRef]
11. Elia, N.; Sougrat, R.; Spurlin, T.A.; Hurley, J.H.; Lippincott-Schwartz, J. Dynamics of endosomal sorting complex required for transport (ESCRT) machinery during cytokinesis and its role in abscission. *Proc. Natl. Acad. Sci. USA* **2011**, *108*, 4846–4851. [CrossRef]
12. Inoue, T.; Tsai, B. How viruses use the endoplasmic reticulum for entry, replication, and assembly. *Cold Spring Harb. Perspect. Biol.* **2013**, *5*, a013250. [CrossRef] [PubMed]
13. Fujii, K.; Munshi, U.M.; Ablan, S.D.; Demirov, D.G.; Soheilian, F.; Nagashima, K.; Stephen, A.G.; Fisher, R.J.; Freed, E.O. Functional role of Alix in HIV-1 replication. *Virology* **2009**, *391*, 284–292. [CrossRef]
14. Vincent, O.; Rainbow, L.; Tilburn, J.; Arst, H.N., Jr.; Peñalva, M.A. YPXL/I is a protein interaction motif recognized by aspergillus PalA and its human homologue, AIP1/Alix. *Mol. Cell. Biol.* **2003**, *23*, 1647–1655. [CrossRef] [PubMed]
15. Pornillos, O.; Higginson, D.S.; Stray, K.M.; Fisher, R.D.; Garrus, J.E.; Payne, M.; He, G.P.; Wang, H.E.; Morham, S.G.; Sundquist, W.I. HIV Gag mimics the Tsg101-recruiting activity of the human Hrs protein. *J. Cell Biol.* **2003**, *162*, 425–434. [CrossRef] [PubMed]
16. Votteler, J.; Sundquist, W.I. Virus budding and the ESCRT pathway. *Cell Host Microbe* **2013**, *14*, 232–241. [CrossRef]
17. Sette, P.; Nagashima, K.; Piper, R.C.; Bouamr, F. Ubiquitin conjugation to Gag is essential for ESCRT-mediated HIV-1 budding. *Retrovirology* **2013**, *10*, 79. [CrossRef]
18. Bendjennat, M.; Saffarian, S. The Race against Protease Activation Defines the Role of ESCRTs in HIV Budding. *PLoS Pathog.* **2016**, *12*, e1005657. [CrossRef]

19. Zhai, Q.; Landesman, M.B.; Robinson, H.; Sundquist, W.I.; Hill, C.P. Identification and structural characterization of the ALIX-binding late domains of simian immunodeficiency virus SIVmac239 and SIVagmTan-1. *J. Virol.* **2011**, *85*, 632–637. [CrossRef]
20. Bello, N.F.; Wu, F.; Sette, P.; Dussupt, V.; Hirsch, V.M.; Bouamr, F. Distal leucines are key functional determinants of Alix-binding simian immunodeficiency virus SIV(smE543) and SIV(mac239) type 3 L domains. *J. Virol.* **2011**, *85*, 11532–11537. [CrossRef]
21. Zhai, Q.; Fisher, R.D.; Chung, H.-Y.; Myszka, D.G.; Sundquist, W.I.; Hill, C.P. Structural and functional studies of ALIX interactions with YPXnL late domains of HIV-1 and EIAV. *Nat. Struct. Mol. Biol.* **2008**, *15*, 43–49. [CrossRef] [PubMed]
22. Lee, S.; Joshi, A.; Nagashima, K.; Freed, E.O.; Hurley, J.H. Structural basis for viral late-domain binding to Alix. *Nat. Struct. Mol. Biol.* **2007**, *14*, 194–199. [CrossRef] [PubMed]
23. Mazeaud, C.; Freppel, W.; Chatel-Chaix, L. The Multiples Fates of the Flavivirus RNA Genome During Pathogenesis. *Front. Genet.* **2018**, *9*, 595. [CrossRef] [PubMed]
24. Mahmoudabadi, G.; Phillips, R. A comprehensive and quantitative exploration of thousands of viral genomes. *eLife* **2018**, *7*, e31955. [CrossRef] [PubMed]
25. Zinzula, L.; Tramontano, E. Strategies of highly pathogenic RNA viruses to block dsRNA detection by RIG-I-like receptors: Hide, mask, hit. *Antivir. Res.* **2013**, *100*, 615–635. [CrossRef] [PubMed]
26. Uchida, L.; Espada-Murao, L.A.; Takamatsu, Y.; Okamoto, K.; Hayasaka, D.; Yu, F.; Nabeshima, T.; Buerano, C.C.; Morita, K. The dengue virus conceals double-stranded RNA in the intracellular membrane to escape from an interferon response. *Sci. Rep.* **2014**, *4*, 7395. [CrossRef]
27. Romero-Brey, I.; Bartenschlager, R. Membranous Replication Factories Induced by Plus-Strand RNA Viruses. *Viruses* **2014**, *6*, 2826. [CrossRef] [PubMed]
28. Tabata, K.; Arimoto, M.; Arakawa, M.; Nara, A.; Saito, K.; Omori, H.; Arai, A.; Ishikawa, T.; Konishi, E.; Suzuki, R.; et al. Unique Requirement for ESCRT Factors in Flavivirus Particle Formation on the Endoplasmic Reticulum. *Cell Rep.* **2016**, *16*, 2339–2347. [CrossRef] [PubMed]
29. Chiou, C.T.; Hu, C.A.; Chen, P.H.; Liao, C.L.; Lin, Y.L.; Wang, J.J. Association of Japanese encephalitis virus NS3 protein with microtubules and tumour susceptibility gene 101 (TSG101) protein. *J. Gen. Virol.* **2003**, *84*, 2795–2805. [CrossRef]
30. Mettenleiter, T.C. Herpesvirus Assembly and Egress. *J. Virol.* **2002**, *76*, 1537–1547. [CrossRef]
31. Bigalke, J.M.; Heldwein, E.E. Structural basis of membrane budding by the nuclear egress complex of herpesviruses. *EMBO J.* **2015**, *34*, 2921–2936. [CrossRef]
32. Rixon, F.J. Structure and assembly of herpesviruses. *Semin. Virol.* **1993**, *4*, 135–144. [CrossRef]
33. Whittaker, G.R.; Helenius, A. Nuclear import and export of viruses and virus genomes. *Virology* **1998**, *246*, 1–23. [CrossRef]
34. Klupp, B.G.; Granzow, H.; Mettenleiter, T.C. Primary envelopment of pseudorabies virus at the nuclear membrane requires the UL34 gene product. *J. Virol.* **2000**, *74*, 10063–10073. [CrossRef] [PubMed]
35. Draganova, E.B.; Zhang, J.; Zhou, Z.H.; Heldwein, E.E. Structural basis for capsid recruitment and coat formation during HSV-1 nuclear egress. *eLife* **2020**, *9*, e56627. [CrossRef]
36. Hagen, C.; Dent, K.C.; Zeev-Ben-Mordehai, T.; Grange, M.; Bosse, J.B.; Whittle, C.; Klupp, B.G.; Siebert, C.A.; Vasishtan, D.; Bäuerlein, F.J.B.; et al. Structural Basis of Vesicle Formation at the Inner Nuclear Membrane. *Cell* **2015**, *163*, 1692–1701. [CrossRef]
37. Arii, J.; Watanabe, M.; Maeda, F.; Tokai-Nishizumi, N.; Chihara, T.; Miura, M.; Maruzuru, Y.; Koyanagi, N.; Kato, A.; Kawaguchi, Y. ESCRT-III mediates budding across the inner nuclear membrane and regulates its integrity. *Nat. Commun.* **2018**, *9*, 3379. [CrossRef]
38. Lee, C.-P.; Liu, P.-T.; Kung, H.-N.; Su, M.-T.; Chua, H.-H.; Chang, Y.-H.; Chang, C.-W.; Tsai, C.-H.; Liu, F.-T.; Chen, M.-R. The ESCRT machinery is recruited by the viral BFRF1 protein to the nucleus-associated membrane for the maturation of Epstein-Barr Virus. *PLoS Pathog.* **2012**, *8*, e1002904. [CrossRef]
39. Lee, C.P.; Liu, G.T.; Kung, H.N.; Liu, P.T.; Liao, Y.T.; Chow, L.P.; Chang, L.S.; Chang, Y.H.; Chang, C.W.; Shu, W.C.; et al. The Ubiquitin Ligase Itch and Ubiquitination Regulate BFRF1-Mediated Nuclear Envelope Modification for Epstein-Barr Virus Maturation. *J. Virol.* **2016**, *90*, 8994–9007. [CrossRef]
40. Briggs, J.A.G.; Riches, J.D.; Glass, B.; Bartonova, V.; Zanetti, G.; Kräusslich, H.G. Structure and assembly of immature HIV. *Proc. Natl. Acad. Sci. USA* **2009**, *106*, 11090. [CrossRef] [PubMed]
41. Pak, A.J.; Grime, J.M.A.; Sengupta, P.; Chen, A.K.; Durumeric, A.E.P.; Srivastava, A.; Yeager, M.; Briggs, J.A.G.; Lippincott-Schwartz, J.; Voth, G.A. Immature HIV-1 lattice assembly dynamics are regulated by scaffolding from nucleic acid and the plasma membrane. *Proc. Natl. Acad. Sci. USA* **2017**, *114*, E10056. [CrossRef] [PubMed]
42. Sundquist, W.I.; Kräusslich, H.-G. HIV-1 assembly, budding, and maturation. *Cold Spring Harb. Perspect. Med.* **2012**, *2*, a006924. [CrossRef] [PubMed]
43. Demirov, D.G.; Orenstein, J.M.; Freed, E.O. The late domain of human immunodeficiency virus type 1 p6 promotes virus release in a cell type-dependent manner. *J. Virol.* **2002**, *76*, 105–117. [CrossRef]
44. Flower, T.G.; Takahashi, Y.; Hudait, A.; Rose, K.; Tjahjono, N.; Pak, A.J.; Yokom, A.L.; Liang, X.; Wang, H.-G.; Bouamr, F.; et al. A helical assembly of human ESCRT-I scaffolds reverse-topology membrane scission. *Nat. Struct. Mol. Biol.* **2020**, *27*, 570–580. [CrossRef]
45. Pornillos, O.; Ganser-Pornillos, B.K. Maturation of retroviruses. *Curr. Opin. Virol.* **2019**, *36*, 47–55. [CrossRef] [PubMed]
46. Bleck, M.; Itano, M.S.; Johnson, D.S.; Thomas, V.K.; North, A.J.; Bieniasz, P.D.; Simon, S.M. Temporal and spatial organization of ESCRT protein recruitment during HIV-1 budding. *Proc. Natl. Acad. Sci. USA* **2014**, *111*, 12211. [CrossRef] [PubMed]

47. Ku, P.-I.; Bendjennat, M.; Ballew, J.; Landesman, M.B.; Saffarian, S. ALIX is recruited temporarily into HIV-1 budding sites at the end of gag assembly. *PLoS ONE* **2014**, *9*, e96950. [CrossRef]

48. Gupta, S.; Bendjennat, M.; Saffarian, S. Abrogating ALIX Interactions Results in Stuttering of the ESCRT Machinery. *Viruses* **2020**, *12*, 1032. [CrossRef]

49. Strack, B.; Calistri, A.; Craig, S.; Popova, E.; Göttlinger, H.G. AIP1/ALIX Is a Binding Partner for HIV-1 p6 and EIAV p9 Functioning in Virus Budding. *Cell* **2003**, *114*, 689–699. [CrossRef]

50. Bertin, A.; de Franceschi, N.; de la Mora, E.; Maity, S.; Alqabandi, M.; Miguet, N.; di Cicco, A.; Roos, W.H.; Mangenot, S.; Weissenhorn, W.; et al. Human ESCRT-III polymers assemble on positively curved membranes and induce helical membrane tube formation. *Nat. Commun.* **2020**, *11*, 2663. [CrossRef]

51. Ladinsky, M.S.; Kieffer, C.; Olson, G.; Deruaz, M.; Vrbanac, V.; Tager, A.M.; Kwon, D.S.; Bjorkman, P.J. Electron tomography of HIV-1 infection in gut-associated lymphoid tissue. *PLoS Pathog.* **2014**, *10*, e1003899. [CrossRef] [PubMed]

52. Gui, D.; Gupta, S.; Xu, J.; Zandi, R.; Gill, S.; Huang, I.C.; Rao, A.L.; Mohideen, U. A novel minimal in vitro system for analyzing HIV-1 Gag-mediated budding. *J. Biol. Phys.* **2015**, *41*, 135–149. [CrossRef] [PubMed]

53. Novikova, M.; Adams, L.J.; Fontana, J.; Gres, A.T.; Balasubramaniam, M.; Winkler, D.C.; Kudchodkar, S.B.; Soheilian, F.; Sarafianos, S.G.; Steven, A.C.; et al. Identification of a Structural Element in HIV-1 Gag Required for Virus Particle Assembly and Maturation. *mBio* **2018**, *9*, e01567-18. [CrossRef] [PubMed]

54. Morita, E.; Sandrin, V.; McCullough, J.; Katsuyama, A.; Baci Hamilton, I.; Sundquist, W.I. ESCRT-III protein requirements for HIV-1 budding. *Cell Host Microbe* **2011**, *9*, 235–242. [CrossRef]

55. Booth, A.M.; Fang, Y.; Fallon, J.K.; Yang, J.-M.; Hildreth, J.E.K.; Gould, S.J. Exosomes and HIV Gag bud from endosome-like domains of the T cell plasma membrane. *J. Cell Biol.* **2006**, *172*, 923–935. [CrossRef] [PubMed]

56. Keller, P.W.; Adamson, C.S.; Heymann, J.B.; Freed, E.O.; Steven, A.C. HIV-1 Maturation Inhibitor Bevirimat Stabilizes the Immature Gag Lattice. *J. Virol.* **2011**, *85*, 1420–1428. [CrossRef]

57. Strickland, M.; Ehrlich, L.S.; Watanabe, S.; Khan, M.; Strub, M.-P.; Luan, C.-H.; Powell, M.D.; Leis, J.; Tjandra, N.; Carter, C.A. Tsg101 chaperone function revealed by HIV-1 assembly inhibitors. *Nat. Commun.* **2017**, *8*, 1391. [CrossRef]

58. VerPlank, L.; Bouamr, F.; LaGrassa, T.J.; Agresta, B.; Kikonyogo, A.; Leis, J.; Carter, C.A. Tsg101, a homologue of ubiquitin-conjugating (E2) enzymes, binds the L domain in HIV type 1 Pr55Gag. *Proc. Natl. Acad. Sci. USA* **2001**, *98*, 7724. [CrossRef]

59. Fuchs, W.; Klupp, B.G.; Granzow, H.; Osterrieder, N.; Mettenleiter, T.C. The interacting UL31 and UL34 gene products of pseudorabies virus are involved in egress from the host-cell nucleus and represent components of primary enveloped but not mature virions. *J. Virol.* **2002**, *76*, 364–378. [CrossRef] [PubMed]

60. Bigalke, J.M.; Heuser, T.; Nicastro, D.; Heldwein, E.E. Membrane deformation and scission by the HSV-1 nuclear egress complex. *Nat. Commun.* **2014**, *5*, 4131. [CrossRef]

61. Arii, J.; Takeshima, K.; Maruzuru, Y.; Koyanagi, N.; Kato, A.; Kawaguchi, Y. Roles of the Interhexamer Contact Site for Hexagonal Lattice Formation of the Herpes Simplex Virus 1 Nuclear Egress Complex in Viral Primary Envelopment and Replication. *J. Virol.* **2019**, *93*, e00498-19. [CrossRef]

62. Yang, B.; Liu, X.J.; Yao, Y.; Jiang, X.; Wang, X.Z.; Yang, H.; Sun, J.Y.; Miao, Y.; Wang, W.; Huang, Z.L.; et al. WDR5 Facilitates Human Cytomegalovirus Replication by Promoting Capsid Nuclear Egress. *J. Virol.* **2018**, *92*, e00207-18. [CrossRef]

63. Crump, C.M.; Yates, C.; Minson, T. Herpes Simplex Virus Type 1 Cytoplasmic Envelopment Requires Functional Vps4. *J. Virol.* **2007**, *81*, 7380–7387. [CrossRef]

64. Dussupt, V.; Javid, M.P.; Abou-Jaoudé, G.; Jadwin, J.A.; de La Cruz, J.; Nagashima, K.; Bouamr, F. The nucleocapsid region of HIV-1 Gag cooperates with the PTAP and LYPXnL late domains to recruit the cellular machinery necessary for viral budding. *PLoS Pathog.* **2009**, *5*, e1000339. [CrossRef] [PubMed]

65. Popov, S.; Popova, E.; Inoue, M.; Göttlinger, H.G. Human Immunodeficiency Virus Type 1 Gag Engages the Bro1 Domain of ALIX/AIP1 through the Nucleocapsid. *J. Virol.* **2008**, *82*, 1389–1398. [CrossRef] [PubMed]

66. Pawliczek, T.; Crump, C.M. Herpes Simplex Virus Type 1 Production Requires a Functional ESCRT-III Complex but Is Independent of TSG101 and ALIX Expression. *J. Virol.* **2009**, *83*, 11254–11264. [CrossRef] [PubMed]

67. Loret, S.; Guay, G.; Lippé, R. Comprehensive Characterization of Extracellular Herpes Simplex Virus Type 1 Virions. *J. Virol.* **2008**, *82*, 8605–8618. [CrossRef] [PubMed]

68. Lee, H.H.; Elia, N.; Ghirlando, R.; Lippincott-Schwartz, J.; Hurley, J.H. Midbody targeting of the ESCRT machinery by a noncanonical coiled coil in CEP55. *Science* **2008**, *322*, 576–580. [CrossRef]

69. Webb, B.; Sali, A. Protein structure modeling with MODELLER. *Methods Mol. Biol.* **2014**, *1137*, 1–15. [CrossRef]

70. Kelley, L.A.; Mezulis, S.; Yates, C.M.; Wass, M.N.; Sternberg, M.J.E. The Phyre2 web portal for protein modeling, prediction and analysis. *Nat. Protoc.* **2015**, *10*, 845–858. [CrossRef]

71. Tan, A.; Pak, A.J.; Morado, D.R.; Voth, G.A.; Briggs, J.A.G. Immature HIV-1 assembles from Gag dimers leaving partial hexamers at lattice edges as potential substrates for proteolytic maturation. *Proc. Natl. Acad. Sci. USA* **2021**, *118*, e2020054118. [CrossRef]

72. Gottwein, E.; Bodem, J.; Müller, B.; Schmechel, A.; Zentgraf, H.; Kräusslich, H.-G. The Mason-Pfizer monkey virus PPPY and PSAP motifs both contribute to virus release. *J. Virol.* **2003**, *77*, 9474–9485. [CrossRef]

73. Stange, A.; Mannigel, I.; Peters, K.; Heinkelein, M.; Stanke, N.; Cartellieri, M.; Göttlinger, H.; Rethwilm, A.; Zentgraf, H.; Lindemann, D. Characterization of prototype foamy virus gag late assembly domain motifs and their role in particle egress and infectivity. *J. Virol.* **2005**, *79*, 5466–5476. [CrossRef] [PubMed]

74. Caballe, A.; Martin-Serrano, J. ESCRT machinery and cytokinesis: The road to daughter cell separation. *Traffic* **2011**, *12*, 1318–1326. [CrossRef] [PubMed]

75. Wollert, T.; Yang, D.; Ren, X.; Lee, H.H.; Im, Y.J.; Hurley, J.H. The ESCRT machinery at a glance. *J. Cell Sci.* **2009**, *122*, 2163–2166. [CrossRef]

76. Arii, J.; Maeda, F.; Maruzuru, Y.; Koyanagi, N.; Kato, A.; Mori, Y.; Kawaguchi, Y. ESCRT-III controls nuclear envelope deformation induced by progerin. *Sci. Rep.* **2020**, *10*, 18877. [CrossRef] [PubMed]

77. Skowyra, M.L.; Schlesinger, P.H.; Naismith, T.V.; Hanson, P.I. Triggered recruitment of ESCRT machinery promotes endolysosomal repair. *Science* **2018**, *360*, eaar5078. [CrossRef] [PubMed]

78. Orvedahl, A.; Alexander, D.; Tallóczy, Z.; Sun, Q.; Wei, Y.; Zhang, W.; Burns, D.; Leib, D.A.; Levine, B. HSV-1 ICP34.5 confers neurovirulence by targeting the Beclin 1 autophagy protein. *Cell Host Microbe* **2007**, *1*, 23–35. [CrossRef] [PubMed]

79. Kyei, G.B.; Dinkins, C.; Davis, A.S.; Roberts, E.; Singh, S.B.; Dong, C.; Wu, L.; Kominami, E.; Ueno, T.; Yamamoto, A.; et al. Autophagy pathway intersects with HIV-1 biosynthesis and regulates viral yields in macrophages. *J. Cell Biol.* **2009**, *186*, 255–268. [CrossRef]

80. Goldman, R.D.; Shumaker, D.K.; Erdos, M.R.; Eriksson, M.; Goldman, A.E.; Gordon, L.B.; Gruenbaum, Y.; Khuon, S.; Mendez, M.; Varga, R.; et al. Accumulation of mutant lamin A causes progressive changes in nuclear architecture in Hutchinson-Gilford progeria syndrome. *Proc. Natl. Acad. Sci. USA* **2004**, *101*, 8963–8968. [CrossRef]

81. Dittmer, T.A.; Misteli, T. The lamin protein family. *Genome Biol.* **2011**, *12*, 222. [CrossRef]

82. Stierlé, V.; Couprie, J.; Östlund, C.; Krimm, I.; Zinn-Justin, S.; Hossenlopp, P.; Worman, H.J.; Courvalin, J.-C.; Duband-Goulet, I. The Carboxyl-Terminal Region Common to Lamins A and C Contains a DNA Binding Domain. *Biochemistry* **2003**, *42*, 4819–4828. [CrossRef]

83. Piekarowicz, K.; Machowska, M.; Dratkiewicz, E.; Lorek, D.; Madej-Pilarczyk, A.; Rzepecki, R. The effect of the lamin A and its mutants on nuclear structure, cell proliferation, protein stability, and mobility in embryonic cells. *Chromosoma* **2017**, *126*, 501–517. [CrossRef] [PubMed]

84. Shimi, T.; Butin-Israeli, V.; Adam, S.A.; Goldman, R.D. Nuclear lamins in cell regulation and disease. *Cold Spring Harb. Symp. Quant. Biol.* **2010**, *75*, 525–531. [CrossRef] [PubMed]

85. Shimi, T.; Pfleghaar, K.; Kojima, S.-i.; Pack, C.-G.; Solovei, I.; Goldman, A.E.; Adam, S.A.; Shumaker, D.K.; Kinjo, M.; Cremer, T.; et al. The A- and B-type nuclear lamin networks: Microdomains involved in chromatin organization and transcription. *Genes Dev.* **2008**, *22*, 3409–3421. [CrossRef] [PubMed]

86. Gonzalo, S.; Kreienkamp, R. DNA repair defects and genome instability in Hutchinson-Gilford Progeria Syndrome. *Curr. Opin. Cell Biol.* **2015**, *34*, 75–83. [CrossRef]

87. Noda, A.; Mishima, S.; Hirai, Y.; Hamasaki, K.; Landes, R.D.; Mitani, H.; Haga, K.; Kiyono, T.; Nakamura, N.; Kodama, Y. Progerin, the protein responsible for the Hutchinson-Gilford progeria syndrome, increases the unrepaired DNA damages following exposure to ionizing radiation. *Genes Environ.* **2015**, *37*, 13. [CrossRef]

88. Parchure, A.; Munson, M.; Budnik, V. Getting mRNA-Containing Ribonucleoprotein Granules Out of a Nuclear Back Door. *Neuron* **2017**, *96*, 604–615. [CrossRef]

89. Matias, N.R.; Mathieu, J.; Huynh, J.R. Abscission is regulated by the ESCRT-III protein shrub in Drosophila germline stem cells. *PLoS Genet.* **2015**, *11*, e1004653. [CrossRef]

90. Speese, S.D.; Ashley, J.; Jokhi, V.; Nunnari, J.; Barria, R.; Li, Y.; Ataman, B.; Koon, A.; Chang, Y.T.; Li, Q.; et al. Nuclear envelope budding enables large ribonucleoprotein particle export during synaptic Wnt signaling. *Cell* **2012**, *149*, 832–846. [CrossRef]

91. Eikenes, Å.H.; Malerød, L.; Christensen, A.L.; Steen, C.B.; Mathieu, J.; Nezis, I.P.; Liestøl, K.; Huynh, J.-R.; Stenmark, H.; Haglund, K. ALIX and ESCRT-III Coordinately Control Cytokinetic Abscission during Germline Stem Cell Division In Vivo. *PLoS Genet.* **2015**, *11*, e1004904. [CrossRef] [PubMed]

92. Davey, N.E.; Travé, G.; Gibson, T.J. How viruses hijack cell regulation. *Trends Biochem. Sci.* **2011**, *36*, 159–169. [CrossRef] [PubMed]

93. Chen, J.J.; Nathaniel, D.L.; Raghavan, P.; Nelson, M.; Tian, R.; Tse, E.; Hong, J.Y.; See, S.K.; Mok, S.A.; Hein, M.Y.; et al. Compromised function of the ESCRT pathway promotes endolysosomal escape of tau seeds and propagation of tau aggregation. *J. Biol. Chem.* **2019**, *294*, 18952–18966. [CrossRef] [PubMed]

 *viruses*

Erratum

Erratum: Rose et al. When in Need of an ESCRT: The Nature of Virus Assembly Sites Suggests Mechanistic Parallels between Nuclear Virus Egress and Retroviral Budding. *Viruses* 2021, *13*, 1138

Kevin M. Rose [1,*], Stephanie J. Spada [2], Vanessa M. Hirsch [2] and Fadila Bouamr [2]

[1] Department of Molecular and Cell Biology, California Institute for Quantitative Biosciences, University of California—Berkeley, Berkeley, CA 94720, USA

[2] Laboratory of Molecular Microbiology, National Institute of Allergy and Infectious Diseases, National Institutes of Health, Bethesda, Rockville, MD 20894, USA; stephanie.spada@nih.gov (S.J.S.); vhirsch@niaid.nih.gov (V.M.H.); bouamrf@mail.nih.gov (F.B.)

* Correspondence: kevin_rose@berkeley.edu

Citation: Rose, K.M.; Spada, S.J.; Hirsch, V.M.; Bouamr, F. Erratum: Rose et al. When in Need of an ESCRT: The Nature of Virus Assembly Sites Suggests Mechanistic Parallels between Nuclear Virus Egress and Retroviral Budding. *Viruses* 2021, *13*, 1138. *Viruses* **2021**, *13*, 1705. https://doi.org/10.3390/v13091705

Received: 14 July 2021
Accepted: 27 July 2021
Published: 27 August 2021

Publisher's Note: MDPI stays neutral with regard to jurisdictional claims in published maps and institutional affiliations.

The authors wish to make the following erratum to this paper [1]:
The published version of Figure 1 was missing labels. The correct Figure 1 is listed below:

Figure 1. The respective roles of ALIX and ESCRT-I in the sorting of membranous cargo. Upon internalization, ubiquitylated cargo is detected by ALIX (**left**) and ESCRT-I (**right**) for compartmentalization into intraluminal vesicles that are destined for degradation via the late endosome. Both ALIX and ESCRT-I contain ubiquitin binding domains that facilitate this first step. Unlike ESCRT-I, ALIX possesses an ESCRT-III binding domain that allows for the direct recruitment of ESCRT-III and VPS4, the machinery required for sealing of cargo within intraluminal vesicles and abscising these vesicles from the endosomal membrane. In a similar fashion, the ESCRT-I component TSG101 binds ubiquitylated cargo, while the VPS28 component can recruit ESCRT-III through ESCRT-II which also binds ubiquitylated cargo as well as phospho-inositol lipids.

The list of authors was incorrect and did not accurately reflect the contributions of individuals that contributed to the overall work. It is changed to:

Kevin M. Rose [1,*], Stephanie J. Spada [2], Vanessa M. Hirsch [2] and Fadila Bouamr [2]

[1] Department of Molecular and Cell Biology, California Institute for Quantitative Biosciences, University of California—Berkeley, Berkeley, CA 94720, USA

[2] Laboratory of Molecular Microbiology, National Institute of Allergy and Infectious Diseases, National Institutes of Health, Bethesda, Rockville, MD 20894, USA; stephanie.spada@nih.gov (S.J.S.); vhirsch@niaid.nih.gov (V.M.H.); bouamrf@mail.nih.gov (F.B.)
* Correspondence: kevin_rose@berkeley.edu

The Author Contribution section was changed to: S.J.S. conceptualized nuclear envelope mechanisms. F.B. co-wrote the manuscript. K.M.R., V.M.H.—writing. All authors have read and agreed to the published version of the manuscript.

The authors would like to apologize for any inconvenience caused to the readers by these changes.

Reference

1. Rose, K.M.; Spada, S.J.; Hirsch, V.M.; Bouamr, F. When in Need of an ESCRT: The Nature of Virus Assembly Sites Suggests Mechanistic Parallels between Nuclear Virus Egress and Retroviral Budding. *Viruses* **2021**, *13*, 1138. [CrossRef] [PubMed]

viruses

Review

Herpesvirus Nuclear Egress across the Outer Nuclear Membrane

Richard J. Roller [1] and David C. Johnson [2,*]

[1] Microbiology & Immunology, University of Iowa, Iowa City, IA 52242, USA; richard-roller@uiowa.edu
[2] Molecular Microbiology & Immunology, Oregon Health & Science University, Portland, OR 97239, USA
* Correspondence: johnsoda@ohsu.edu

Abstract: Herpesvirus capsids are assembled in the nucleus and undergo a two-step process to cross the nuclear envelope. Capsids bud into the inner nuclear membrane (INM) aided by the nuclear egress complex (NEC) proteins UL31/34. At that stage of egress, enveloped virions are found for a short time in the perinuclear space. In the second step of nuclear egress, perinuclear enveloped virions (PEVs) fuse with the outer nuclear membrane (ONM) delivering capsids into the cytoplasm. Once in the cytoplasm, capsids undergo re-envelopment in the Golgi/trans-Golgi apparatus producing mature virions. This second step of nuclear egress is known as de-envelopment and is the focus of this review. Compared with herpesvirus envelopment at the INM, much less is known about de-envelopment. We propose a model in which de-envelopment involves two phases: (i) fusion of the PEV membrane with the ONM and (ii) expansion of the fusion pore leading to release of the viral capsid into the cytoplasm. The first phase of de-envelopment, membrane fusion, involves four herpes simplex virus (HSV) proteins: gB, gH/gL, gK and UL20. gB is the viral fusion protein and appears to act to perturb membranes and promote fusion. gH/gL may also have similar properties and appears to be able to act in de-envelopment without gB. gK and UL20 negatively regulate these fusion proteins. In the second phase of de-envelopment (pore expansion and capsid release), an alpha-herpesvirus protein kinase, US3, acts to phosphorylate NEC proteins, which normally produce membrane curvature during envelopment. Phosphorylation of NEC proteins reverses tight membrane curvature, causing expansion of the membrane fusion pore and promoting release of capsids into the cytoplasm.

Keywords: de-envelopment; phosphorylation; nuclear envelopment complex; membrane fusion; hemi-fusion

Citation: Roller, R.J.; Johnson, D.C. Herpesvirus Nuclear Egress across the Outer Nuclear Membrane. *Viruses* **2021**, *13*, 2356. https://doi.org/10.3390/v13122356

Academic Editor: Donald M. Coen

Received: 12 October 2021
Accepted: 16 November 2021
Published: 24 November 2021

Publisher's Note: MDPI stays neutral with regard to jurisdictional claims in published maps and institutional affiliations.

1. Introduction

Herpesviruses construct their nucleocapsids and fill them with DNA in the nucleus. These large virus particles then face the difficult challenge of crossing the nuclear envelope (NE), a structure that is not designed to allow large objects to pass. Decades ago, there was controversy, with one camp suggesting that herpes simplex virus (HSV) acquires an envelope by budding into the inner nuclear membrane (INM) and that these enveloped particles then acquire a second membrane at the outer nuclear membrane (ONM) [1]. These vesicles were thought to then ferry the enveloped virions through the cytoplasm to the plasma membrane. Others suggested that herpesvirus capsids can squeeze through impaired nuclear pores despite the large size of these particles [2]. A third model, originally proposed by Stackpole [3], suggested that herpesviruses cross the NE via a two-step process involving capsid envelopment at the INM followed by de-envelopment at the ONM (see Figure 1). The vast majority of the evidence now supports this third model, often known as envelopment/de-envelopment (reviewed in [4]). In the first step of this pathway, viral proteins disrupt the nuclear lamina then nucleocapsids interact with the INM that is wrapped around the capsids as enveloped particles bud into the space between the INM and ONM, also known as the perinuclear space. This envelopment step involves the well-characterized nuclear envelopment complex (NEC) proteins HSV UL31 and UL34, which have been extensively reviewed [4–7]. Enveloped particles found in the perinuclear space are known as perinuclear virions (PEVs). In the second step of nuclear egress, the envelopes of PEVs fuse with the ONM so that capsids with tegument proteins bound to their surfaces disengage from the ONM and move into the cytoplasm. Later these capsids acquire another membrane by budding into the Golgi or trans-Golgi network followed by delivery to the plasma membrane.

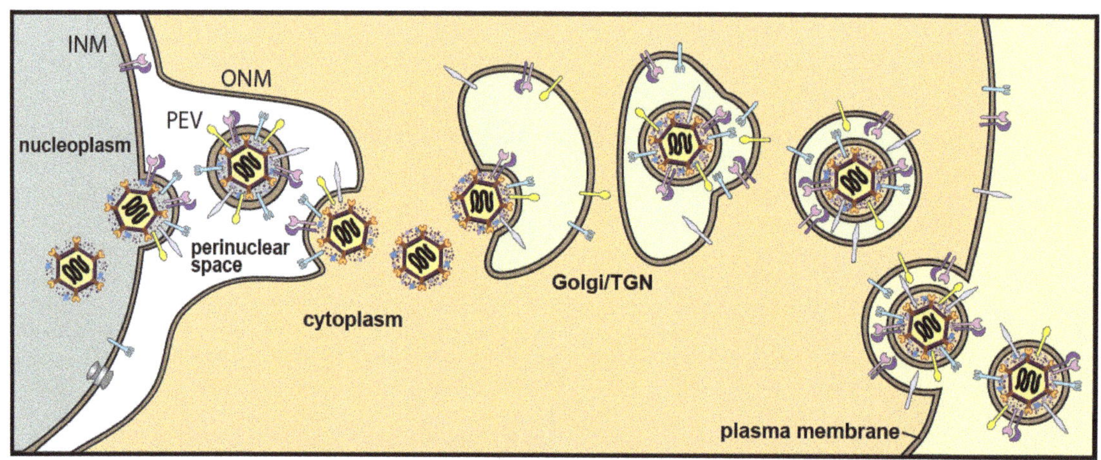

Figure 1. Envelopment/de-envelopment model of herpesvirus egress from cells. Capsids in the nucleus acquire an envelope from the INM by budding into the perinuclear space. The PEV envelope fuses with the ONM delivering capsids into the cytoplasm. These capsids bud into cytoplasmic membranes. Vesicles containing mature virions fuse with the plasma membrane delivering virions to the extracellular space.

Capsid de-envelopment at the ONM consists of at least two phases. The first phase involves fusion of the PEV membrane with the ONM. This process may involve an initial phase in which the lipids in outer leaflet of the two membranes mix to produce hemi-fusion, followed by mixing of the lipids in the inner leaflets of the membranes to produce full membrane fusion and a fusion pore between PEV and the cytoplasm. The second, easily overlooked phase of this process involves expansion of the fusion pore and release of the capsid into the cytoplasm. Small fusion pores between membranes can theoretically resolve either by expansion or closure of the pore [8]. With herpesviruses, we have the example of the NEC proteins that have the capacity to alter the curvature of membranes in order to promote envelopment, the opposite of de-envelopment [4,6]. Reversing this process of curving membranes, which involves removing or altering the NEC present in PEV, might lead to expansion of the fusion pore and capsid release into the cytoplasm.

While there is extensive mechanistic information about the first step of nuclear egress, envelopment, much less is known about the viral machinery that promotes the second step, de-envelopment. In this review, we describe what is known about herpesvirus egress across the ONM, with much of the information from studies of herpes simplex virus (HSV). Given that this is such a fundamentally important and basic process, we assume that other herpesvirus families may also utilize similar mechanisms for this egress. However, we hasten to acknowledge that molecular mechanisms involved in de-envelopment remain sketchy, and thus there is ample conjecture, as well as some theoretical models attempting to explain puzzles represented by certain observations. Most of what is known about de-envelopment comes from studies of HSV and the related alpha-herpesvirus pseudorabies virus (PRV), and our review focuses on these two viruses. It would seem that such a basic process of virus egress should involve similar proteins in beta- and gamma-herpesviruses, although there are major differences between HSV and PRV. Moreover, little to nothing is known about de-envelopment with beta- and gamma-herpesviruses. The review is broken down into two sections. The first focuses on four membrane proteins: gB, gH/gL, gK and UL20, which apparently participate in the membrane fusion phase of de-envelopment. The second section focuses on the second phase of de-envelopment involving the US3 protein, a threonine/serine kinase that phosphorylates NEC proteins so as to potentially reverse effects of NEC proteins and promote fusion pore expansion and capsid entry into the cytoplasm.

2. HSV Membrane Proteins That Promote De-Envelopment

2.1. HSV Glycoproteins gB and gH/gL in Fusion of PEVs with the ONM

A role for herpes simplex virus (HSV) glycoproteins in de-envelopment at the ONM was suggested by the observations that a mutant lacking two glycoproteins, gB and gH, exhibited defects in this step [9]. Enveloped virions accumulated in the perinuclear space or in membrane vesicles containing enveloped virions known as herniations (Figure 2). Loss of both gB and gH in HaCaT human keratinocytes produced large herniations, extensions of the INM extending into the nucleoplasm containing enveloped virions (Figure 2A) [9]. Most of the total enveloped particles in these cells were present as PEVs or in herniations, and there were substantial reductions in cell surface and cytoplasmic particles. In monkey Vero cells, there were fewer herniations and instead more extensive accumulation of enveloped virions in the perinuclear space which was often ballooned (Figure 2B).

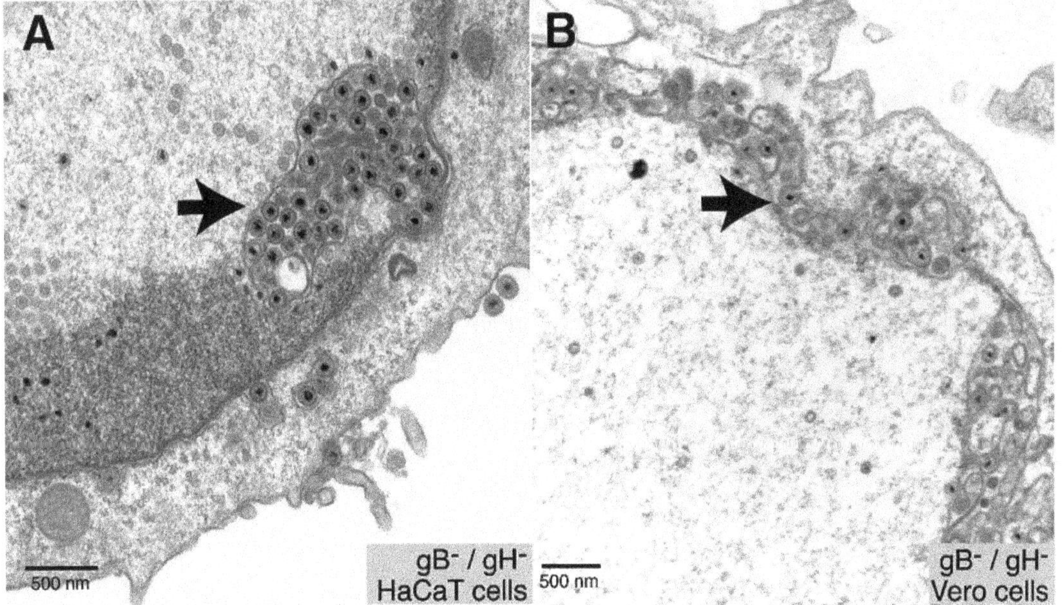

Figure 2. Herniations and PEVs. (**A**) HaCaT keratinocytes infected with an HSV gB⁻/gH⁻ double mutant display herniations (arrow). These vesicle membranes are continuous with the INM. (**B**) Vero cells infected with the HSV gB⁻/gH⁻ mutant display perinuclear enveloped particles (PEVs) (arrow).

Both gH and gB are components of the mature virus particle and are essential for the membrane fusion events that initiate virus infection (reviewed in [10,11]). It should be noted that the gH glycoprotein is found in a complex with a second smaller glycoprotein gL that is not membrane anchored [12,13]. Without gL, HSV gH is not folded normally, does not leave the endoplasmic reticulum, and is not incorporated into the mature virion envelope. Observations with the gB⁻/gH⁻ double mutant suggested that gB and gH/gL act in a manner that bears similarities to their role in virus entry, where these proteins mediate fusion of the virion envelope with host membranes. We describe data in the section below suggesting that both gB and gH/gL can affect lipid mixing and, perhaps, promote membrane fusion. Thus, in the absence of both gH/gL and gB, fusion between the virion envelope and the ONM is blocked, and enveloped particles accumulate in the perinuclear space or back up into the nucleoplasm in herniations. Consistent with this model, both gB and gH/gL were detected in nuclear membranes and in PEVs by immunofluorescence and immunoEM [9]. Others have also shown intense staining of the NE and ER by gB-specific antibodies [14].

Mutations in the gB fusion loops abolished the capacity of gB to function in de-envelopment when expressed in an HSV recombinant unable to express gH [15]. gB fusion loops are thought to play a direct role in membrane fusion by inserting hydrophobic peptide sequences into cellular membranes [16]. HSV recombinants expressing gB with any one of four fusion loop mutations (W174R, W174Y, Y179K and A261D) were unable to enter cells. In addition, all four fusion loop mutants expressed in viruses lacking gH accumulated PEVs, and there were fewer enveloped virions on cell surfaces [15]. These results support the hypothesis that gB functions directly in membrane fusion to promote nuclear egress, rather than in other processes such as binding receptors or other viral proteins. The observations also decreased the likelihood that the original HSV gB⁻/gH⁻ null mutant possessed gross defects in the virion envelope making it unable to cross the NE.

The herniations observed with the HSV gB⁻/gH⁻ mutant appeared similar to those previously observed in cells infected with HSV US3-null mutants [17,18]. HSV US3 is a viral serine/threonine kinase and is reviewed in the second section of this paper. US3 directly phosphorylated the gB cytoplasmic (CT) domain in in vitro assays, as well as in extracts of cells [19,20]. Deletion of gB in the context of a US3-null virus did not add substantially to defects in nuclear egress. The majority of the US3-dependent phosphorylation involved amino acid T887 in the cytoplasmic domain of gB, which is present in a motif similar to that recognized by US3 in other proteins. HSV recombinants expressing a gB substitution T887A or a gB truncated at residue 886 and lacking gH accumulated in the perinuclear space and in herniations [19]. These observations supported the conclusion that phosphorylation of the gB CT domain is important for gB-mediated fusion with the ONM. US3 is incorporated into the tegument layer (between the capsid and envelope) of HSV virions, and this close proximity to the gB CT tail might lead to phosphorylation and triggering of gB⁻ mediated fusion. Alternatively, US3 phosphorylation of gB might control the incorporation of gB into the PEV membrane. Related to this possibility, PEV particles that accumulate with an HSV US3-null mutant were characterized by cryo-electron microscopy and found to contain few glycoprotein spikes [21].

2.2. Caveats and Puzzles Associated with Observations That gB and gH/gL Promote De-Envelopment

The observations that gB and gH/gL participate in de-envelopment bear similarities to the process of virus entry into cells. However, there is a big difference. Models for entry of HSV and essentially all herpesviruses suggest that various forms of gH/gL interact with gB to cause gB-mediated cell fusion (reviewed in [10,22,23]. However, evidence for direct interactions between these two glycoproteins is rather thin for most herpesviruses (see discussion in [24]). HSV virus entry begins with glycoprotein gD binding to cellular receptors, then gD apparently interacts with and alters gH/gL [25]. Following this, gH/gL is said to interact with gB, which is clearly a membrane fusion protein based on its structure. Pre- and post-fusion structures of gB are similar to the pre- and post-fusion forms of the vesicular stomatitis virus G protein, a type III fusion protein that promotes VSV entry [26–28]. HSV mutants lacking any one of gD, gH/gL or gB are unable to enter cells [13,29–31]. In contrast, it was necessary to mutate both gB and gH to substantially reduce nuclear egress. However, it should be noted that the numbers of PEVs observed with a gB⁻ mutant (expressing gH/gL) were about triple those observed with w.t. HSV in HaCaT cells, though much more major defects (30X more PEVs) were observed with gB$^-$/gH$^-$ double mutant [9]. These results suggest that gH/gL, without gB, can mediate some level of de-envelopment.

To explain these differences, there is some evidence that gH/gL might interact directly with membranes. There were reports that an alpha-helical peptide domain of gH (in the region of a.a. 626–644) can interact with membranes and peptides can inhibit cell–cell fusion and virus entry [32–34]. Other studies involving the entire gH/gL protein supported the hypothesis that gH/gL can induce membrane hemi-fusion, which was first described as the first step in influenza virus HA-mediated entry fusion [35]. Hemi-fusion involves mixing of the outer leaflets of two different membranes, but not the mixing of the inner leaflets of these membranes (see Figure 3). Subramanian et al. used assays involving the transfer of the lipid ganglioside GM1, which was present in the outer leaflet of the plasma membrane of Vero cells, into the surface membranes of CHO cells, which do not express GM1 [36]. Expression of gD (to bring membranes close) and gH/gL in Vero cells produced hemi-fusion. However, complete fusion of cells, detected by transfer of soluble GFP from the cytoplasm of Vero cells into CHO cells, was produced by expression of gD, gH/gL and gB. These data produce strong support for the conclusion that gH/gL can interact directly with membranes, perturbing the lipids sufficiently so that the outer leaflets of two opposing membranes mix with one another. This membrane mixing activity of gH/gL may be sufficient to promote de-envelopment at the ONM.

It is tempting to think of de-envelopment fusion as analogous to virus entry and therefore likely to be mediated by fusion proteins embedded in the PEV envelope that are activated in response to interactions with some component of the ONM. This seems most likely given how gB and gH/gL function in virus entry, and these proteins might also promote de-envelopment in this manner. However, it is also possible that the fusion machinery (gB and gH/gL or other proteins) might reside in the ONM and be activated by interaction with some component of the PEV envelope. Current evidence does not distinguish between these two possibilities. Moreover, there is no information on which cellular or viral proteins might trigger the viral fusion machinery.

A second puzzle relates to observations that another herpesvirus does not appear to rely on gB and gH/gL for de-envelopment. All herpesviruses express gB and gH/gL proteins, and thus it might seem likely that all might use these glycoproteins in this fundamental process. Nevertheless, porcine pseudorabies virus (PRV) mutants lacking gB and gH, gB and gD, gD and gH or gH and gL all showed no defects in virus egress [37]. In addition, immunoEM studies involving both monoclonal and polyclonal antibodies failed to detect PRV glycoproteins in the INM or in PEV. These PRV immunoEM studies differ from the immunoEM results from several laboratories that demonstrated HSV glycoproteins gB, gD, gH/gL and gM present in the INM and perinuclear virions [9,38,39], and gD was observed in PEVs purified from infected cells and subjected to Western blotting [40]. Moreover, HSV gB and gH/gL display intense immunofluorescence in nuclear and ER membranes, more intense than that observed in the plasma membrane [9,14]. These differences between HSV and PRV must be considered in light of other major differences in egress pathways of egress for these viruses. For example, PRV mutant lacking gM and gE/gI showed major defects in secondary envelopment, i.e., unenveloped virions accumulated in the cytoplasm [41]. In contrast, an HSV mutant lacking both gM and gE was only marginally compromised in virus replication [42]. Moreover, an HSV mutant lacking both gB and gD was found to be defective in secondary envelopment [43], but that was not the case with a PRV gB$^-$/gD$^-$ double mutant [37]. There are also major differences in how PRV and HSV egress into neuronal axons (reviewed in [44]).

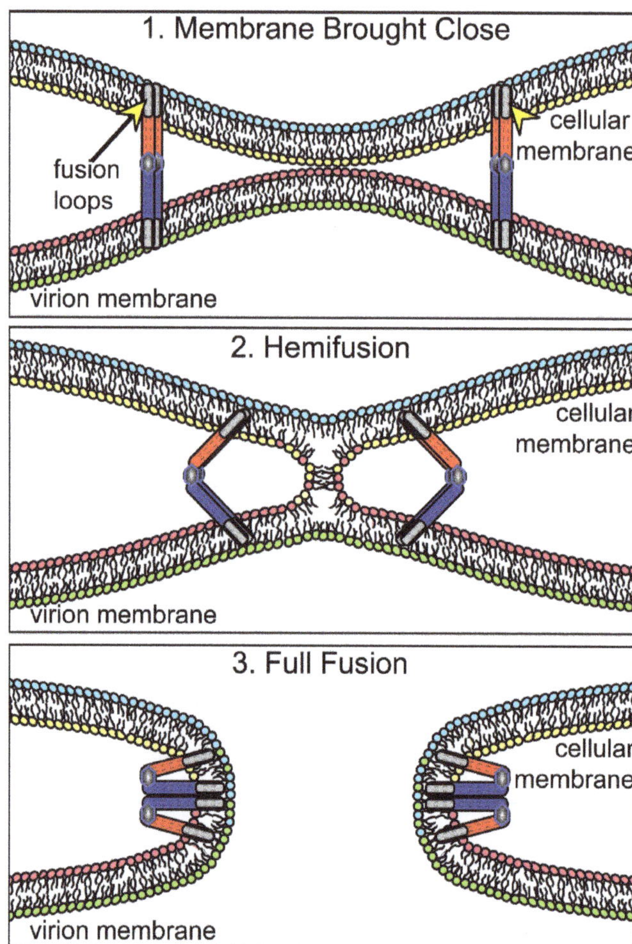

Figure 3. Model of virion envelope fusion with cellular membranes. (**1**) The virion envelope is brought close to a cellular membrane by the action of viral fusion proteins (illustrated to represent any viral fusion protein). Fusion loops in the fusion protein insert into the cellular membrane. (**2**) The viral fusion protein, acting like a hinge, pulls the two membranes close together, promoting hemi-fusion, i.e., mixing of the lipids in the outer leaflets. (**3**) Continued folding back of the fusion proteins produces mixing of lipids in the inner leaflets of the two membranes creating a fusion pore.

2.3. gK and UL20, Other Membrane Proteins That May Regulate De-Envelopment Fusion

HSV-1 gK is a ~40 kDa glycosylated multi-pass membrane protein that is the encoded by the UL53 gene [45]. gK is essential for virus replication, and gK null mutants exhibited defects in secondary envelopment, shown by accumulation of unenveloped or partially enveloped capsids in the cytoplasm and far fewer cell surface virions [46,47]. Point mutations in gK represent, by far, the most frequent genetic mutations that produce the syncytial phenotype in which infected cells fuse extensively in early stages of infection [48–53]. This cell–cell fusion is apparently mediated by gB, the viral fusion protein [54]. It had long been hypothesized that gK might negatively regulate the capacity of gB to fuse membranes [55–58].

The involvement of gK in nuclear egress was indicated by the surprising observations that overexpression of gK in stably transfected cells followed by infection of wild-type HSV resulted in extensive accumulation of enveloped virus particles in the perinuclear space [46]. There were also very few virus particles in the cytoplasm or on cell surfaces in these cells. Consistent with an involvement in nuclear egress, gK displayed subcellular localization consistent with a function in nuclear egress. gK peptide-specific antibodies showed that gK was not on cell surfaces, but accumulated extensively in the ER and NE [56]. Moreover, gK N-linked oligosaccharides remained in the high mannose form, consistent with an inability to be transported to the Golgi apparatus and suggesting that most gK is retained in the ER and NE [56]. Subsequent studies of gK with insertions of a 14 a.a. epitope produced evidence that some gK reached cell surfaces [59]. Unfortunately, these studies did not compare to the anti-gK antibodies used in the earlier studies, and thus it is possible that the insertion of 14 a.a. into gK altered its traffic in cells. In contrast to the overexpression of gK, the loss of gK by deletion caused defects in secondary envelopment [46,47]. Given the hypothesis that HSV gK negatively regulates the capacity of gB to fuse membranes, the observations of marked accumulation of PEVs with overexpression of gK supports a model in which gK reduces gB-mediated membrane fusion between the virus and the ONM. However, it is not clear whether overexpression of gK also impairs gH/gL in the de-envelopment process.

HSV UL20 may be another player in the de-envelopment process. UL20 is a ~24 kDa multi-pass transmembrane protein, but, unlike gK, UL20 is not glycosylated [60,61]. UL20 also extensively accumulated in the NE and ER, with little of the protein present on cell surface membranes [61]. UL20 interacted with gK, and the two proteins affect the traffic of one another in transfected cells [62,63]. Early studies involving an HSV UL20 deletion mutant described marked accumulation of enveloped virions in the perinuclear space [64]. However, this UL20 mutant did not just contain a simple deletion of the entire UL20 open reading frame, but instead the virus expressed a C-terminal fragment of the UL20 protein fused in frame to the UL20.5 protein [65]. The UL20.5 protein also accumulates within nuclear membranes [66]. Subsequently, an HSV deletion mutant lacking all UL20 sequences and without effects on the UL20.5 gene did not exhibit these defects in nuclear egress [65]. Instead, this UL20 mutant showed defects in secondary envelopment, so that cytoplasmic capsids accumulated in the cytoplasm, similar to the defects seen with the HSV gK- null mutant [46]. A PRV UL20 mutant also showed defects in cytoplasmic assembly, accumulating vesicles containing numerous enveloped virions [67]. Therefore, the retention of HSV enveloped virions in the perinuclear space observed with the first HSV UL20 mutant [64] was apparently due to expression of the UL20–UL20.5 fusion protein, not loss of UL20. Nevertheless, the phenotype of this UL20/UL20.5 fusion protein appears similar to that of gK overexpression, causing reduced de-envelopment.

There are also important relationships between gK and UL20 and gB. Syn mutations most frequently affect the UL53 (gK) gene, but mutations altering the gB cytoplasmic domain or the UL20 protein also produce the syncytial phenotype [13,48–53]. Moreover, there was a report that gK and UL20 interact with gB [58]. In addition, coexpression of UL20 and gK blocks cell–cell fusion mediated by transfecting cells with gD, gH/gL and gB [68]. These results, when coupled with observations of gK overexpression and the UL20–UL20.5 fusion protein, support the hypothesis that gK and UL20 negatively regulate gB in de-envelopment fusion.

2.4. Summary of the Roles of HSV Membrane Proteins gB, gH/gL, gK and UL20 in Nuclear Egress

Mutants lacking both gB and gH accumulated enveloped particles in the perinuclear space or herniations. gB proteins with mutations in the fusion loops or in cytoplasmic residues phosphorylated by US3 when combined with a gH-null mutant also displayed these defects. To explain the observations that both gB and gH must be deleted in order to substantially reduce de-envelopment, it appears that both gB and gH/gL may have the capacity to cause lipid mixing. Related to the role of gB in de-envelopment, there are two other membrane proteins gK and UL20 that are thought to interact with gB and might negatively regulate gB in membrane fusion in cells. All three of UL20, gK and gB are mutated in HSV syn mutants. It seems more than coincidence that all three of these proteins are involved in cell–cell fusion and mutant forms of these proteins can alter de-envelopment. It also makes ample sense that herpesviruses construct fusion machinery including gB and gH/gL in order to enter cells and could also use this machinery for the similar process of fusing the virion envelope with the ONM. That said, there is also good evidence that there are other mechanisms by which HSV and other herpesviruses cross the ONM (see Section 3 below).

3. US3-Mediated Phosphorylation of NEC Proteins in De-Envelopment

The Role of US3 in De-Envelopment

The first indication that alpha-herpesvirus proteins promote de-envelopment involved a PRV mutant lacking that US3 gene that accumulated PEVs [69]. These enveloped virions were not distributed throughout the perinuclear space but were found in herniations of the inner nuclear membrane [17,69,70], similar to the herniations observed later with gB⁻/gH⁻ double mutants [9] (see examples in Figure 2). US3 mutations in other, distantly related alpha-herpesviruses also produced similar defects in de-envelopment [71–74]. Therefore, pUS3 is important for efficient de-envelopment of alpha-herpesvirus capsids at the ONM.

US3 is a serine/threonine protein kinase and is conserved among the alpha-herpesviruses, but not in beta- and gamma-herpesviruses. It has been reported to phosphorylate numerous viral and cellular substrates in the infected cell and to regulate infected cell properties as diverse as protection from apoptosis, promotion of translation and inhibition of antigen presentation (reviewed in [72]). The de-envelopment function of HSV pUS3 requires its kinase activity, since point mutations that ablate kinase activity have the same de-envelopment phenotype as a US3 deletion [17]. This suggests that pUS3 functions in de-envelopment by phosphorylation of other protein substrates. US3 is a structural component of the PEVs, but whether its de-envelopment function requires incorporation into the virion is not known [75].

The critical question about US3 function in de-envelopment is the identity of the relevant protein substrate(s). The available data are consistent with two non-exclusive hypotheses. The first hypothesis involves US3 phosphorylation of gB, and perhaps gH/gL, so that this machinery, which causes fusion of the PEV and ONM membranes, is given increased fusogenic activity by virtue of phosphorylation. Evidence of this hypothesis is summarized above.

The second hypothesis is that pUS3 phosphorylation regulates the reversal of the tight membrane curvature brought about by the presence of the NEC (UL31/UL34 proteins), so that there is an expanded fusion pore and release of unenveloped capsids into the cytoplasm (see Figure 4). Evidence in favor of this second hypothesis comes from the observation that mutation of pUS3 phosphorylation sites on one of the NEC proteins produces a de-envelopment defect that is similar to that seen with US3 mutants [18]. HSV-1 US3 phosphorylates both UL34 and UL31 [17,18,76], but phosphorylation of UL34 is not conserved in PRV [70]. HSV-1 UL34 is phosphorylated in its flexible C-terminal stalk and UL31 in its N-terminal domain [17,18,76]. The function of UL34 phosphorylation in HSV-1 is unclear, since mutation of the phosphorylation site has no effect on virus replication or nuclear egress [17]. In contrast, phosphorylation of UL31 at US3-specific phosphorylation sites is required for efficient virus replication and nuclear egress. Replacement of UL31 phosphorylated residues with alanines resulted in a de-envelopment defect and diminished single-step replication similar to that seen for US3 mutant viruses [18]. Replacement of the phosphorylated residues with the phosphomimetic glutamic acid resulted in a replication defect, and a nuclear egress defect that precedes de-envelopment, i.e., capsids accumulated in the nucleus and accumulation of PEVs, was not observed. These observations suggest that UL31 is a critical substrate for the de-envelopment function of US3. UL31 might participate directly in de-envelopment in a manner regulated by pUS3 or, alternatively, phosphorylated UL31 might recruit some other de-envelopment factor. All alpha-herpesviruses encode UL31 homologs with serine or threonine residues in the context of basic residues, motifs that are similar or identical to HSV and PRV US3 phosphorylation motifs.

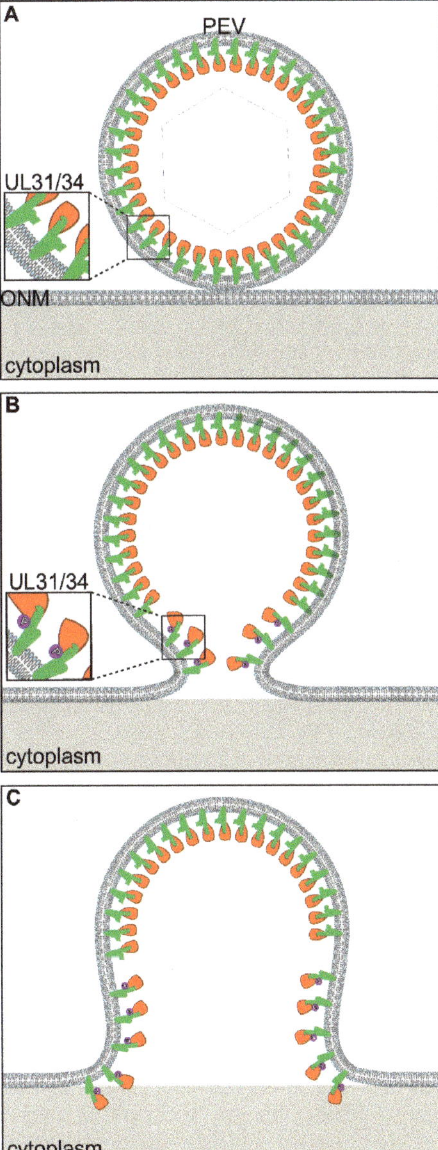

Figure 4. Model for fusion pore expansion by phosphorylation of the NEC. (**A**) Membrane curvature in the PEV is induced and maintained by interactions between NEC heterodimers embedded in the lipid bilayer. (**B**) Upon opening of a fusion pore, pore expansion can be driven by US3-mediated phosphorylation of UL31 N-terminal domain near the membrane proximal region of the heterodimer. (**C**) UL31/UL34 continues to be altered in conformation and moves into the cellular membrane further relaxing membrane curvature and allowing expansion of the fusion pore.

One mechanism by which US3-mediated phosphorylation of UL31 might promote de-envelopment involves altering the capacity of the NEC to induce and maintain membrane curvature. By releasing the UL31/34 imposed curvature of the virion envelope that is imposed at the INM, the fusion pore may expand during de-envelopment and allow release of the capsid into the cytoplasm. To understand this better it is worth reviewing how UL31 and UL34 mediate membrane curvature during envelope at the INM. UL31 and UL34 expressed in transfected cells (without other viral proteins) promote vesicle formation in the perinuclear space [77]. The crystal structures of UL31/UL34 heterodimers suggest assembly into hexameric arrays similar to those observed in vesicles formed by UL31 and UL34 in vitro [78–82]. Mutations in UL31 and UL34 that disrupt hexamer formation depress membrane budding in vitro and produce a budding defect in infected cells [78]. This suggests that association of UL31/UL34 heterodimers into hexameric arrays is important for membrane budding, perhaps by inducing lipid ordering that favors curvature [83]. Evidence that pUS3 regulates the association state of UL31/UL34 heterodimers comes from observations that US3 mutants form NEC aggregates in the absence of capsid budding and that these aggregates correspond to deformations of the nuclear membrane and areas of tight membrane curvature [84]. Thus, the NEC can produce tight curvature of the INM, and this is regulated by pUS3. Our model illustrated in Figure 4 would suggest that reversal of the NEC-imposed curvature at the ONM during de-envelopment is necessary to permit expansion of the fusion pore so that the capsid can be released into the cytoplasm. Without reversal of NEC-imposed curvature, the fusion pore may close, regenerating a fully enveloped PEV. US3 would function in this model by phosphorylating UL31 to alter the structure of UL31/34, thereby reversing membrane curvature.

The HSV UL21 protein also plays a significant role in de-envelopment as shown by the accumulation of PEVs in cells infected with a UL21 deletion virus [85]. Indeed, the PEV accumulations appear quite similar to those seen in US3 mutant-infected cells, and mutation of UL21 also causes formation of punctate NEC aggregates in the nuclear envelope similar to those seen with US3 mutants [85,86]. These observations suggest that UL21 and US3 are on the same pathway for promoting de-envelopment or have overlapping functions. The situation, however, is complex. UL21 functions as a phosphatase adapter, recruiting protein phosphatase 1 (PP1) by way of a sequence motif that is conserved across the alpha-herpesviruses [86]. Deletion of UL21 or mutation of its phosphatase adapter motif results in increased phosphorylation of some US3 substrates, including UL31 [86]. Furthermore, mutation of the UL21 PP1 adapter motif results in a virus replication defect that can be partially suppressed by mutations in US3 that may affect its catalytic activity [86]. These results suggest that, in some circumstances, US3 and UL21 activities are opposed to one another. How these results might be reconciled is not clear at this point, but it is possible that UL21 may function differently in the envelopment and de-envelopment steps of nuclear egress. UL21 associates with intranuclear capsids [87,88], and Benedyk et al. have suggested that capsid-associated UL21 might locally promote dephosphorylation of UL31 and thereby promote NEC assembly and virion budding into the perinuclear space [86]. US3 phosphorylation of UL31 in areas where capsids are not docked might prevent NEC self-association and inhibit capsid-less budding events. At the de-envelopment step, UL21 and US3 might cooperate in a way that may not involve the phosphatase adapter activity of UL21. However, these results again highlight the important role of phosphorylation and, perhaps, de-phosphorylation in regulating de-envelopment.

4. Models for HSV De-Envelopment

The observations reviewed here suggest a consensus model for HSV de-envelopment in which gB and gH/gL can each independently promote the membrane fusion phase of de-envelopment and pUS3 both regulates the de-envelopment activity of gB and promotes the fusion pore expansion/capsid release phase. The two multi-span proteins gK and UL20 may also contribute to regulating gB, and perhaps gH/gL, in this process. It is important to note, however, that neither loss of both gB and gH nor loss of US3 completely blocks

HSV nuclear egress; there were still significant quantities of cell surfaces virions [17,19]. With respect to gB and gH/gL, in Vero cells, wild-type HSV produced 85% cell surface and 5% PEVs, while the gB$^-$/gH$^-$ mutant produced 20% cell surface and 59% PEVs. Thus, there were approximately 4–5 fold reductions in cell surface particles and 10 fold increases in PEVs. With loss of US3 there were defects in growth that are far more modest than those associated with loss of either UL31 or UL34. This suggests that nuclear egress proceeds relatively efficiently even without US3 function [17,70,89–93]. Indeed, the fraction of the US3 deletion growth defect that is due to its function in nuclear egress is small, since adding a US3 inactivating mutation onto an NEC deletion gives a further growth defect similar in magnitude to the US3 mutation on a wild-type background [84]. Therefore, there appear to be viral or cellular proteins, in addition to US3 and gB and gH/gL, that promote de-envelopment. These other viral or cellular proteins might also account for the de-envelopment observed with the PRV gB$^-$/gH$^-$ double mutant [37].

EM images of cells infected with wild-type HSV show relatively few PEVs; instead, the majority of enveloped particles accumulate on cell surfaces. Thus, the de-envelopment process is rapid. However, in US3$^-$ or gB$^-$/gH$^-$ mutant virus-infected cells, there is substantial accumulation of PEVs over time, even though some virus reaches the cytoplasm and cell surfaces. This delay continues until long after infection so that more and more PEVs accumulate and herniations appear to represent a major log jam, as PEVs back up into the nucleoplasm. Our model to explain these observations suggests that the other putative viral or cellular de-envelopment factors can function at earlier stages of virus replication, but these factors are overwhelmed by the large numbers of HSV particles produced in the nucleus at late stages of infection. By this model HSV gB, gH/gL and US3 expedite the de-envelopment process, for which other viral and cellular factors are insufficient.

As far as we are aware, there have not been descriptions of cellular machinery that act to fuse NE membranes. Fusion of the inner and outer nuclear membranes appears not to be required for dissolution of the NE during mitosis. Rather, nuclear membranes are subsumed into the ER, which then vesiculates (reviewed in [94]). The formation of the de-envelopment fusion pore is topologically analogous to the formation of fenestrations in sheets of endoplasmic reticulum and to the creation of nuclear pores in the interphase nucleus, but the mechanism by which these fenestrations and pores are formed is currently unknown [94,95]. Intracellular transport vesicles fuse with acceptor molecules such as the plasma membrane, but the machinery involved, including Rab and SNARE proteins, are on the wrong face of the membrane to apply to virus particles or the ONM [96]. For fusion pore expansion and capsid release, it should be noted that the consensus sequence motif for pUS3 phosphorylation, RRRXS/T, overlaps that of cellular kinases including protein kinase A, Akt and protein kinase C [89,97–99]. Thus, it is possible that the US3-mediated phosphorylation can be assumed by cellular kinases, which are lacking in quantities or are less efficient. A similar picture may be proposed for gB and gH/gL, and other viral or cellular proteins may promote de-envelopment fusion.

Funding: D.C.J. was supported by NIH grants R01 EY018755 and R21 EY029082. R.R was supported by NIH grants R21 AI148831 and R21 AI153683.

Institutional Review Board Statement: Not applicable.

Informed Consent Statement: Not applicable.

Acknowledgments: We are most grateful to Tiffani Howard for the illustration in Figure 1 and to Grayson DuRaine for the illustrations in Figures 2–4.

Conflicts of Interest: The authors declare no conflict of interest.

References

1. Campadelli-Fiume, G.; Farabegoli, F.; Di Gaeta, S.; Roizman, B. Origin of unenveloped capsids in the cytoplasm of cells infected with herpes simplex virus 1. *J. Virol.* **1991**, *65*, 1589–1595. [CrossRef]
2. Wild, P.; Engels, M.; Senn, C.; Tobler, K.; Ziegler, U.; Schraner, E.M.; Loepfe, E.; Ackermann, M.; Mueller, M.; Walther, P. Impairment of nuclear pores in bovine herpesvirus 1-infected MDBK cells. *J. Virol.* **2005**, *79*, 1071–1083. [CrossRef]
3. Stackpole, C.W. Herpes-type virus of the frog renal adenocarcinoma. I. Virus development in tumor transplants maintained at low temperature. *J. Virol.* **1969**, *4*, 75–93. [CrossRef]
4. Draganova, E.B.; Thorsen, M.K.; Heldwein, E.E. Nuclear Egress. *Curr. Issues Mol. Biol.* **2021**, *41*, 125–170. [CrossRef] [PubMed]
5. Arii, J. Host and Viral Factors Involved in Nuclear Egress of Herpes Simplex Virus 1. *Viruses* **2021**, *13*, 754. [CrossRef] [PubMed]
6. Draganova, E.B.; Zhang, J.; Zhou, Z.H.; Heldwein, E.E. Structural basis for capsid recruitment and coat formation during HSV-1 nuclear egress. *Elife* **2020**, *9*, e56627. [CrossRef] [PubMed]
7. Crump, C. Virus Assembly and Egress of HSV. *Adv. Exp. Med. Biol.* **2018**, *1045*, 23–44. [CrossRef] [PubMed]
8. Akimov, S.A.; Volynsky, P.E.; Galimzyanov, T.R.; Kuzmin, P.I.; Pavlov, K.V.; Batishchev, O.V. Pore formation in lipid membrane I: Continuous reversible trajectory from intact bilayer through hydrophobic defect to transversal pore. *Sci. Rep.* **2017**, *7*, 12152. [CrossRef]
9. Farnsworth, A.; Wisner, T.W.; Webb, M.; Roller, R.; Cohen, G.; Eisenberg, R.; Johnson, D.C. Herpes simplex virus glycoproteins gB and gH function in fusion between the virion envelope and the outer nuclear membrane. *Proc. Natl. Acad. Sci. USA* **2007**, *104*, 10187–10192. [CrossRef] [PubMed]
10. Connolly, S.A.; Jardetzky, T.S.; Longnecker, R. The structural basis of herpesvirus entry. *Nat. Rev. Microbiol.* **2021**, *19*, 110–121. [CrossRef]
11. Vollmer, B.; Grunewald, K. Herpesvirus membrane fusion—A team effort. *Curr. Opin. Struct. Biol.* **2020**, *62*, 112–120. [CrossRef]
12. Hutchinson, L.; Browne, H.; Wargent, V.; Davis-Poynter, N.; Primorac, S.; Goldsmith, K.; Minson, A.C.; Johnson, D.C. A novel herpes simplex virus glycoprotein, gL, forms a complex with glycoprotein H (gH) and affects normal folding and surface expression of gH. *J. Virol.* **1992**, *66*, 2240–2250. [CrossRef]
13. Roop, C.; Hutchinson, L.; Johnson, D.C. A mutant herpes simplex virus type 1 unable to express glycoprotein L cannot enter cells, and its particles lack glycoprotein H. *J. Virol.* **1993**, *67*, 2285–2297. [CrossRef] [PubMed]
14. Imai, T.; Sagou, K.; Arii, J.; Kawaguchi, Y. Effects of phosphorylation of herpes simplex virus 1 envelope glycoprotein B by Us3 kinase in vivo and in vitro. *J. Virol.* **2010**, *84*, 153–162. [CrossRef]
15. Wright, C.C.; Wisner, T.W.; Hannah, B.P.; Eisenberg, R.J.; Cohen, G.H.; Johnson, D.C. Fusion between perinuclear virions and the outer nuclear membrane requires the fusogenic activity of herpes simplex virus gB. *J. Virol.* **2009**, *83*, 11847–11856. [CrossRef] [PubMed]
16. Hannah, B.P.; Heldwein, E.E.; Bender, F.C.; Cohen, G.H.; Eisenberg, R.J. Mutational evidence of internal fusion loops in herpes simplex virus glycoprotein B. *J. Virol.* **2007**, *81*, 4858–4865. [CrossRef]
17. Ryckman, B.J.; Roller, R.J. Herpes simplex virus type 1 primary envelopment: UL34 protein modification and the US3-UL34 catalytic relationship. *J. Virol.* **2004**, *78*, 399–412. [CrossRef]
18. Mou, F.; Wills, E.; Baines, J.D. Phosphorylation of the U(L)31 protein of herpes simplex virus 1 by the U(S)3-encoded kinase regulates localization of the nuclear envelopment complex and egress of nucleocapsids. *J. Virol.* **2009**, *83*, 5181–5191. [CrossRef] [PubMed]
19. Wisner, T.; Wright, C.; Kato, A.; Kawaguchi, Y.; Mou, F.; Baines, J.; Roller, R.; Johnson, D. Herpesvirus gB-induced fusion between the virion envelope and outer nuclear membrane during virus egress is regulated by the viral US3 kinase. *J. Virol.* **2009**, *83*, 3115–3126. [CrossRef]
20. Kato, A.; Arri, J.; Shiratori, I.; Akashi, H.; Arase, H.; Kawaguchi, Y. Herpes simplex virus 1 protein kinease US3 phosphorylates viral envelope glycoprotein B and regulates its expression on the cell surface. *J. Virol.* **2009**, *83*, 250–261. [CrossRef]
21. Newcomb, W.W.; Fontana, J.; Winkler, D.C.; Cheng, N.; Heymann, J.B.; Steven, A.C. The Primary Enveloped Virion of Herpes Simplex Virus 1: Its Role in Nuclear Egress. *mBio* **2017**, *8*, e00825-17. [CrossRef] [PubMed]
22. Nishimura, M.; Mori, Y. Entry of betaherpesviruses. *Adv. Virus Res.* **2019**, *104*, 283–312. [CrossRef] [PubMed]
23. Cairns, T.M.; Connolly, S.A. Entry of Alphaherpesviruses. *Curr. Issues Mol. Biol.* **2021**, *41*, 63–124. [CrossRef] [PubMed]
24. Vanarsdall, A.; Howard, P.; Wisner, T.; Johnson, D. Human Cytomegalovirus gH/gL Forms a Stable Complex with the Fusion Protein gB in Virions. *PLoS Pathog.* **2016**, *12*, e1005564. [CrossRef]
25. Cairns, T.M.; Atanasiu, D.; Saw, W.T.; Lou, H.; Whitbeck, J.C.; Ditto, N.T.; Bruun, B.; Browne, H.; Bennett, L.; Wu, C.; et al. Localization of the Interaction Site of Herpes Simplex Virus Glycoprotein D (gD) on the Membrane Fusion Regulator, gH/gL. *J. Virol.* **2020**, *94*, e00983-20. [CrossRef]
26. Heldwein, E.; Lou, H.; Bender, F.; Cohen, G.; Eisenberg, R.; Harrison, S. Crystal structure of glycoprotein B from herpes simplex virus 1. *Science* **2006**, *313*, 217–220. [CrossRef]
27. Vollmer, B.; Prazak, V.; Vasishtan, D.; Jefferys, E.E.; Hernandez-Duran, A.; Vallbracht, M.; Klupp, B.G.; Mettenleiter, T.C.; Backovic, M.; Rey, F.A.; et al. The prefusion structure of herpes simplex virus glycoprotein B. *Sci. Adv.* **2020**, *6*, eabc1726. [CrossRef] [PubMed]
28. Liu, Y.; Heim, K.P.; Che, Y.; Chi, X.; Qiu, X.; Han, S.; Dormitzer, P.R.; Yang, X. Prefusion structure of human cytomegalovirus glycoprotein B and structural basis for membrane fusion. *Sci. Adv.* **2021**, *7*, eabf3718. [CrossRef]

29. Cai, W.H.; Gu, B.; Person, S. Role of glycoprotein B of herpes simplex virus type 1 in viral entry and cell fusion. *J. Virol.* **1988**, *62*, 2596–2604. [CrossRef] [PubMed]

30. Ligas, M.W.; Johnson, D.C. A herpes simplex virus mutant in which glycoprotein D sequences are replaced by beta-galactosidase sequences binds to but is unable to penetrate into cells. *J. Virol.* **1988**, *62*, 1486–1494. [CrossRef]

31. Forrester, A.; Farrell, H.; Wilkinson, G.; Kaye, J.; Davis-Poynter, N.; Minson, T. Construction and properties of a mutant of herpes simplex virus type 1 with glycoprotein H coding sequences deleted. *J. Virol.* **1992**, *66*, 341–348. [CrossRef]

32. Galdiero, S.; Vitiello, M.; D'Isanto, M.; Falanga, A.; Collins, C.; Raieta, K.; Pedone, C.; Browne, H.; Galdiero, M. Analysis of synthetic peptides from heptad-repeat domains of herpes simplex virus type 1 glycoproteins H and B. *J. Gen. Virol.* **2006**, *87*, 1085–1097. [CrossRef] [PubMed]

33. Galdiero, S.; Falanga, A.; Vitiello, M.; Raiola, L.; Fattorusso, R.; Browne, H.; Pedone, C.; Isernia, C.; Galdiero, M. Analysis of a membrane interacting region of herpes simplex virus type 1 glycoprotein H. *J. Biol. Chem.* **2008**, *283*, 29993–30009. [CrossRef]

34. Galdiero, S.; Falanga, A.; Vitiello, M.; D'Isanto, M.; Cantisani, M.; Kampanaraki, A.; Benedetti, E.; Browne, H.; Galdiero, M. Peptides containing membrane-interacting motifs inhibit herpes simplex virus type 1 infectivity. *Peptides* **2008**, *29*, 1461–1471. [CrossRef] [PubMed]

35. Kemble, G.W.; Danieli, T.; White, J.M. Lipid-anchored influenza hemagglutinin promotes hemifusion, not complete fusion. *Cell* **1994**, *76*, 383–391. [CrossRef]

36. Subramanian, R.P.; Geraghty, R.J. Herpes simplex virus type 1 mediates fusion through a hemifusion intermediate by sequential activity of glycoproteins D, H, L, and B. *Proc. Natl. Acad. Sci. USA* **2007**, *104*, 2903–2908. [CrossRef]

37. Klupp, B.; Altenschmidt, J.; Granzow, H.; Fuchs, W.; Mettenleiter, T.C. Glycoproteins required for entry are not necessary for egress of pseudorabies virus. *J. Virol.* **2008**, *82*, 6299–6309. [CrossRef] [PubMed]

38. Torrisi, M.R.; Di Lazzaro, C.; Pavan, A.; Pereira, L.; Campadelli-Fiume, G. Herpes simplex virus envelopment and maturation studied by fracture label. *J. Virol.* **1992**, *66*, 554–561. [CrossRef]

39. Baines, J.D.; Wills, E.; Jacob, R.J.; Pennington, J.; Roizman, B. Glycoprotein M of herpes simplex virus 1 is incorporated into virions during budding at the inner nuclear membrane. *J. Virol.* **2007**, *81*, 800–812. [CrossRef]

40. Padula, M.E.; Sydnor, M.L.; Wilson, D.W. Isolation and preliminary characterization of herpes simplex virus 1 primary enveloped virions from the perinuclear space. *J. Virol.* **2009**, *83*, 4757–4765. [CrossRef]

41. Brack, A.R.; Dijkstra, J.M.; Granzow, H.; Klupp, B.G.; Mettenleiter, T.C. Inhibition of virion maturation by simultaneous deletion of glycoproteins E, I, and M of pseudorabies virus. *J. Virol.* **1999**, *73*, 5364–5372. [CrossRef]

42. Browne, H.; Bell, S.; Minson, T. Analysis of the requirement for glycoprotein m in herpes simplex virus type 1 morphogenesis. *J. Virol.* **2004**, *78*, 1039–1041. [CrossRef]

43. Johnson, D.C.; Wisner, T.W.; Wright, C.C. Herpes simplex virus glycoproteins gB and gD function in a redundant fashion to promote secondary envelopment. *J. Virol.* **2011**, *85*, 4910–4926. [CrossRef] [PubMed]

44. DuRaine, G.; Johnson, D.C. Anterograde transport of alpha-herpesviruses in neuronal axons. *Virology* **2021**, *559*, 65–73. [CrossRef] [PubMed]

45. Hutchinson, L.; Goldsmith, K.; Snoddy, D.; Ghosh, H.; Graham, F.L.; Johnson, D.C. Identification and characterization of a novel herpes simplex virus glycoprotein, gK, involved in cell fusion. *J. Virol.* **1992**, *66*, 5603–5609. [CrossRef]

46. Hutchinson, L.; Johnson, D.C. Herpes simplex virus glycoprotein K promotes egress of virus particles. *J. Virol.* **1995**, *69*, 5401–5413. [CrossRef] [PubMed]

47. Jayachandra, S.; Baghian, A.; Kousoulas, K.G. Herpes simplex virus type 1 glycoprotein K is not essential for infectious virus production in actively replicating cells but is required for efficient envelopment and translocation of infectious virions from the cytoplasm to the extracellular space. *J. Virol.* **1997**, *71*, 5012–5024. [CrossRef] [PubMed]

48. Ruyechan, W.T.; Morse, L.S.; Knipe, D.M.; Roizman, B. Molecular genetics of herpes simplex virus. II. Mapping of the major viral glycoproteins and of the genetic loci specifying the social behavior of infected cells. *J. Virol.* **1979**, *29*, 677–697. [CrossRef]

49. Read, G.S.; Person, S.; Keller, P.M. Genetic studies of cell fusion induced by herpes simplex virus type 1. *J. Virol.* **1980**, *35*, 105–113. [CrossRef]

50. Bond, V.C.; Person, S. Fine structure physical map locations of alterations that affect cell fusion in herpes simplex virus type 1. *Virology* **1984**, *132*, 368–376. [CrossRef]

51. Little, S.P.; Schaffer, P.A. Expression of the syncytial (syn) phenotype in HSV-1, strain KOS: Genetic and phenotypic studies of mutants in two syn loci. *Virology* **1981**, *112*, 686–702. [CrossRef]

52. Dolter, K.E.; Ramaswamy, R.; Holland, T.C. Syncytial mutations in the herpes simplex virus type 1 gK (UL53) gene occur in two distinct domains. *J. Virol.* **1994**, *68*, 8277–8281. [CrossRef] [PubMed]

53. Pertel, P.E.; Spear, P.G. Modified entry and syncytium formation by herpes simplex virus type 1 mutants selected for resistance to heparin inhibition. *Virology* **1996**, *226*, 22–33. [CrossRef]

54. Turner, A.; Bruun, B.; Minson, T.; Browne, H. Glycoproteins gB, gD, and gHgL of herpes simplex virus type 1 are necessary and sufficient to mediate membrane fusion in a Cos cell transfection system. *J. Virol.* **1998**, *72*, 873–875. [CrossRef] [PubMed]

55. Hutchinson, L.; Graham, F.L.; Cai, W.; Debroy, C.; Person, S.; Johnson, D.C. Herpes simplex virus (HSV) glycoproteins B and K inhibit cell fusion induced by HSV syncytial mutants. *Virology* **1993**, *196*, 514–531. [CrossRef]

56. Hutchinson, L.; Roop-Beauchamp, C.; Johnson, D.C. Herpes simplex virus glycoprotein K is known to influence fusion of infected cells, yet is not on the cell surface. *J. Virol.* **1995**, *69*, 4556–4563. [CrossRef]

57. Avitabile, E.; Lombardi, G.; Campadelli-Fiume, G. Herpes simplex virus glycoprotein K, but not its syncytial allele, inhibits cell-cell fusion mediated by the four fusogenic glycoproteins, gD, gB, gH, and gL. *J. Virol.* **2003**, *77*, 6836–6844. [CrossRef]

58. Chouljenko, V.N.; Iyer, A.V.; Chowdhury, S.; Kim, J.; Kousoulas, K.G. The herpes simplex virus type 1 UL20 protein and the amino terminus of glycoprotein K (gK) physically interact with gB. *J. Virol.* **2010**, *84*, 8596–8606. [CrossRef]

59. Foster, T.P.; Alvarez, X.; Kousoulas, K.G. Plasma membrane topology of syncytial domains of herpes simplex virus type 1 glycoprotein K (gK): The UL20 protein enables cell surface localization of gK but not gK-mediated cell-to-cell fusion. *J. Virol.* **2003**, *77*, 499–510. [CrossRef]

60. McGeoch, D.J.; Dalrymple, M.A.; Davison, A.J.; Dolan, A.; Frame, M.C.; McNab, D.; Perry, L.J.; Scott, J.E.; Taylor, P. The complete DNA sequence of the long unique region in the genome of herpes simplex virus type 1. *J. Gen. Virol.* **1988**, *69*, 1531–1574. [CrossRef]

61. Ward, P.L.; Campadelli-Fiume, G.; Avitabile, E.; Roizman, B. Localization and putative function of the UL20 membrane protein in cells infected with herpes simplex virus 1. *J. Virol.* **1994**, *68*, 7406–7417. [CrossRef] [PubMed]

62. Foster, T.P.; Melancon, J.M.; Olivier, T.L.; Kousoulas, K.G. Herpes simplex virus type 1 glycoprotein K and the UL20 protein are interdependent for intracellular trafficking and trans-Golgi network localization. *J. Virol.* **2004**, *78*, 13262–13277. [CrossRef]

63. Foster, T.P.; Chouljenko, V.N.; Kousoulas, K.G. Functional and physical interactions of the herpes simplex virus type 1 UL20 membrane protein with glycoprotein K. *J. Virol.* **2008**, *82*, 6310–6323. [CrossRef]

64. Baines, J.D.; Ward, P.L.; Campadelli-Fiume, G.; Roizman, B. The UL20 gene of herpes simplex virus 1 encodes a function necessary for viral egress. *J. Virol.* **1991**, *65*, 6414–6424. [CrossRef] [PubMed]

65. Foster, T.P.; Melancon, J.M.; Baines, J.D.; Kousoulas, K.G. The herpes simplex virus type 1 UL20 protein modulates membrane fusion events during cytoplasmic virion morphogenesis and virus-induced cell fusion. *J. Virol.* **2004**, *78*, 5347–5357. [CrossRef]

66. Ward, P.L.; Taddeo, B.; Markovitz, N.S.; Roizman, B. Identification of a novel expressed open reading frame situated between genes U(L)20 and U(L)21 of the herpes simplex virus 1 genome. *Virology* **2000**, *266*, 275–285. [CrossRef] [PubMed]

67. Fuchs, W.; Klupp, B.G.; Granzow, H.; Mettenleiter, T.C. The UL20 gene product of pseudorabies virus functions in virus egress. *J. Virol.* **1997**, *71*, 5639–5646. [CrossRef]

68. Avitabile, E.; Lombardi, G.; Gianni, T.; Capri, M.; Campadelli-Fiume, G. Coexpression of UL20p and gK inhibits cell-cell fusion mediated by herpes simplex virus glycoproteins gD, gH-gL, and wild-type gB or an endocytosis-defective gB mutant and downmodulates their cell surface expression. *J. Virol.* **2004**, *78*, 8015–8025. [CrossRef] [PubMed]

69. Wagenaar, F.; Pol, J.M.; Peeters, B.; Gielkens, A.L.; de Wind, N.; Kimman, T.G. The US3-encoded protein kinase from pseudorabies virus affects egress of virions from the nucleus. *J. Gen. Virol.* **1995**, *76*, 1851–1859. [CrossRef]

70. Klupp, B.G.; Granzow, H.; Mettenleiter, T.C. Effect of the pseudorabies virus US3 protein on nuclear membrane localization of the UL34 protein and virus egress from the nucleus. *J. Gen. Virol.* **2001**, *82*, 2363–2371. [CrossRef]

71. Deng, L.; Wang, M.; Cheng, A.; Yang, Q.; Wu, Y.; Jia, R.; Chen, S.; Zhu, D.; Liu, M.; Zhao, X.; et al. The Pivotal Roles of US3 Protein in Cell-to-Cell Spread and Virion Nuclear Egress of Duck Plague Virus. *Sci. Rep.* **2020**, *10*, 7181. [CrossRef] [PubMed]

72. Kato, A.; Kawaguchi, Y. Us3 Protein Kinase Encoded by HSV: The Precise Function and Mechanism on Viral Life Cycle. *Adv. Exp. Med. Biol.* **2018**, *1045*, 45–62. [CrossRef] [PubMed]

73. Proft, A.; Spiesschaert, B.; Izume, S.; Taferner, S.; Lehmann, M.J.; Azab, W. The Role of the Equine Herpesvirus Type 1 (EHV-1) US3-Encoded Protein Kinase in Actin Reorganization and Nuclear Egress. *Viruses* **2016**, *8*, 275. [CrossRef]

74. Schumacher, D.; Tischer, B.K.; Trapp, S.; Osterrieder, N. The protein encoded by the US3 orthologue of Marek's disease virus is required for efficient de-envelopment of perinuclear virions and involved in actin stress fiber breakdown. *J. Virol.* **2005**, *79*, 3987–3997. [CrossRef]

75. Reynolds, A.E.; Wills, E.G.; Roller, R.J.; Ryckman, B.J.; Baines, J.D. Ultrastructural localization of the herpes simplex virus type 1 UL31, UL34, and US3 proteins suggests specific roles in primary envelopment and egress of nucleocapsids. *J. Virol.* **2002**, *76*, 8939–8952. [CrossRef] [PubMed]

76. Purves, F.C.; Spector, D.; Roizman, B. U_L34, the target of the herpes simplex virus U_s3 protein kinase, is a membrane protein which in its unphosphorylated state associates with novel phosphoproteins. *J. Virol.* **1992**, *66*, 4295–4303. [CrossRef]

77. Klupp, B.G.; Granzow, H.; Fuchs, W.; Keil, G.M.; Finke, S.; Mettenleiter, T.C. Vesicle formation from the nuclear membrane is induced by coexpression of two conserved herpesvirus proteins. *Proc. Natl. Acad. Sci. USA* **2007**, *104*, 7241–7246. [CrossRef]

78. Bigalke, J.M.; Heldwein, E.E. Structural basis of membrane budding by the nuclear egress complex of herpesviruses. *EMBO J.* **2015**, *34*, 2921–2936. [CrossRef]

79. Bigalke, J.M.; Heuser, T.; Nicastro, D.; Heldwein, E.E. Membrane deformation and scission by the HSV-1 nuclear egress complex. *Nat. Commun.* **2014**, *5*, 4151. [CrossRef]

80. Hagen, C.; Dent Kyle, C.; Zeev-Ben-Mordehai, T.; Grange, M.; Bosse Jens, B.; Whittle, C.; Klupp Barbara, G.; Siebert, C.A.; Vasishtan, D.; Bäuerlein Felix, J.B.; et al. Structural Basis of Vesicle Formation at the Inner Nuclear Membrane. *Cell* **2015**, *163*, 1692–1701. [CrossRef]

81. Lye, M.F.; Sharma, M.; El Omari, K.; Filman, D.J.; Schuermann, J.P.; Hogle, J.M.; Coen, D.M. Unexpected features and mechanism of heterodimer formation of a herpesvirus nuclear egress complex. *EMBO J.* **2015**, *34*, 2937–2952. [CrossRef]

82. Walzer, S.A.; Egerer-Sieber, C.; Sticht, H.; Sevvana, M.; Hohl, K.; Milbradt, J.; Muller, Y.A.; Marschall, M. Crystal Structure of the Human Cytomegalovirus pUL50-pUL53 Core Nuclear Egress Complex Provides Insight into a Unique Assembly Scaffold for Virus-Host Protein Interactions. *J. Biol. Chem.* **2015**, *290*, 27452–27458. [CrossRef]

83. Thorsen, M.K.; Lai, A.; Lee, M.W.; Hoogerheide, D.P.; Wong, G.C.L.; Freed, J.H.; Heldwein, E.E. Highly Basic Clusters in the Herpes Simplex Virus 1 Nuclear Egress Complex Drive Membrane Budding by Inducing Lipid Ordering. *mBio* **2021**, *12*, e0154821. [CrossRef] [PubMed]

84. Bahnamiri, M.M.; Roller, R.J. Mechanism of Nuclear Lamina Disruption and the Role of pUS3 in HSV-1 Nuclear Egress. *J. Virol.* **2021**, *95*, e02432-20. [CrossRef]

85. Gao, J.; Finnen, R.L.; Sherry, M.R.; Le Sage, V.; Banfield, B.W. Differentiating the Roles of UL16, UL21, and Us3 in the Nuclear Egress of Herpes Simplex Virus Capsids. *J. Virol.* **2020**, *94*, e00738-20. [CrossRef] [PubMed]

86. Benedyk, T.H.; Muenzner, J.; Connor, V.; Han, Y.; Brown, K.; Wijesinghe, K.J.; Zhuang, Y.; Colaco, S.; Stoll, G.A.; Tutt, O.S.; et al. pUL21 is a viral phosphatase adaptor that promotes herpes simplex virus replication and spread. *PLoS Pathog.* **2021**, *17*, e1009824. [CrossRef] [PubMed]

87. de Wind, N.; Wagenaar, F.; Pol, J.; Kimman, T.; Berns, A. The pseudorabies virus homology of the herpes simplex virus UL21 gene product is a capsid protein which is involved in capsid maturation. *J. Virol.* **1992**, *66*, 7096–7103. [CrossRef] [PubMed]

88. Takakuwa, H.; Goshima, F.; Koshizuka, T.; Murata, T.; Daikoku, T.; Nishiyama, Y. Herpes simplex virus encodes a virion-associated protein which promotes long cellular processes in over-expressing cells. *Genes Cells* **2001**, *6*, 955–966. [CrossRef]

89. Purves, F.C.; Longnecker, R.M.; Leader, D.P.; Roizman, B. Herpes simplex virus 1 protein kinase is encoded by open reading frame US3 which is not essential for virus growth in cell culture. *J. Virol.* **1987**, *61*, 2896–2901. [CrossRef]

90. Chang, Y.E.; Van Sant, C.; Krug, P.W.; Sears, A.E.; Roizman, B. The null mutant of the U_L31 gene of herpes simplex virus 1: Construction and phenotype of infected cells. *J. Virol.* **1997**, *71*, 8307–8315. [CrossRef]

91. Fuchs, W.; Klupp, B.G.; Granzow, H.; Osterrieder, N.; Mettenleiter, T.C. The interacting UL31 and UL34 gene products of pseudorabies virus are involved in egress from the host-cell nucleus and represent components of primary enveloped but not mature virions. *J. Virol.* **2002**, *76*, 364–378. [CrossRef]

92. Klupp, B.G.; Granzow, H.; Mettenleiter, T.C. Primary envelopment of pseudorabies virus at the nuclear membrane requires the UL34 gene product. *J. Virol.* **2000**, *74*, 10063–10073. [CrossRef]

93. Roller, R.; Zhou, Y.; Schnetzer, R.; Ferguson, J.; Desalvo, D. Herpes simplex virus type 1 U_L34 gene product is required for viral envelopment. *J. Virol.* **2000**, *74*, 117–129. [CrossRef]

94. De Magistris, P.; Antonin, W. The Dynamic Nature of the Nuclear Envelope. *Curr. Biol.* **2018**, *28*, R487–R497. [CrossRef] [PubMed]

95. Puhka, M.; Joensuu, M.; Vihinen, H.; Belevich, I.; Jokitalo, E. Progressive sheet-to-tubule transformation is a general mechanism for endoplasmic reticulum partitioning in dividing mammalian cells. *Mol. Biol. Cell* **2012**, *23*, 2424–2432. [CrossRef] [PubMed]

96. Benetti, L.; Roizman, B. Herpes simplex virus protein kinase US3 activates and functionally overlaps protein kinase A to block apoptosis. *Proc. Natl. Acad. Sci. USA* **2004**, *101*, 9411–9416. [CrossRef] [PubMed]

97. Daikoku, T.; Yamashita, Y.; Tsurumi, T.; Maeno, K.; Nishiyama, Y. Purification and biochemical characterization of the protein kinase encoded by the US3 gene of herpes simplex virus type 2. *Virology* **1993**, *197*, 685–694. [CrossRef]

98. Frame, M.C.; Purves, F.C.; McGeoch, D.J.; Marsden, H.S.; Leader, D.P. Identification of the herpes simplex virus protein kinase as the product of virus gene US3. *J. Gen.Virol.* **1987**, *68*, 2699–2704. [CrossRef]

99. Jahn, R.; Lang, T.; Sudhoff, T.C. Membrane fusion. *Cell* **2003**, *112*, 519–533. [CrossRef]

Review

Human Cytomegalovirus Egress: Overcoming Barriers and Co-Opting Cellular Functions

Veronica Sanchez [1,*] and William Britt [1,2]

1 Department of Pediatrics, University of Alabama School of Medicine, Birmingham, AL 35294, USA; wbritt@uabmc.edu
2 Department of Microbiology, University of Alabama School of Medicine, Birmingham, AL 35294, USA
* Correspondence: veronicasanchez@uabmc.edu

Abstract: The assembly of human cytomegalovirus (HCMV) and other herpesviruses includes both nuclear and cytoplasmic phases. During the prolonged replication cycle of HCMV, the cell undergoes remarkable changes in cellular architecture that include marked increases in nuclear size and structure as well as the reorganization of membranes in cytoplasm. Similarly, significant changes occur in cellular metabolism, protein trafficking, and cellular homeostatic functions. These cellular modifications are considered integral in the efficient assembly of infectious progeny in productively infected cells. Nuclear egress of HCMV nucleocapsids is thought to follow a pathway similar to that proposed for other members of the herpesvirus family. During this process, viral nucleocapsids must overcome structural barriers in the nucleus that limit transit and, ultimately, their delivery to the cytoplasm for final assembly of progeny virions. HCMV, similar to other herpesviruses, encodes viral functions that co-opt cellular functions to overcome these barriers and to bridge the bilaminar nuclear membrane. In this brief review, we will highlight some of the mechanisms that define our current understanding of HCMV egress, relying heavily on the current understanding of egress of the more well-studied α-herpesviruses, HSV-1 and PRV.

Keywords: human cytomegalovirus; nuclear egress; virus assembly; virus–cell interactions

Citation: Sanchez, V.; Britt, W. Human Cytomegalovirus Egress: Overcoming Barriers and Co-Opting Cellular Functions. *Viruses* **2022**, *14*, 15. https://doi.org/10.3390/v14010015

Academic Editor: Donald M. Coen

Received: 8 October 2021
Accepted: 16 December 2021
Published: 22 December 2021

Publisher's Note: MDPI stays neutral with regard to jurisdictional claims in published maps and institutional affiliations.

1. Introduction

Herpesviruses follow a complex assembly pathway in which viral DNA is first replicated and encapsidated in the nucleus, followed by the acquisition of a mature virion envelope within cytoplasmic membranes. The nuclear membrane, the protein lattice known as the lamina underlying the nuclear membrane, and chromatin represent significant barriers to the egress of subviral particles from the nucleus. To breach these physical barriers, herpesviruses have evolved mechanisms that include both viral and cellular functions to modify nuclear structures, such as modification of the nuclear lamina that allows access of capsids to the inner nuclear membrane (INM) and facilitates budding into the perinuclear space (PNS). While several components of the viral nuclear egress complex (NEC) that mediate the process of the capsid nuclear exit are conserved between members of the herpesvirus family, cellular functions that contribute to nuclear egress appear to vary among the α, β, and γ subfamilies presumably because of unique host cell responses that follow infection by viruses from each family. Even though different members of the herpesvirus family utilize different cellular functions in transit of their capsids from the nucleus to the cytoplasm, the conservation of the core components of the nuclear egress machinery of these viruses makes the core NEC an attractive target for the development of antiviral drugs to inhibit herpesvirus replication [1].

2. Human Cytomegalovirus Replication Cycle

Like other herpesviruses, viral DNA replication occurs in the nucleus, while acquisition of the mature envelope occurs in the cytoplasm, but unlike α-herpesviruses, the

HCMV life cycle is protracted and progeny virions are produced over a period of days, as shown in Figure 1. Nonetheless, major steps of the nuclear phase of herpesvirus infections are similar between the different virus families. After entry of HCMV into the host cell, the viral genomes are deposited into the nucleus at PML nuclear domains and circularize. Expression of IE (immediate-early) and E (early) genes is initiated, and viral proteins accumulate at these viral pre-replication sites to stimulate rolling circle replication of the input genomes. As infection proceeds and depending on the multiplicity of infection, cells can contain several replication foci that coalesce into larger replication compartments (RC) containing the replication machinery and early-late proteins required for encapsidation of the viral genome. Concurrent with expansion of the RCs, host DNA is marginalized, and nuclear morphology is altered [2–6] (Figure 1). Although herpesvirus infection-induced changes in nuclear size have been noted by many investigators, advances in microscopy have facilitated accurate measurements of the increases in nuclear volume and overall surface area and of changes in substructures such as the nuclear envelope. For example, Aho et al. measured an average nuclear volume of 260 μm^3 in HSV-infected cells versus 170 μm^3 in uninfected controls [5]. The surface area of the nuclei of HSV-infected cells also increased to 260 μm^2 compared to 176 μm^2 for uninfected cells [5]. In HCMV-infected cells, nuclei are markedly larger than uninfected cells, and the surface area of the nucleus is further increased by the presence of NE infoldings [4,7,8].

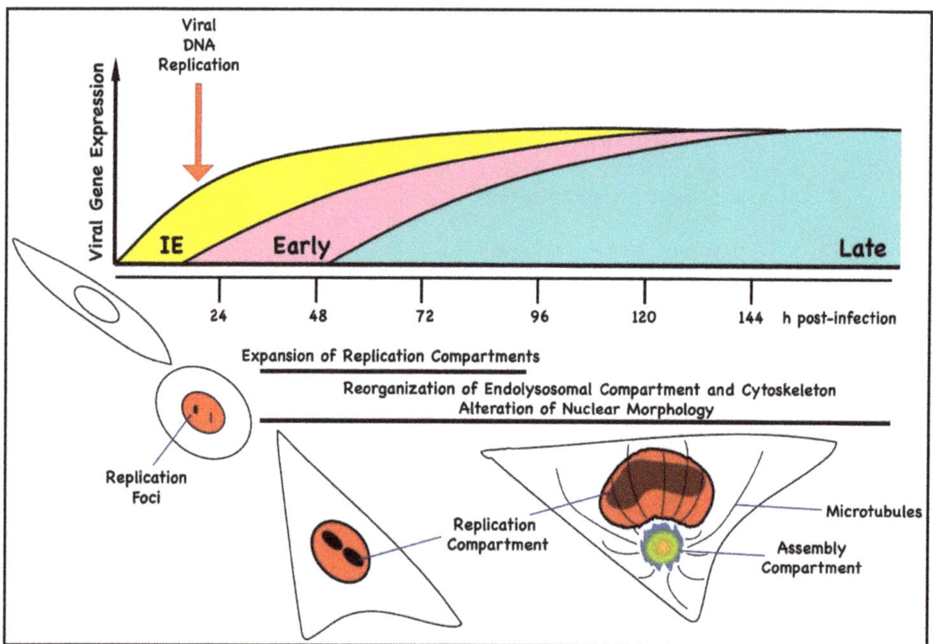

Figure 1. Timeline of viral gene expression and morphological changes in HCMV-infected cells. In the top panel, the kinetics of viral gene is shown. Immediate-early gene (yellow) expression is observed 1 h following infection of permissive human fibroblasts and within 24 h pi, cell rounding is observed, concurrent with the expression of early gene (pink) products. Small replication foci containing IE and E proteins can be detected at this time point. By 48 h pi, cells begin to flatten and enlarge, and nuclear replication compartments can be seen along with expression of late gene (blue) products, including virion structural proteins. In the lower panel, the morphological changes that take place during a permissive infection of human fibroblast cell are illustrated. At 72 h pi, the enlarged nuclei begin to adopt the characteristic kidney shape, and the juxtanuclear assembly compartment (AC) becomes readily detected with microtubules emanating from the Golgi-derived MTOC.

Immediate-early (IE) viral gene products can be detected within 8–12 h post-infection (h pi) in nuclear pre-replication foci after HCMV infection of permissive cells in G_0/G_1 [9–11] (Figure 1). Viral DNA replication begins by 24 h pi and plateaus by 72–96 h pi (Figure 1). The expression of virion structural genes is an early-late event, and many of the protein products of these viral genes can be detected by 48 h pi. Remodeling of the nuclear lamina and changes in nuclear morphology are detected during a similar time interval after infection. HCMV induces major reorganization of intracellular membranes and dysregulates membrane trafficking to generate a juxtanuclear membrane compartment that is required for efficient cytoplasmic envelopment and virus assembly [6,12–15]. The morphogenesis of the juxtanuclear site of cytoplasmic envelopment, also known as the assembly compartment (AC), begins at approximately 24 h pi when depolymerization of actin and changes in Golgi morphology are observed [12,13,15,16]. Virus particles exiting the nucleus acquire their mature envelope by budding into AC vesicles that contain virus-encoded glycoproteins and that subsequently fuse with the plasma membrane, releasing virions into the extracellular space. The morphogenesis and function of the AC in virus assembly are areas of intense investigation and beyond the scope of this review; however, there are several structural features of the AC that are relevant to a discussion of nuclear egress [6]. First, the AC serves as a Golgi-nucleated microtubule organizing center [16]. The microtubules that extend from the AC become acetylated, a modification that imparts stability. Acetylation of the microtubules is associated with nuclear rotation, a phenomenon often associated with mechanical stress during cell migration and also during preparation for cell division [16]. Lastly, activity of the AC MTs and their associated motors is required for changes in nuclear morphology observed in HCMV-infected cells [6,16,17].

Overview of Nuclear Egress of HCMV

Following viral DNA replication and packaging of newly formed capsids, genome-containing capsids must transit from RC and cross the nuclear envelope through a process of envelopment at the inner nuclear membrane (INM) and de-envelopment at the outer nuclear membrane (ONM) to reach sites of tegumentation and final envelopment in the cytoplasm (Figure 2). In addition to the nuclear membrane, several other barriers restrict the egress of capsids from the viral replication centers to the cytoplasmic AC, including the nuclear lamina and chromatin (Figure 2). Modification of these nuclear structures during herpesvirus egress are briefly reviewed below.

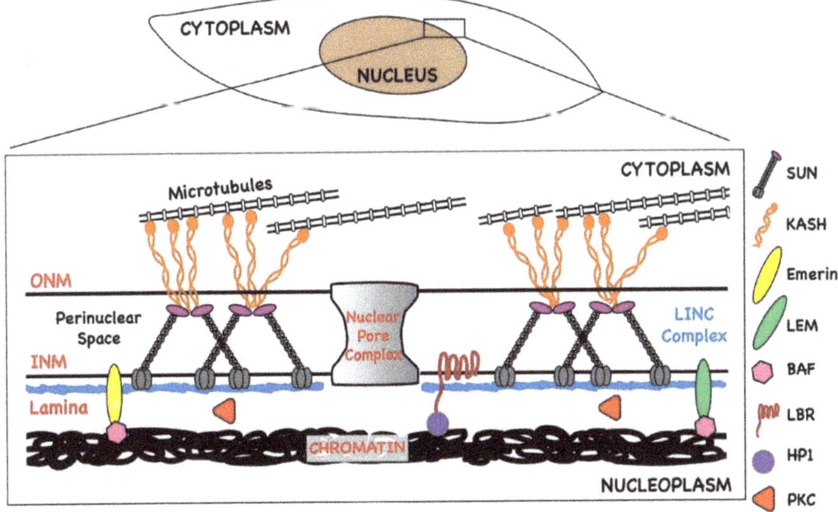

Figure 2. Organization of nuclear envelope in normal cells. The nuclear envelope consists of a double

bilayer, the inner nuclear membrane (INM) and outer nuclear membrane (ONM). These membranes are connected at nuclear pores, a protein complex that permits diffusion and active transport of molecules across the envelope. The space between the ONM and INM, the perinuclear space (PNS), is traversed by the LINC (linker of nucleoskeleton and cytoskeleton) complex that is composed of KASH domain proteins (nesprins) anchored in the ONM and SUN proteins anchored in the INM. The lamina underlies the INM and is composed of intermediate filament proteins known as the lamins and is connected to the INM by proteins including the lamin B receptor (LBR) and LEM-domain proteins (Lap2-emerin-Man1). The lamina maintains nuclear shape and stability and serves as a scaffold for other nuclear proteins. BAF (Barrier-to-autointegration factor) binds LEM-domain proteins and chromatin. Phosphorylation of BAF during mitosis leads to its release from chromatin. Similarly, HP1 (heterochromatin protein 1) is a component of heterochromatin that links condensed chromatin to the INM through LBR.

3. Interphase Nuclear Architecture

The nuclear envelope consists of a double bilayer, the INM and the ONM, as well as the underlying lamina (Figure 2). The ONM is contiguous with the endoplasmic reticulum (ER), thus allowing lateral diffusion of membrane-bound proteins destined for localization within nuclear membranes. In addition, since the ER is the main site of phospholipid synthesis, the continuity of the ONM with the ER allows for nuclear membrane expansion by lateral diffusion of lipids, although local synthesis and storage of lipids at the INM have been described [18]. In fact, local synthesis of phospholipids has been shown to contribute to the formation of NM invaginations known as the nucleoplasmic reticulum (NR) [19]. One potential function proposed for the NR is to facilitate communication from the nuclear periphery to inner domains of the nucleus. Local phospholipid accumulation in the INM has also been shown to be regulated by host cell functions contained within the endosomal sorting complex required for transport III (ESCRT-III), a complex of proteins that has a critical role in vesicle formation, abscission during cytokinesis, and the budding of enveloped viruses [20,21]. Proteins within this complex have been shown to contribute to the regulation of INM proliferation and repair during mitosis, resealing nuclear ruptures in migrating cells, and, most recently, nuclear egress of herpesvirus particles [20–23].

Nuclear permeability is maintained by nuclear pores that serve as junctions between the INM and ONM (Figure 2). Pore complexes are large, multi-subunit structures 110 MDa in size (~800 Å wide) that allow nucleocytoplasmic transit of macromolecules such as RNAs and proteins in an energy-dependent manner [24]. Smaller molecules (<40 kDa) can diffuse through pore channels without ATP consumption. The intermembrane or perinuclear space (PNS) between the INM and ONM spans approximately 20–40 nM and contains components of the LINC (linker of nucleoskeleton and cytoskeleton) complex (Figure 2). LINC complexes function as mechanosensors relaying signals to and from the nuclear environment and cytoplasm [25]. They are composed of a trimer of SUN-domain (Sad1 and UNC-84) proteins anchored in the INM that are in turn bound to a trimer of nesprins (KASH domain proteins) anchored in the ONM. Thus, the LINC complex extends from the INM and PNS into the cytoplasm (Figure 2). Nesprins bind to cytoskeletal elements such as tubulin, and thus the LINC complex provides a bridge to the karyoskeleton both structurally and for intracellular signaling. For example, the LINC complex has been implicated in the regulation of transcription in response to changes in the extracellular matrix [26]. The LINC complex is also important for nuclear positioning [27–30].

Together, the lamina, the nuclear pores, the LINC complex, spectrin, actin, and the LEM-domain proteins, which are involved in tethering chromatin to the NE, all contribute to maintaining nuclear architecture in interphase cells [31,32] (Figure 2). Regulation of components of the nucleoskeleton effect changes in nuclear structure and function during the cell cycle, during stress responses, and in response to extracellular cues that trigger cell migration. The dynamic nature of nuclear architecture is essential for homeostasis and genome maintenance. Like other viruses that replicate in the nucleus, herpesviruses take advantage of the plasticity of nuclear structure to create an environment that favors the replication of the viral genome over that of the host without compromising the integrity of the nuclear compartment prematurely in order to maximize virus production.

4. Herpesviruses Circumvent Nuclear Barriers

Perhaps the most well understood step in the nuclear egress of α, β, and γ herpesviruses is the transit of the viral capsid through the INM into the perinuclear space (PNS) and in particular, the role of viral proteins in this step of the assembly pathway. Budding of membrane-bound capsids from the INM into the PNS, alternatively termed primary envelopment, has been most been well studied in the α-herpesviruses herpes simplex virus (HSV) and pseudorabies virus (PRV). Primary envelopment of α-herpesviruses is dependent on the interaction between a heterodimer of pUL31, a nuclear phosphoprotein, and pUL34, a type II membrane protein inserted into the INM, both of which are required for budding of the capsid into the PNS [33,34]. Together these viral proteins constitute the nuclear egress complex (NEC). Interactions between these core NEC proteins and formation of the heterodimer have been argued to lead to the recruitment of other viral proteins, including viral kinases and host cell proteins that are required for efficient passage through the INM. The NEC is conserved in all families of herpesviruses. In the prototypic β-herpesvirus, HCMV, the NEC consists of the pUL53 and pUL50, orthologs of the HSV pUL31 and pUL34 respectively, and a heterodimer of HCMV pUL53 and pUL50 has also been shown to contribute to the recruitment of viral and cellular proteins that contribute to nuclear egress [35,36]. Similarly, the NEC of the γ-herpesvirus EBV has been identified and is represented by the HSV pUL31 and pUL34 orthologs, BFLF2 and BFRF1, respectively [37–39]. A further description of the function of the HCMV NEC will be provided in greater detail in the following sections.

In the case of PRV, the interaction between the orthologs of HSV pUL31 and pUL34 is sufficient for vesiculation of the INM and formation of particles resembling L particles of α-herpesviruses in transfected cells in the absence of expression of other PRV encoded proteins, arguing that interactions between the NEC and the PRV capsid are not required for budding of particles at the INM [40]. Furthermore, purified pUL31 and pUL34 of HSV and the PRV orthologues can vesiculate membranes in cell-free systems, strongly suggesting that the minimal NEC of α-herpesviruses is sufficient for budding from lipid-containing membranes [40–43]. Although these observations could be interpreted as evidence that other viral and host cell functions are non-essential for herpesvirus nuclear egress, multiple studies have shown that efficient nuclear egress leading to WT levels of virus replication requires additional viral proteins together with the viral NEC. Furthermore, it could be argued that cellular proteins can functionally complement or even partially replace viral functions encoded by non-NEC viral proteins. Consistent with this possibility is the finding that although the deletion of the PRV-encoded protein kinase pUS3 led to large invaginations of the INM and to an increased number of primary enveloped capsids in the PNS in PRV-infected cells, this deletion had a marginal effect on the production of infectious virus [34,44], suggesting that other redundant functions compensate for the loss of US3. Consistent with this possibility, protein kinase C (PKC) was argued to disrupt the nuclear lamina during nuclear egress of murine CMV, whereas more recent findings have argued for a more prominent role of the conserved herpesvirus protein kinase (CHPK) of HCMV, pUL97, in this step in nuclear egress [45,46]. A number of other host cell functions have been proposed to contribute to budding of the herpesvirus capsid at the INM, including emerin and p53, and deletion or decreased expression of these cellular functions in some cases has been shown to result in a quantifiable change in nuclear egress and the production of infectious virus (Table 1) [35,47,48]. Thus, current data suggest that herpesviruses have usurped physiologic processes to facilitate efficient nuclear egress and, ultimately, virus replication. In the following sections, we will attempt to update the current understanding of the role of cellular and viral proteins in HCMV nuclear egress, often in the context of the more well-studied and well-understood models of the nuclear egress of α-herpesviruses.

Table 1. Viral and cellular factors involved in nuclear egress.

	Role during Infection	Citation
VIRAL NUCLEAR EGRESS COMPLEX		
pUL50/53	Core NEC, forms hexameric ring structure that can deform membranes and perform membrane scission	[36,49–54]
pUL93/pUL77	Capsid vertex components 1/2, potential adaptor proteins binding to core NEC, required for encapsidation	[55,56]
pUL97	Conserved herpesvirus protein kinase, phosphorylates lamins and NEC components, possibly host proteins; disrupts lamin polymerization	[46,57–59]
NEC-DIRECTED TRANSIT THROUGH INM		
Lamins	Phosphorylated by viral and possibly cellular kinases; focal disruption of lamin network required for egress	[46,57,60–64]
WDR5	Required for efficient egress by unknown mechanism; may remodel chromatin and regulate nuclear architecture	[65]
Protein kinase C	Host kinase that phosphorylates lamina and disrupts lamin polymerization	[45,61]
p32	Recruited to INM by interactions with NEC; binds lamin B receptor; recruits UL97 to NEC	[62]
Nuclear actin/Myosin Va	Required for efficient transit of capsids to INM	[66,67]
ESCRT-III	Required for efficient egress in α-herpesviruses; repairs nuclear membrane ruptures	[21,23,68]
p53	Regulates expression of core NEC protein UL53	[47,48]
Emerin	Polarized expression in HCMV-infected cells; interacts with NEC; reduced expression disrupts AC formation	[8,35]
TRANSIT FROM THE PNS THOUGH THE ONM		
Viral gB	In α-herpesvirus infected cells, reported to be required for membrane fusion at ONM	[69]
Torsin A	ATPase; may regulate LINC complex; deletion causes accumulation of capsids in PNS in α-herpesvirus infected cells	[70]
LINC COMPLEX	Sun2 levels downregulated by infection; Sun1 localization polarized, levels may be downregulated; reduction in expression correlated with dilation of PNS; expression of Sun dominant-negative proteins reduces virus yield	[4,8,71]

4.1. Nuclear Egress: Budding of Newly Assembled HCMV Capsids into the PNS

Role of Virus-Encoded Proteins

Several detailed descriptions of the transit of the α-herpesvirus capsid through the INM have been published, and because studies have shown that HCMV follows a similar overall pathway for budding into the PNS, only a brief description of this process will be provided with features unique to the transit of HCMV into the PNS highlighted. As noted previously, the HCMV-encoded proteins of pUL50 and pUL53 are the core components of the HCMV NEC, and deletion of either viral gene is lethal; thus, both are classified as essential viral genes [72]. Neither viral protein has been convincingly detected in the mature extracellular virion by mass spectrometry; however, early studies using immunoelectron microscopy suggested that cytoplasmic-enveloped particles contained pUL53 [49]. pUL53 is a 367 aa nuclear phosphoprotein with a defined monopartite nuclear localization signal

(NLS) that results in its localization in the nucleus in the absence of other viral proteins [50]. pUL53 has been shown to interact with pUL50 as well as a number of other proteins, including pUL97 [35,46,57,60]. pUL50 is a 397 aa type II membrane protein anchored in the INM with the C-terminal 16 aa in the PNS [61]. Of note, HCMV pUL50 is approximately 125 aa larger than the HSV-1 ortholog and 50 aa larger than the EBV ortholog. pUL50 does not express an identified NLS and is thought to localize to the INM following insertion into the ER membranes and translocation through the membranes bridged by nuclear pores that are contiguous with the INM [50]. Expression of pUL50 in the absence of other viral proteins resulted in its localization in the INM [50]. Deletion of the transmembrane domain of pUL50 redirected its expression to the cytoplasm, and expression of this pUL50 mutant with pUL53 results in its localization in the nucleus, not the INM, consistent with the requirement of the TM domain for its INM localization [46,50,73]. Both biochemical and structural studies have described in considerable detail the interaction between pUL50 and pUL53 that leads to the formation of a heterodimer that constitutes the NEC of HCMV [51,52]. Structural studies of the NEC from both α- and β-herpesviruses have described a heterodimer with a hook-into-groove type interaction [43,51–54,74,75]. The groove component of this heterodimer encoded by pUL50 consists of four helical segments and an anti-parallel β-sheet sandwich [51,53]. The hook component encoded by pUL53 contains two consecutive α-helices within a short strand of residues (aa 59–87) in the N-terminus [51,53]. Although this structure is highly conserved for all NEC thus far described, it is interesting that the primary sequence of the core NEC proteins is not highly conserved between different herpesviruses, including contact residues within the hook-into-groove interfaces [75].

Structures of the NEC complex of HSV derived by cryoelectron microscopy revealed that the pUL31/pUL34 hetrodimer forms a hexameric ring-like structure in the presence of membranes [41]. The oligomerization of the pUL31/pUL34 heterodimer into the hexameric ring is required for egress as mutations that disrupt the oligomerization also block egress [41,76–78]. A similar ring-like structure of the heterodimer of HCMV pUL50/pUL53 has also been described for the HCMV NEC [51]. This ring-like structure consists of a pUL50 hexamer with pUL53 located between individual pUL50 molecules, thus contacting two pUL50 molecules but not in contact with adjacent pUL53 molecules [51]. The hexameric structure of the NEC appears highly conserved in herpesviruses, thus suggesting a common function of this structure in this large family of viruses possibly acting as a scaffold for the association and/or recruitment of other viral and cellular proteins that could facilitate the budding process. Although the hexameric ring-like structure of the NEC heterodimers has been confirmed using both cell-free systems and infected cells, the planar structure of the hexameric ring has also presented a conundrum for mechanisms of budding at the INM, a process that includes membrane deformation resulting in the generation of spherical enveloped capsids in the PNS [42]. Recently, this question was further addressed in a study of nuclear egress of HSV using cryoelectron microscopy. These investigators demonstrated that HSV pUL25 interacts with membrane-associated pUL31, leading to the formation of NEC pentagons that are proposed to secure the NEC to the vertices of the capsid [42]. Based on these findings, these authors have proposed that irregular incorporation of pUL25 containing NEC pentagons into the hexameric NEC lattice can disrupt the hexameric ring, thus permitting the formation of an icosahedral-like structure surrounding the capsid and the budding of a spherical structure into the PNS [42].

The interaction between pUL50 and pUL53 has been shown to result in concentration of the heterodimer in the INM, a key step in nuclear egress [50]. In addition, more recent findings have argued that the interaction between pUL50 and pUL53 could also facilitate the recruitment of other viral and cellular proteins that contribute to nuclear egress to the NEC [79]. Several host proteins were identified using co-immune precipitation protocols, and the analysis of the NEC by mass spectrometry identified additional viral and host proteins associated with the NEC [35]. The CHPK of HCMV, pUL97, is the most consistently detected viral protein associated with the NEC of HCMV and is thought to interact specifically with pUL53. As discussed in more detail in the following sections, pUL97 has been shown to have a critical role in egress secondary to its role in phosphorylation of the nuclear lamina leading to disruption of this barrier [46,57,62].

More recent studies have identified a number of HCMV capsid components associated with pUL53 in co-immunoprecipitation assays [79]. Furthermore, the association of pUL53 with intranuclear capsids was confirmed using immunoelectron microscopy [79]. These observations led these investigators to propose that intact capsids could interact with pUL53 prior to the interaction between pUL53 and pUL50, thus providing a central role of UL53 in the recruitment of capsids to budding sites on the INM [79]. The interactions between pUL53 and intranuclear capsids is thought to require a viral adaptor protein(s), a postulate based on studies of the role in egress of the α-herpesviruses capsid-associated proteins, pUL17 and pUL25 [80–83]. Findings from studies in PRV and HSV have proposed that the pUL25/pUL17 heterodimer serves as an adaptor between capsids and the NEC, thus facilitating targeting of capsids and subsequent budding from the INM [82,84]. More recently, specific residues on pUL31 required for capsid binding to the NEC have been identified, suggesting the likelihood of interactions between pUL25, the capsid, and the NEC [85]. Mutations of these residues in pUL31 resulted in mutant viruses with reduced virus yield, increased numbers of capsids in the nucleus, and empty vesicles in the PNS [85]. However, it should be noted that other roles in assembly have been proposed for these α-herpesvirus proteins, including a role of encapsidation for PRV pUL17 [83]. Interestingly, pUL25 of HSV has been reported to be required for genome encapsidation, whereas the PRV pUL25 is not required for the encapsidation of newly replicated viral DNA [40,86].

Candidates for the function of a viral adaptor protein linking the capsid to the NEC in HCMV include pUL77 (HSV UL25 ortholog) and pUL93 (HSV UL17 ortholog). Both proteins are expressed with early late kinetics and interact to form a pUL77/pUL93 heterodimer [55,56]. Imaging studies have indicated that both proteins were localized with capsid proteins and, perhaps, intranuclear capsids [55]. Co-immunoprecipitation assays demonstrated interactions between pUL77 and capsid proteins MCP (pUL86), SCP (pUL49a), mCP (pUL85), and mCP-BP (pUL46) [55]. In contrast to reports that described preferential binding of the α-herpesviruses pUL25/pUL17 heterodimer to intranuclear C capsids as a mechanism to enrich primary envelopment of infectious capsids during nuclear egress, neither HCMV pUL77 nor pUL93 exhibited preferential binding for intranuclear A, B, or C capsids [55,87,88]. Nonetheless, both pUL77 and pUL93 were tightly associated with capsids in extracellular virions [55]. Unit length genomic DNA could not be detected in cells infected with mutant viruses in which either UL77 or UL93 was deleted, suggesting that these proteins were required for viral DNA encapsidation [55].

Lastly, virion proteins could directly regulate the function of components of the NEC, thus providing another level of regulation of egress. As noted previously, deletions of the α-herpesviruses protein kinase, pUS3, have been shown to alter the capsid egress with the resulting accumulation of NEC-coated particles in the PNS; however, whether similar phenotypes would be observed following mutation of proteins that regulate pUS3 function such as has been suggested for the CHPK of α-herpesviruses, UL13, is less well understood [89–94]. Although HSV pUL13 has been shown to phosphorylate HSV pUS3 in vivo and in vitro, the importance of this pUL13-mediated post-translational modification of pUS3 in phosphorylation of components of the NEC of HSV is unclear [94]. More recently, pUL21 of HSV has been shown to play a role in the regulation of phosphorylation of pUS3 as well as proteins of the NEC, and it has been shown that the phenotype of pUL21 deletion mutant viruses could be secondary to the hyperphosphorylation of components of the NEC and defective primary envelopment at the INM [95]. In the case of HCMV, pUL97 has been shown to phosphorylate both pUL50 and pUL53 and to have autophosphorylating activity [58,59]. However, it remains unknown if other HCMV proteins also regulate the functions of pUL97 required for efficient nuclear egress.

4.2. Role of Cellular Proteins in Nuclear Egress of HCMV: Modification of Cellular Barriers to Capsid Budding into the PNS

4.2.1. Chromatin as a Barrier to Nuclear Egress

In the context of viruses that replicate in the nucleus of interphase cells, cellular chromatin represents a significant barrier to the virus in terms of competing for the availability of nucleotides and other factors required for replication and egress of subviral particles from the nucleus. The replication of herpesviruses causes marginalization of the cellular chromatin towards the nuclear periphery as RC expand [2,3,5,8]. Importantly, nascent capsids must cross this barrier to access the INM.

A cellular protein that has been reported to be involved in nuclear egress and potentially in the regulation of INM infoldings is WDR5 (WD repeat domain 5 or WD repeat-containing protein 5) [65]. This nuclear protein is highly conserved and is a member of SET1/MLL1 complexes that methylate histone H3, thus regulating chromatin structure and accessibility to the transcriptional machinery [96–98]. Like the nuclear lamina, chromatin is a recognized factor in nuclear plasticity, and WDR5 has been implicated in the modulation of nuclear morphology related to chromatin structure independent of its role in transcription [98]. H3K4 methylation by WDR5-containing complexes causes chromatin decompaction, making the nucleus less rigid and thereby facilitating nuclear deformation in response to environmental cues [97,98]. Based on these and other observations, it could be speculated that a linkage between epigenetic marks and nuclear deformation is required for optimal nuclear egress of herpesviruses. In the context of HCMV infection, WDR5 has been implicated in primary and secondary envelopment [65]. In cells depleted of WRD5 by siRNA, the distribution of the NEC on the rim of the INM in infected cells was heterogenous in comparison to control cells with a reduced number of pUL53-positive nuclear puncta and fewer cytoplasmic virions compared to control cells [65]. Increased expression of WDR5 during infection could promote decompaction of cellular chromatin as it is pushed to the nuclear periphery by the expansion of the viral replication compartment, thereby reducing the nuclear barrier imposed by chromatin and permitting movement of capsid to sites of budding on the INM [65,99–101]. Studies by Procter et al. described the polarized accumulation of condensed histone H3K9me3 host DNA at a site proximal to the AC, while active H3K4me3-marked DNA appeared to be distributed uniformly across the nucleus relative to the AC [8]. These data suggest that the area of the nucleus immediately adjacent to the AC is rigid as a result of the underlying chromatin structure, while the remainder of the nuclear membrane is more flexible as a result of chromatin decompaction. From this, it is inferred that the nuclear membrane closest to the AC would represent a significant barrier to capsid egress, as exit would be inhibited by condensed chromatin. However, it should be noted that Buchkovich et al. demonstrated that the area of the nucleus proximal to the AC was permeable to high molecular weight dextran at late times post-infection, consistent with their observations from EM studies that the ONM was not intact in that location [4]. These authors speculated that the increased permeability at this area of the nucleus would accelerate nuclear egress and provide a mechanism for vectorial transport of capsids to the AC. One way to reconcile these data is to propose that the accumulation of condensed histone H3K9me3-marked chromatin helps to maintain the integrity of the NE next to the AC that is otherwise compromised by the reorganization of cellular membranes and the cytoskeleton during infection. Lastly, data from α-herpesvirus infected cells support a model in which capsids diffuse to the INM through channels or corrals in the chromatin [5,102] (Figure 3). Thus, it is possible that the formation of such channels in the marginalized cellular DNA also occurs in HCMV-infected cells, including the chromatin located closest to the AC. Capsid diffusion may occur in tandem with more directed types for trafficking mediated by nuclear actin and myosin Va, which is discussed below [66,67].

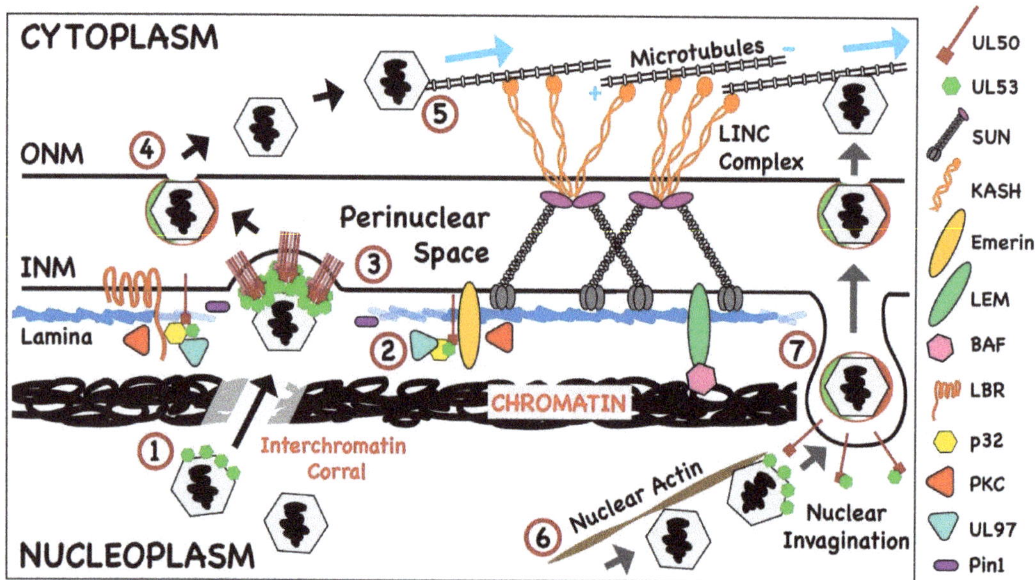

Figure 3. Modification of nuclear structures that allow HCMV nuclear egress. The structure of nuclear membrane described in Figure 2 is shown with modification of nuclear structure that could act as barriers for egress of HCMV capsids. (1) Newly formed capsids decorated with pUL53 are shown transiting chromatin through an interchromatin corral; (2) disruption of the nuclear lamina by host cell proteins such as Pin 1 and the viral protein kinase, UL97; (3) loss of this barrier followed by heterodimerization of the core NEC of HCMV (pUL53 and pUL50) at the INM promotes primary envelopment of capsids at the INM. (4) Following primary envelopment at the INM and transit through the PNS, the enveloped capsid then fuses with the ONM and undergoes deenvelopment, releasing the particle into the cytoplasm; (5) following acquisition of the inner tegument proteins, the tegumented capsid is transported on MT to sites of cytoplasmic assembly. (6) Alternatively, the capsid could be transported by nuclear actin to sites of budding from the INM; (7) following interactions between accessory viral proteins and pUL53 of the NEC, heterodimerization of pUL53 and pUL50 leading to primary envelopment at a nuclear invagination generated through disruption of the nuclear lamina as described above. Following primary envelopment, transit through the PNS space and fusion with the ONM, the capsid is released from ONM and trafficked to sites of virus assembly in the cytoplasm as described above.

4.2.2. Modifications of Nuclear Lamins and HCMV Nuclear Egress

The plasticity of the nucleus is controlled in large part by lamin polymerization. In mammalian cells, there are three lamin genes encoding lamins A and C, through alternative splicing, and lamins B1 and B2, which are distinct proteins that share about 60% identity [103]. All lamins share a similar structure consisting of a central rod domain flanked by an N-terminal head and C-terminal tail. In A- and B-type lamins but not lamin C, the tail contains a CaaX-motif that is the site of farnesylation and extensive processing, which results in membrane targeting. All domains of the lamins are required for polymerization that involves the head-to-tail polymers of dimers and subsequent assembly into protofilaments and intermediate filaments. Importantly, lamins are targets for several types of post-translational modifications including phosphorylation, acetylation, addition of lipid moieties (farnesylation and myristoylation), and proteolytic cleavage. These modifications, particularly phosphorylation, regulate polymerization of the lamina and protein–protein, as well as protein–DNA, interactions [104]. Several kinases target the lamins including protein kinases A and C, cyclin-dependent kinase 1 (Cdk1), and mitogen-activated protein kinase (MAPK), and because phosphorylation is reversible, this modification plays a key role in controlling the structure and function of the nuclear lamina; PP1 and PP2A mediate the dephosphorylation of lamins upon mitotic exit, allowing for lamin polymerization after cell division [103].

As discussed earlier, studies of the mechanism of MCMV capsid egress suggested that host proteins played a critical role in remodeling the nuclear membrane during nuclear egress (Figure 3). One such observation detailed the phosphorylation of the nuclear lamina by the protein kinase C (PKC) during nuclear egress of MCMV [45]. Subsequent studies provided compelling data that the CHPK of HCMV, pUL97, is responsible for the phosphorylation of lamins A/C and that this viral protein kinase mimics the activity of the host cell mitotic kinase, Cdk1 [46,57]. Consistent with this interpretation is the finding that UL97 phosphorylates Ser22 of lamin A/C, which is also a Cdk1 consensus site [57]. Although protein kinase C has been implicated in disruption of the lamina in cells infected with MCMV, Sharma et al. could not demonstrate recruitment of PKC to the lamina in cells infected with HCMV [46]. Moreover, a pan-PKC inhibitor did not block dissolution of the lamina in HCMV-infected cells [105]. More recent studies have described additional cellular proteins that are associated with the NEC and/or proposed to have a role in HCMV nuclear egress [35,36,65] (Table 1). For example, Pin 1 has been proposed to play a role in the disruption of the lamina in HCMV-infected cells [63]. Phosphorylation of lamins at consensus Cdk1 S/TP sites recruits isomerases that catalyze the conversion of proline from the *cis* to *trans* conformation, a modification that introduces kinks into the secondary structure of lamin that can then alter protein interactions and lamin polymerization [104]. Pin1 levels have been reported to increase with early kinetics during HCMV infection [63]. In addition, Pin1 has been shown to co-precipitate with lamin A in lysates from infected cells and localize to the nuclear envelope late in infection in association with the HCMV CHPK, pUL97 [63] (Figure 3). In the absence of Pin1, disruption of the nuclear lamina by pUL97 is decreased, suggesting that Pin1 participates in the modification of the nuclear lamina during HCMV infection [64].

4.2.3. Nuclear Actin and Nuclear Egress

Along with microtubules, actin filaments are considered major components of the cytoskeleton; however, nuclear actin, which represents about 20% of the cellular pool of actin, is less well characterized than its cytoplasmic counterpart. In the nucleus, actin has multiple roles including regulation of chromatin, transcription, and movement of chromosomes. The role of nuclear actin in herpesvirus-infected cells remains unresolved, but data suggest that nuclear actin may contribute to nuclear expansion and capsid motility in a virus- and cell-type dependent manner.

A role for nuclear actin in the herpesvirus replication cycle was first suggested by Forest et al. who reported that intranuclear trafficking of capsids was temperature- and energy-dependent, as well as sensitive to an inhibitor of actin polymerization, latrunculin A (LatA) [106]. Capsid movement was also reduced in the presence of 2,3-butanedione monoxime (BDM), a putative inhibitor of the actin-associated molecular motors, myosins, further suggesting an important role for actin in intranuclear capsid trafficking, although the mechanism(s) of action of BDM remains unclear [106]. Interestingly, capsid movement was not affected by another inhibitor of actin polymerization, cytochalasin D (CytoD), which has a mode of action distinct from LatA, which was attributed at the time to differences between the nuclear and cytoplasmic actin networks [106]. Later work by Simpson-Holley et al. demonstrated that the treatment of HSV-1 infected cells with LatA but not CytoD led to significant alteration in nuclear size and distribution and expansion of viral replication compartments relative to control infected cells, suggesting that nuclear actin acts as a scaffold supporting the changes in nuclear architecture observed during infection [107]. Importantly, these authors reported that even though LatA inhibited nuclear expansion during infection, focal disruption of the lamina caused by infection was still observed in the presence of the actin polymerization inhibitor; moreover, LatA treatment did not reduce virus yields at the concentrations used [107].

Both of the studies described above suggested that nuclear actin plays a role during the herpesvirus life cycle in mediating nuclear expansion and capsid transport; however, these assertions have been challenged by results from other studies. Feierbach et al. were the first to report induction of nuclear actin filaments in neurons infected with the α-herpesviruses PRV and HSV-1 [108]. In addition, these authors reported the association of capsids with nuclear actin filaments, which were found to be required for formation of capsid assembly domains termed assemblons [108]. Capsids were also found to colocalize with myosin V, suggesting a role for actin filaments in capsid transport [108]. Later work by Bosse et al. contradicted this hypothesis by demonstrating that intranuclear capsid trafficking of HSV-1, murine cytomegalovirus, and MHV68 capsids occurs in primary murine fibroblasts (MEFs) that do not contain discernable nuclear actin filaments after infection [109]. Moreover, inhibitors of actin polymerization, including LatA, did not affect intranuclear capsid motility in murine fibroblasts [109]. Thus, the authors concluded that intranuclear capsid motility is not dependent on nuclear actin in MEFs and other cells in which nuclear actin filaments are not induced by infection [109]. In a subsequent publication, these authors suggested that capsid motility occurs by diffusion [102].

In contrast to these studies, Wilkie et al. demonstrated a role for nuclear actin in egress in the context of HCMV infection [66] (Figure 3). These authors demonstrated that nuclear actin filaments were induced in cells infected with HCMV AD169 by 6 h p.i. and that their formation required IE gene expression [66]. HSV-1 did not induce nuclear actin in the same cells, suggesting that induction of nuclear actin filaments is cell type- and herpesvirus-specific. The accumulation of nuclear actin filaments continued until late times post infection in HCMV-infected cells, and the filaments were shown to extend from the viral RC to the nuclear rim. These observations suggested that the actin filaments could play a role in the trafficking of nascent capsids to the NEC. Depolymerization of nuclear actin by treatment of cells with LatA at late times post-infection reduced the number of capsids in the cytoplasm and decreased viral titer, again suggesting that nuclear actin plays a role in nuclear egress [66]. Importantly, the concentrations of LatA that were used in these studies to alter virus assembly were also shown to lead to depolymerization of nuclear actin [66]. Electron microscopy revealed that in the absence of nuclear actin, capsids were more often associated with the RC, and fewer were observed near the nuclear rim [66]. Taken together, these data suggested that nuclear actin could play a significant role in the movement of capsids from the RC to the NE (Figure 3). This model was supported the finding that myosin Va, a molecular motor associated with nuclear actin, was also required for nuclear egress [67]. Importantly, these authors showed colocalization between capsids and myosin Va by EM and of the major capsid protein, myosin Va, and nuclear actin filaments by immunofluorescence [67]. Similar to the findings that were obtained when inhibitors of nuclear actin polymerization were used, inhibition of myosin Va activity with siRNA or expression of a dominant negative allele decreased the production of infectious virus, the accumulation of cytoplasmic capsids, and the trafficking for capsids from the RC to the NE [67]. These results strongly implicate nuclear actin and myosin Va in nuclear trafficking and egress of HCMV capsids in infected human fibroblasts. As mentioned above, although nuclear actin has been shown to be important in RC distribution and expansion in HCMV- and HSV-1-infected cells [107], the data from other studies of this herpesvirus support a model in which capsids diffuse to the NE rather than using nuclear actin in their trafficking itinerary within the nucleus [5,102].

4.2.4. ESCRT-III and Nuclear Egress

Although several studies have reported that overexpression or inhibition of various cellular functions resulted in decreased virus replication, the impact on virus yield was often less than an order of magnitude, suggesting redundancies in the contribution of individual host cell functions to nuclear egress of herpesviruses. In contrast to these findings, a striking phenotype in the nuclear egress of HSV has been described in infected cells following knockdowns of members of the endosomal sorting complex required for transport (ESCRT) [23]. This well-studied cellular complex plays a role in a wide spectrum of cytoplasmic membrane remodeling and membrane scission events. Relevant to this review, ESCRT-III has also been shown to contribute to budding processes from the nuclear envelope, including repair of the nuclear membrane during cytokinesis and maintenance of the nuclear membrane integrity in interphase cells [20,22,110]. Because components of the ESCRT-III system play a critical role in maintaining the structure and function of the INM, it is not surprising that INM deformation and rupture during herpesvirus egress likely would elicit this cellular response [111,112]. Previously, the ESCRT-III pathway has been shown to play a role in EBV nuclear egress [37]. This study suggested that recruitment of an adaptor of ESCRT-III, ALIX, by the pUL34 EBV ortholog BFRF1 to the INM rim was required for efficient EBV egress [37]. More recent studies have shown that depletion of ALIX reduced recruitment of the HSV NEC to the INM and decreased virus yield; however, the latter finding could also be consistent with the role of ESCRT-III in cytoplasmic envelopment [23]. Importantly, this same report described an increase in primary enveloped capsids in the nucleoplasm and the accumulation of partially enveloped capsids in the PNS as compared to control cells, strongly suggesting a role for ESCRT-III in the nuclear egress of HSV-1 [23]. Deletion of a component of ESCRT-III, the charged multivesicular body protein 4B (CHMP4B), in HeLa cells followed by siRNA knockdowns of CHMP4A and CHMP4C and infection with HSV-1 resulted in a similar phenotype as deletion of the ESCRT-III adaptor ALIX [23]. In this latter experiment, the yield of infectious virus fell by about $3 \log_{10}$, and consistent with the initial observation, there was an accumulation of enveloped capsids associated with INM invaginations in the nucleoplasm, suggesting a failure of capsids to successfully transit through the INM [23]. Consistent with these findings, a similar phenotype was described when a dominant negative VPS4 ATPase was used to alter ESCRT-III function in HSV infected cells [23]. Lastly, a recent study from the same group of investigators provided additional evidence for the role of the ESCRT-III complex in nuclear egress by demonstrating that infection with a replication competent HSV-1 containing a mutation in the disordered domain of pUL34 led to the altered localization of the NEC-infected cells, loss of pUL34 colocalization with CHMP4B of the ESCRT-III complex, and the accumulation of enveloped capsids in the PNS [68]. In contrast, studies in HCMV have thus far failed to provide convincing evidence for a role of the ESCRT-III pathway in nuclear egress, although it is important to note that studies to specifically determine the role of the ESCRT-III in HCMV nuclear egress have not been accomplished presumably because of the experimental challenge to deconvolute the impact of ESCRT-III functions on nuclear and cytoplasmic phases of virus assembly. For example, in one study, inhibition of ESCRT-III function by overexpression of a dominant negative VSP4a inhibited virus production, presumably through blocking cytoplasmic envelopment of HCMV [113]. However, definitive evidence excluding the potential role of the expression of the dominant negative VPS4 on nuclear egress was not explored [113]. A more definitive series of experiments using dominant negative constructs to inhibit ESCRT-III function that also controlled for the effects of overexpression of these dominant negative proteins on cell viability failed to demonstrate a significant impact on envelopment and the release of infectious HCMV, a finding, as noted above, that was inconsistent with results from similar experiments in studies of HSV-1 [23,114,115]. Although the results from various studies of the role of the ESCRT-III complex in nuclear egress appear contradictory, the experimental approaches used in these studies vary considerably, ranging from dominant negative mutants, siRNA depletions, CRISPR-derived knockout cells, and recombinant viruses. Yet, a summation of the data from studies in α and γ herpesviruses strongly suggests a role of ESCRT-III in nuclear egress and likely illustrates another example of herpesviruses co-opting a normal cellular function for more efficient replication.

4.3. Nuclear Egress: Transit of Newly Assembled HCMV Capsids from the PNS through the ONM to the Cytoplasm

4.3.1. Role of Viral Proteins

Although an extensive amount of literature describing the primary envelopment of herpesvirus capsids at the INM and movement into PNS is available, the transit of the enveloped capsid through the PNS and ONM to the cytoplasm of the infected cells, including that of HCMV, remains understudied and largely undefined. As discussed previously, the deletion of the α-herpesvirus protein kinase, pUS3, has been reported to result in the accumulation of enveloped capsids in the PNS [44,89]. Deletion of HSV gB and gH led to a significant defect in nuclear egress that was characterized by accumulation of enveloped particles in the PNS and herniations of the INM, suggesting that deletion of these viral glycoproteins that constitute the fusion machinery of herpesviruses resulted in the failure of enveloped capsids to fuse with the ONM [116]. Subsequently, a similar nuclear egress phenotype was observed in cells infected with HSV mutant virus in which only the fusion loops of gB were mutated [69]. These findings strongly argued that HSV gB, which has been shown to require expression of gH/gL for the fusion of incoming virions in permissive cells, was required for fusion at the ONM of primary enveloped capsids during nuclear egress. However, similar phenotypes have not been observed in PRV mutant viruses with deletions of homologous envelope glycoproteins [117]. To date, the differences between the nuclear egress phenotypes of these two closely related α-herpesviruses following deletion of essential glycoproteins have not been resolved. In the case of HCMV, there are no conclusive data relevant to its transit through the PNS and ONM to the cytoplasm.

4.3.2. Role of Cellular Proteins

During formation of the nuclear pores complex (NPC) in interphase cells, fusion between the INM and ONM has been proposed to take place by an inside-out evagination of the INM followed by approximation of the INM and ONM and subsequent fusion [118,119]. Although components of ESCRT-III have been implicated in this process, other cellular proteins have also been shown to have a role in NPC formation. Two of these, Torsin A and Torsin B, both members of the AAA+ ATPase superfamily of proteins, have been studied in the context of herpesvirus egress [70,120]. These proteins are localized to the PNS in interphase cells, and similar to ESCRT-III, the Torsins have been proposed to have a role in maintaining the nuclear envelope. Mutations in the gene encoding Torsin A, in both humans and in a transgenic mouse model of a human disease associated with loss of Torsin A function, led to altered NEC structure and blebbing of the INM [121,122]. A modest decrease in HSV replication in cells overexpressing Torsin A has been reported [70]. More recently, deletion of both Torsin A and Torsin B resulted in decreased PRV replication early in infection and, more interestingly, an accumulation of primary enveloped capsids in the PNS [120]. These enveloped particles were not associated with the ONM but appeared to remain attached to the INM [120]. These most recent findings are provocative in that retention of primary enveloped particles in the PNS suggested that Torsins could contribute to the transit through the PNS and de-envelopment at the ONM.

Torsin A has also been shown to play an important role in the assembly of the LINC complexes, possibly through its interactions with the nesprins 1–3 and SUN proteins [120]. As described previously, the canonical LINC complex consists of a trimer of KASH domain proteins (nesprins), which are inserted in the ONM, and a trimer of SUN proteins that reside in the INM and act as a receptor for the C-terminal KASH domain of the nesprin (Figure 2) [26,30,123]. Recent studies show that the LINC complex forms 6:6 branched assemblies consisting of three KASH domain protein dimers and two SUN domain protein trimers [124]. Crosslinking of cytoskeletal elements by nesprins in branched assemblies could allow for more robust distribution of mechanical forces along the nuclear envelope and transduction of signals from the cytoplasm and external environment.

Virus yield in PRV-infected cells overexpressing dominant negative forms SUN1 and SUN2 was decreased, and perhaps more interestingly, primary enveloped capsids accumulated in the PNS and in the ER [71]. This latter observation provided evidence that the altered function of the LINC complex could limit nuclear egress of PRV and at least partially phenocopy the loss of Torsin A in PRV-infected cells [71,120]. These authors also reported that overexpression of these dominant negative forms of SUN1 and SUN2 led to a widening of the perinuclear space. The functions of Torsins in LINC complex dynamics, maintenance of the perinuclear space, and membrane fusion events during NPC assembly fusion in normal cells all suggest that these proteins could also have a role in herpesvirus transit through the PNS and ONM during nuclear egress.

In the context of HCMV infection, Buchkovich et al. described a decrease in the steady state levels of the LINC proteins Sun1 and Sun2, which was correlated with an increase in the thickness of the PNS [4]. This finding is reminiscent of results from the knockdown of SUN protein expression in PRV-infected cells described above [71,120]. More recent studies have provided evidence of a potential function(s) of the LINC complex in the context of infection-induced nuclear rotation, a cellular response that has also been reported to occur during cell migration and secondary to mechanical stress [8,27,28]. These authors observed the accumulation of Sun1 at the nuclear membrane in close proximity to the cytoplasmic AC at late times post-infection, similar to the localization of condensed chromatin [8]. Interestingly, these authors previously reported that nuclear rotation in infected cells preceded the maturation of the cytoplasmic AC [16], a finding that could be interpreted as evidence that nuclear and cytoplasmic phases of HCMV assembly were coordinated through nuclear-cytoplasmic signals, a proposed function of the LINC complex [8]. With the exception of these observations on the potential contribution of the LINC complex in HCMV nuclear egress, there are few, if any, other published data available that have related the activity of cellular proteins to the transit of HCMV through the PNS and ONM.

5. Summary

There has been considerable progress in defining the mechanism of egress of herpesviruses from the nucleus, including the role of individual virus-encoded proteins in egress. However, the contribution of cellular proteins to nuclear egress remains less well understood and an area of active investigation. Several cellular proteins and protein complexes have been shown to play a role in nuclear egress, and in some cases, specific function(s) in egress have been assigned. Moreover, the contribution of cellular proteins to egress in many cases appears redundant; thus, it has been difficult to establish robust phenotypes following conventional loss-of-function approaches used to study individual proteins. This redundancy in function can likely be accounted for by a similar redundancy in function of these cellular proteins in the homeostasis of uninfected interphase cells. In addition, efficient egress could also require combinatorial functions of several cellular proteins, thus further limiting conventional approaches commonly used for the definition of the function of a single cellular protein. In the case of HCMV, efficient nuclear egress is almost certainly dependent on the contribution of cellular responses to the infection if nuclear events during egress parallel the remarkable modifications of the host cell architecture, organelles, and intracellular trafficking pathways during HCMV replication. Defining the coordinated activities of host cell proteins and viral proteins during egress will provide an opportunity to extend current understanding of the role of cellular proteins in trafficking of macromolecules in and out of the nucleus.

Author Contributions: V.S. and W.B. contributed to the analysis of the published literature and writing of this manuscript. V.S. and W.B. declare they have no conflicts of interest with the publication of this manuscript. All authors have read and agreed to the published version of the manuscript.

Funding: This work was supported by the NIH (R01AI035602 and 1R01AI120619) to W.B.

Institutional Review Board Statement: Not applicable.

Informed Consent Statement: Not applicable.

Data Availability Statement: Not applicable.

Conflicts of Interest: The authors declare no conflict of interest.

References

1. Kicuntod, J.; Alkhashrom, S.; Häge, S.; Diewald, B.; Müller, R.; Hahn, F.; Lischka, P.; Sticht, H.; Eichler, J.; Marschall, M. Properties of Oligomeric Interaction of the Cytomegalovirus Core Nuclear Egress Complex (NEC) and Its Sensitivity to an NEC Inhibitory Small Molecule. *Viruses* **2021**, *13*, 462. [CrossRef]
2. Monier, K.; Armas, J.C.G.; Etteldorf, S.; Ghazal, P.; Sullivan, K. Annexation of the interchromosomal space during viral infection. *Nat. Cell Biol.* **2000**, *2*, 661–665. [CrossRef]
3. Aho, V.; Mäntylä, E.; Ekman, A.; Hakanen, S.; Mattola, S.; Chen, J.-H.; Weinhardt, V.; Ruokolainen, V.; Sodeik, B.; Larabell, C.; et al. Quantitative Microscopy Reveals Stepwise Alteration of Chromatin Structure during Herpesvirus Infection. *Viruses* **2019**, *11*, 935. [CrossRef] [PubMed]
4. Buchkovich, N.J.; Maguire, T.G.; Alwine, J.C. Role of the Endoplasmic Reticulum Chaperone BiP, SUN Domain Proteins, and Dynein in Altering Nuclear Morphology during Human Cytomegalovirus Infection. *J. Virol.* **2010**, *84*, 7005–7017. [CrossRef] [PubMed]
5. Aho, V.; Myllys, M.; Ruokolainen, V.; Hakanen, S.; Mäntylä, E.; Virtanen, J.; Hukkanen, V.; Kühn, T.; Timonen, J.; Mattila, K.; et al. Chromatin organization regulates viral egress dynamics. *Sci. Rep.* **2017**, *7*, 3692. [CrossRef] [PubMed]
6. Alwine, J.C. The Human Cytomegalovirus Assembly Compartment: A Masterpiece of Viral Manipulation of Cellular Processes That Facilitates Assembly and Egress. *PLoS Pathog.* **2012**, *8*, e1002878. [CrossRef]
7. Villinger, C.; Neusser, G.; Kranz, C.; Walther, P.; Mertens, T. 3D Analysis of HCMV Induced-Nuclear Membrane Structures by FIB/SEM Tomography: Insight into an Unprecedented Membrane Morphology. *Viruses* **2015**, *7*, 5686–5704. [CrossRef]
8. Procter, D.J.; Furey, C.; Garza-Gongora, A.G.; Kosak, S.T.; Walsh, D. Cytoplasmic control of intranuclear polarity by human cytomegalovirus. *Nat. Cell Biol.* **2020**, *587*, 109–114. [CrossRef]
9. Sanchez, V.; Clark, C.L.; Yen, J.Y.; Dwarakanath, R.; Spector, D.H. Viable Human Cytomegalovirus Recombinant Virus with an Internal Deletion of the IE2 86 Gene Affects Late Stages of Viral Replication. *J. Virol.* **2002**, *76*, 2973–2989. [CrossRef] [PubMed]
10. Sanchez, V.; McElroy, A.K.; Yen, J.; Tamrakar, S.; Clark, C.L.; Schwartz, R.A.; Spector, D.H. Cyclin-Dependent Kinase Activity Is Required at Early Times for Accurate Processing and Accumulation of the Human Cytomegalovirus UL122-123 and UL37 Immediate-Early Transcripts and at Later Times for Virus Production. *J. Virol.* **2004**, *78*, 11219–11232. [CrossRef]
11. Salvant, B.S.; Fortunato, E.A.; Spector, D.H. Cell Cycle Dysregulation by Human Cytomegalovirus: Influence of the Cell Cycle Phase at the Time of Infection and Effects on Cyclin Transcription. *J. Virol.* **1998**, *72*, 3729–3741. [CrossRef] [PubMed]
12. Sanchez, V.; Greis, K.D.; Sztul, E.; Britt, W.J. Accumulation of Virion Tegument and Envelope Proteins in a Stable Cytoplasmic Compartment during Human Cytomegalovirus Replication: Characterization of a Potential Site of Virus Assembly. *J. Virol.* **2000**, *74*, 975–986. [CrossRef] [PubMed]
13. Das, S.; Vasanji, A.; Pellett, P.E. Three-Dimensional Structure of the Human Cytomegalovirus Cytoplasmic Virion Assembly Complex Includes a Reoriented Secretory Apparatus. *J. Virol.* **2007**, *81*, 11861–11869. [CrossRef]
14. Hook, L.; Grey, F.; Grabski, R.; Tirabassi, R.; Doyle, T.; Hancock, M.; Landais, I.; Jeng, S.; McWeeney, S.; Britt, W.; et al. Cytomegalovirus miRNAs Target Secretory Pathway Genes to Facilitate Formation of the Virion Assembly Compartment and Reduce Cytokine Secretion. *Cell Host Microbe* **2014**, *15*, 363–373. [CrossRef]
15. Rebmann, G.M.; Grabski, R.; Sanchez, V.; Britt, W.J. Phosphorylation of Golgi Peripheral Membrane Protein Grasp65 Is an Integral Step in the Formation of the Human Cytomegalovirus Cytoplasmic Assembly Compartment. *mBio* **2016**, *7*, e01554-16. [CrossRef] [PubMed]
16. Procter, D.J.; Banerjee, A.; Nukui, M.; Kruse, K.; Gaponenko, V.; Murphy, E.A.; Komarova, Y.; Walsh, D. The HCMV Assembly Compartment Is a Dynamic Golgi-Derived MTOC that Controls Nuclear Rotation and Virus Spread. *Dev. Cell* **2018**, *45*, 83–100.e7. [CrossRef]
17. Indran, S.V.; Ballestas, M.E.; Britt, W.J. Bicaudal D1-Dependent Trafficking of Human Cytomegalovirus Tegument Protein pp150 in Virus-Infected Cells. *J. Virol.* **2010**, *84*, 3162–3177. [CrossRef]
18. Barbosa, A.D.; Sembongi, H.; Su, W.-M.; Abreu, S.; Reggiori, F.; Carman, G.M.; Siniossoglou, S. Lipid partitioning at the nuclear envelope controls membrane biogenesis. *Mol. Biol. Cell* **2015**, *26*, 3641–3657. [CrossRef]
19. Malhas, A.; Goulbourne, C.; Vaux, D. The nucleoplasmic reticulum: Form and function. *Trends Cell Biol.* **2011**, *21*, 362–373. [CrossRef]
20. Vietri, M.; Radulovic, M.; Stenmark, H. The many functions of ESCRTs. *Nat. Rev. Mol. Cell Biol.* **2020**, *21*, 25–42. [CrossRef]
21. Rose, K.M.; Spada, S.J.; Hirsch, V.M.; Bouamr, F. When in Need of an ESCRT: The Nature of Virus Assembly Sites Suggests Mechanistic Parallels between Nuclear Virus Egress and Retroviral Budding. *Viruses* **2021**, *13*, 1138. [CrossRef] [PubMed]
22. Vietri, M.; Stenmark, H.; Campsteijn, C. Closing a gap in the nuclear envelope. *Curr. Opin. Cell Biol.* **2016**, *40*, 90–97. [CrossRef]
23. Arii, J.; Watanabe, M.; Maeda, F.; Tokai-Nishizumi, N.; Chihara, T.; Miura, M.; Maruzuru, Y.; Koyanagi, N.; Kato, A.; Kawaguchi, Y. ESCRT-III mediates budding across the inner nuclear membrane and regulates its integrity. *Nat. Commun.* **2018**, *9*, 3379. [CrossRef]

24. Lin, D.H.; Hoelz, A. The Structure of the Nuclear Pore Complex (An Update). *Annu. Rev. Biochem.* **2019**, *88*, 725–783. [CrossRef]
25. Starr, D.A.; Fridolfsson, H.N. Interactions Between Nuclei and the Cytoskeleton Are Mediated by SUN-KASH Nuclear-Envelope Bridges. *Annu. Rev. Cell Dev. Biol.* **2010**, *26*, 421–444. [CrossRef] [PubMed]
26. Alam, S.G.; Zhang, Q.; Prasad, N.; Li, Y.; Chamala, S.; Kuchibhotla, R.; Kc, B.; Aggarwal, V.; Shrestha, S.; Jones, A.L.; et al. The mammalian LINC complex regulates genome transcriptional responses to substrate rigidity. *Sci. Rep.* **2016**, *6*, 38063. [CrossRef]
27. Gundersen, G.G.; Worman, H.J. Nuclear Positioning. *Cell* **2013**, *152*, 1376–1389. [CrossRef] [PubMed]
28. Maninová, M.; Klímová, Z.; Parsons, J.T.; Weber, M.J.; Iwanicki, M.P.; Vomastek, T. The Reorientation of Cell Nucleus Promotes the Establishment of Front–Rear Polarity in Migrating Fibroblasts. *J. Mol. Biol.* **2013**, *425*, 2039–2055. [CrossRef]
29. Razafsky, D.; Wirtz, D.; Hodzic, D. Nuclear Envelope in Nuclear Positioning and Cell Migration. *Adv. Exp. Med. Biol.* **2014**, *773*, 471–490. [CrossRef]
30. Burke, B. Chain reaction: LINC complexes and nuclear positioning. *F1000Research* **2019**, *8*, 136. [CrossRef]
31. Burla, R.; La Torre, M.; Maccaroni, K.; Verni, F.; Giunta, S.; Saggio, I. Interplay of the nuclear envelope with chromatin in physiology and pathology. *Nucleus* **2020**, *11*, 205–218. [CrossRef]
32. Uhler, C.; Shivashankar, G.V. Regulation of genome organization and gene expression by nuclear mechanotransduction. *Nat. Rev. Mol. Cell Biol.* **2017**, *18*, 717–727. [CrossRef] [PubMed]
33. Johnson, D.C.; Baines, J.D. Herpesviruses remodel host membranes for virus egress. *Nat. Rev. Genet.* **2011**, *9*, 382–394. [CrossRef]
34. Fuchs, W.; Klupp, B.G.; Granzow, H.; Osterrieder, N.; Mettenleiter, T.C. The Interacting UL31 and UL34 Gene Products of Pseudorabies Virus Are Involved in Egress from the Host-Cell Nucleus and Represent Components of Primary Enveloped but Not Mature Virions. *J. Virol.* **2002**, *76*, 364–378. [CrossRef] [PubMed]
35. Milbradt, J.; Kraut, A.; Hutterer, C.; Sonntag, E.; Schmeiser, C.; Ferro, M.; Wagner, S.; Lenac, T.; Claus, C.; Pinkert, S.; et al. Proteomic Analysis of the Multimeric Nuclear Egress Complex of Human Cytomegalovirus. *Mol. Cell. Proteom.* **2014**, *13*, 2132–2146. [CrossRef] [PubMed]
36. Milbradt, J.; Auerochs, S.; Sticht, H.; Marschall, M. Cytomegaloviral proteins that associate with the nuclear lamina: Components of a postulated nuclear egress complex. *J. Gen. Virol.* **2009**, *90*, 579–590. [CrossRef]
37. Lee, C.-P.; Liu, P.-T.; Kung, H.-N.; Su, M.-T.; Chua, H.-H.; Chang, Y.-H.; Chang, C.-W.; Tsai, C.-H.; Liu, F.-T.; Chen, M.-R. The ESCRT Machinery Is Recruited by the Viral BFRF1 Protein to the Nucleus-Associated Membrane for the Maturation of Epstein-Barr Virus. *PLoS Pathog.* **2012**, *8*, e1002904. [CrossRef]
38. Gonnella, R.; Farina, A.; Santarelli, R.; Raffa, S.; Feederle, R.; Bei, R.; Granato, M.; Modesti, A.; Frati, L.; Delecluse, H.-J.; et al. Characterization and Intracellular Localization of the Epstein-Barr Virus Protein BFLF2: Interactions with BFRF1 and with the Nuclear Lamina. *J. Virol.* **2005**, *79*, 3713–3727. [CrossRef]
39. Farina, A.; Feederle, R.; Raffa, S.; Gonnella, R.; Santarelli, R.; Frati, L.; Angeloni, A.; Torrisi, M.R.; Faggioni, A.; Delecluse, H.-J. BFRF1 of Epstein-Barr Virus Is Essential for Efficient Primary Viral Envelopment and Egress. *J. Virol.* **2005**, *79*, 3703–3712. [CrossRef]
40. Klupp, B.G.; Granzow, H.; Fuchs, W.; Keil, G.M.; Finke, S.; Mettenleiter, T.C. Vesicle formation from the nuclear membrane is induced by coexpression of two conserved herpesvirus proteins. *Proc. Natl. Acad. Sci. USA* **2007**, *104*, 7241–7246. [CrossRef]
41. Bigalke, J.M.; Heuser, T.; Nicastro, D.; Heldwein, E.E. Membrane deformation and scission by the HSV-1 nuclear egress complex. *Nat. Commun.* **2014**, *5*, 4131. [CrossRef]
42. Draganova, E.B.; Zhang, J.; Zhou, Z.H.; E Heldwein, E. Structural basis for capsid recruitment and coat formation during HSV-1 nuclear egress. *eLife* **2020**, *9*, e56627. [CrossRef] [PubMed]
43. Hagen, C.; Dent, K.C.; Zeev-Ben-Mordehai, T.; Grange, M.; Bosse, J.B.; Whittle, C.; Klupp, B.G.; Siebert, C.; Vasishtan, D.; Bäuerlein, F.J.; et al. Structural Basis of Vesicle Formation at the Inner Nuclear Membrane. *Cell* **2015**, *163*, 1692–1701. [CrossRef] [PubMed]
44. Wagenaar, F.; Pol, J.M.A.; Peeters, B.; Gielkens, A.L.J.; De Wind, N.; Kimman, T.G. The US3-encoded protein kinase from pseudorabies virus affects egress of virions from the nucleus. *J. Gen. Virol.* **1995**, *76*, 1851–1859. [CrossRef]
45. Muranyi, W.; Haas, J.; Wagner, M.; Krohne, G.; Koszinowski, U.H. Cytomegalovirus Recruitment of Cellular Kinases to Dissolve the Nuclear Lamina. *Science* **2002**, *297*, 854–857. [CrossRef] [PubMed]
46. Sharma, M.; Kamil, J.P.; Coughlin, M.; Reim, N.I.; Coen, D.M. Human Cytomegalovirus UL50 and UL53 Recruit Viral Protein Kinase UL97, Not Protein Kinase C, for Disruption of Nuclear Lamina and Nuclear Egress in Infected Cells. *J. Virol.* **2014**, *88*, 249–262. [CrossRef]
47. Kuan, M.I.; O'Dowd, J.M.; Chughtai, K.; Hayman, I.; Brown, C.J.; Fortunato, E.A. Human Cytomegalovirus nuclear egress and secondary envelopment are negatively affected in the absence of cellular p53. *Virology* **2016**, *497*, 279–293. [CrossRef]
48. I Kuan, M.; O'Dowd, J.M.; Fortunato, E.A. The absence of p53 during Human Cytomegalovirus infection leads to decreased UL53 expression, disrupting UL50 localization to the inner nuclear membrane, and thereby inhibiting capsid nuclear egress. *Virology* **2016**, *497*, 262–278. [CrossRef]
49. Monte, P.D.; Pignatelli, S.; Zini, N.; Maraldi, N.M.; Perret, E.; Prevost, M.C.; Landini, M.P. Analysis of intracellular and intraviral localization of the human cytomegalovirus UL53 protein. *J. Gen. Virol.* **2002**, *83*, 1005–1012. [CrossRef]
50. Schmeiser, C.; Borst, E.; Sticht, H.; Marschall, M.; Milbradt, J. The cytomegalovirus egress proteins pUL50 and pUL53 are translocated to the nuclear envelope through two distinct modes of nuclear import. *J. Gen. Virol.* **2013**, *94*, 2056–2069. [CrossRef]

51. Walzer, S.A.; Egerer-Sieber, C.; Sticht, H.; Sevvana, M.; Hohl, K.; Milbradt, J.; Muller, Y.A.; Marschall, M. Crystal Structure of the Human Cytomegalovirus pUL50-pUL53 Core Nuclear Egress Complex Provides Insight into a Unique Assembly Scaffold for Virus-Host Protein Interactions. *J. Biol. Chem.* **2015**, *290*, 27452–27458. [CrossRef]

52. Leigh, K.; Sharma, M.; Mansueto, M.S.; Boeszoermenyi, A.; Filman, D.; Hogle, J.; Wagner, G.; Coen, D.M.; Arthanari, H. Structure of a herpesvirus nuclear egress complex subunit reveals an interaction groove that is essential for viral replication. *Proc. Natl. Acad. Sci. USA* **2015**, *112*, 9010–9015. [CrossRef] [PubMed]

53. Muller, Y.A.; Häge, S.; Alkhashrom, S.; Höllriegl, T.; Weigert, S.; Dolles, S.; Hof, K.; Walzer, S.A.; Egerer-Sieber, C.; Conrad, M.; et al. High-resolution crystal structures of two prototypical β- and γ-herpesviral nuclear egress complexes unravel the determinants of subfamily specificity. *J. Biol. Chem.* **2020**, *295*, 3189–3201. [CrossRef] [PubMed]

54. Lye, M.F.; Sharma, M.; El Omari, K.; Filman, D.; Schuermann, J.P.; Hogle, J.M.; Coen, D.M. Unexpected features and mechanism of heterodimer formation of a herpesvirus nuclear egress complex. *EMBO J.* **2015**, *34*, 2937–2952. [CrossRef]

55. Borst, E.M.; Bauerfeind, R.; Binz, A.; Stephan, T.M.; Neuber, S.; Wagner, K.; Steinbrück, L.; Sodeik, B.; Roviš, T.L.; Jonjić, S.; et al. The Essential Human Cytomegalovirus Proteins pUL77 and pUL93 are Structural Components Necessary for Viral Genome Encapsidation. *J. Virol.* **2016**, *90*, 5860–5875. [CrossRef] [PubMed]

56. Köppen-Rung, P.; Dittmer, A.; Bogner, E. Intracellular Distribution of Capsid-Associated pUL77 of Human Cytomegalovirus and Interactions with Packaging Proteins and pUL93. *J. Virol.* **2016**, *90*, 5876–5885. [CrossRef]

57. Hamirally, S.; Kamil, J.P.; Ndassa-Colday, Y.M.; Lin, A.J.; Jahng, W.J.; Baek, M.-C.; Noton, S.; Silva, L.A.; Simpson-Holley, M.; Knipe, D.M.; et al. Viral Mimicry of Cdc2/Cyclin-Dependent Kinase 1 Mediates Disruption of Nuclear Lamina during Human Cytomegalovirus Nuclear Egress. *PLoS Pathog.* **2009**, *5*, e1000275. [CrossRef]

58. Sharma, M.; Bender, B.J.; Kamil, J.; Lye, M.F.; Pesola, J.M.; Reim, N.I.; Hogle, J.; Coen, D.M. Human Cytomegalovirus UL97 Phosphorylates the Viral Nuclear Egress Complex. *J. Virol.* **2014**, *89*, 523–534. [CrossRef] [PubMed]

59. He, Z.; He, Y.S.; Kim, Y.; Chu, L.; Ohmstede, C.; Biron, K.K.; Coen, D.M. The human cytomegalovirus UL97 protein is a protein kinase that autophosphorylates on serines and threonines. *J. Virol.* **1997**, *71*, 405–411. [CrossRef]

60. Sonntag, E.; Milbradt, J.; Svrlanska, A.; Strojan, H.; Häge, S.; Kraut, A.; Hesse, A.-M.; Amin, B.; Sonnewald, U.; Couté, Y.; et al. Protein kinases responsible for the phosphorylation of the nuclear egress core complex of human cytomegalovirus. *J. Gen. Virol.* **2017**, *98*, 2569–2581. [CrossRef]

61. Milbradt, J.; Auerochs, S.; Marschall, M. Cytomegaloviral proteins pUL50 and pUL53 are associated with the nuclear lamina and interact with cellular protein kinase C. *J. Gen. Virol.* **2007**, *88*, 2642–2650. [CrossRef]

62. Marschall, M.; Marzi, A.; aus dem Siepen, P.; Jochmann, R.; Kalmer, M.; Auerochs, S.; Lischka, P.; Leis, M.; Stamminger, T. Cellular p32 Recruits Cytomegalovirus Kinase pUL97 to Redistribute the Nuclear Lamina. *J. Biol. Chem.* **2005**, *280*, 33357–33367. [CrossRef] [PubMed]

63. Milbradt, J.; Webel, R.; Auerochs, S.; Sticht, H.; Marschall, M. Novel Mode of Phosphorylation-triggered Reorganization of the Nuclear Lamina during Nuclear Egress of Human Cytomegalovirus. *J. Biol. Chem.* **2010**, *285*, 13979–13989. [CrossRef]

64. Milbradt, J.; Hutterer, C.; Bahsi, H.; Wagner, S.; Sonntag, E.; Horn, A.H.C.; Kaufer, B.B.; Mori, Y.; Sticht, H.; Fossen, T.; et al. The Prolyl Isomerase Pin1 Promotes the Herpesvirus-Induced Phosphorylation-Dependent Disassembly of the Nuclear Lamina Required for Nucleocytoplasmic Egress. *PLoS Pathog.* **2016**, *12*, e1005825. [CrossRef] [PubMed]

65. Yang, B.; Liu, X.-J.; Yao, Y.; Jiang, X.; Wang, X.-Z.; Yang, H.; Sun, J.-Y.; Miao, Y.; Wang, W.; Huang, Z.-L.; et al. WDR5 Facilitates Human Cytomegalovirus Replication by Promoting Capsid Nuclear Egress. *J. Virol.* **2018**, *92*, e00207-18. [CrossRef] [PubMed]

66. Wilkie, A.R.; Lawler, J.; Coen, D.M. A Role for Nuclear F-Actin Induction in Human Cytomegalovirus Nuclear Egress. *mBio* **2016**, *7*, e01254-16. [CrossRef] [PubMed]

67. Wilkie, A.R.; Sharma, M.; Pesola, J.M.; Ericsson, M.; Fernandez, R.; Coen, D.M. A Role for Myosin Va in Human Cytomegalovirus Nuclear Egress. *J. Virol.* **2018**, *92*, e01849-17. [CrossRef] [PubMed]

68. Arii, J.; Takeshima, K.; Maruzuru, Y.; Koyanagi, N.; Nakayama, Y.; Kato, A.; Mori, Y.; Kawaguchi, Y. Role of the arginine cluster in the disordered domain of Herpes Simplex Virus 1 UL34 for the recruitment of ESCRT-III for viral primary envelopment. *J. Virol.* **2021**. [CrossRef]

69. Wright, C.C.; Wisner, T.W.; Hannah, B.P.; Eisenberg, R.J.; Cohen, G.H.; Johnson, D.C. Fusion between Perinuclear Virions and the Outer Nuclear Membrane Requires the Fusogenic Activity of Herpes Simplex Virus gB. *J. Virol.* **2009**, *83*, 11847–11856. [CrossRef]

70. Maric, M.; Shao, J.; Ryan, R.J.; Wong, C.-S.; Gonzalez-Alegre, P.; Roller, R.J. A Functional Role for TorsinA in Herpes Simplex Virus 1 Nuclear Egress. *J. Virol.* **2011**, *85*, 9667–9679. [CrossRef]

71. Klupp, B.G.; Hellberg, T.; Granzow, H.; Franzke, K.; Gonzalez, B.D.; Goodchild, R.E.; Mettenleiter, T.C. Integrity of the Linker of Nucleoskeleton and Cytoskeleton Is Required for Efficient Herpesvirus Nuclear Egress. *J. Virol.* **2017**, *91*, e00330-17. [CrossRef]

72. Dunn, W.; Chou, C.; Li, H.; Hai, R.; Patterson, D.; Stolc, V.; Zhu, H.; Liu, F. Functional profiling of a human cytomegalovirus genome. *Proc. Natl. Acad. Sci. USA* **2003**, *100*, 14223–14228. [CrossRef] [PubMed]

73. Sam, M.D.; Evans, B.T.; Coen, D.M.; Hogle, J.M. Biochemical, Biophysical, and Mutational Analyses of Subunit Interactions of the Human Cytomegalovirus Nuclear Egress Complex. *J. Virol.* **2009**, *83*, 2996–3006. [CrossRef] [PubMed]

74. Bigalke, J.M.; E Heldwein, E. Structural basis of membrane budding by the nuclear egress complex of herpesviruses. *EMBO J.* **2015**, *34*, 2921–2936. [CrossRef]

75. Marschall, M.; Häge, S.; Conrad, M.; Alkhashrom, S.; Kicuntod, J.; Schweininger, J.; Kriegel, M.; Lösing, J.; Tillmanns, J.; Neipel, F.; et al. Nuclear Egress Complexes of HCMV and Other Herpesviruses: Solving the Puzzle of Sequence Coevolution, Conserved Structures and Subfamily-Spanning Binding Properties. *Viruses* **2020**, *12*, 683. [CrossRef]
76. Arii, J.; Takeshima, K.; Maruzuru, Y.; Koyanagi, N.; Kato, A.; Kawaguchi, Y. Roles of the Interhexamer Contact Site for Hexagonal Lattice Formation of the Herpes Simplex Virus 1 Nuclear Egress Complex in Viral Primary Envelopment and Replication. *J. Virol.* **2019**, *93*, e00498-19. [CrossRef]
77. Roller, R.J.; Bjerke, S.L.; Haugo, A.C.; Hanson, S. Analysis of a Charge Cluster Mutation of Herpes Simplex Virus Type 1 UL34 and Its Extragenic Suppressor Suggests a Novel Interaction between pUL34 and pUL31 That Is Necessary for Membrane Curvature around Capsids. *J. Virol.* **2010**, *84*, 3921–3934. [CrossRef]
78. Bjerke, S.L.; Cowan, J.M.; Kerr, J.K.; Reynolds, A.E.; Baines, J.D.; Roller, R.J. Effects of Charged Cluster Mutations on the Function of Herpes Simplex Virus Type 1 U L 34 Protein. *J. Virol.* **2003**, *77*, 7601–7610. [CrossRef]
79. Milbradt, J.; Sonntag, E.; Wagner, S.; Strojan, H.; Wangen, C.; Rovis, T.L.; Lisnic, B.; Jonjic, S.; Sticht, H.; Britt, W.J.; et al. Human Cytomegalovirus Nuclear Capsids Associate with the Core Nuclear Egress Complex and the Viral Protein Kinase pUL97. *Viruses* **2018**, *10*, 35. [CrossRef] [PubMed]
80. Newcomb, W.W.; Homa, F.L.; Brown, J.C. Herpes Simplex Virus Capsid Structure: DNA Packaging Protein UL25 Is Located on the External Surface of the Capsid near the Vertices. *J. Virol.* **2006**, *80*, 6286–6294. [CrossRef]
81. Conway, J.F.; Cockrell, S.K.; Copeland, A.M.; Newcomb, W.W.; Brown, J.C.; Homa, F.L. Labeling and Localization of the Herpes Simplex Virus Capsid Protein UL25 and Its Interaction with the Two Triplexes Closest to the Penton. *J. Mol. Biol.* **2010**, *397*, 575–586. [CrossRef] [PubMed]
82. Klupp, B.G.; Granzow, H.; Keil, G.M.; Mettenleiter, T.C. The Capsid-Associated UL25 Protein of the Alphaherpesvirus Pseudorabies Virus Is Nonessential for Cleavage and Encapsidation of Genomic DNA but Is Required for Nuclear Egress of Capsids. *J. Virol.* **2006**, *80*, 6235–6246. [CrossRef] [PubMed]
83. Klupp, B.G.; Granzow, H.; Karger, A.; Mettenleiter, T.C. Identification, Subviral Localization, and Functional Characterization of the Pseudorabies Virus UL17 Protein. *J. Virol.* **2005**, *79*, 13442–13453. [CrossRef] [PubMed]
84. Yang, K.; Wills, E.; Lim, H.Y.; Zhou, Z.H.; Baines, J.D. Association of Herpes Simplex Virus pUL31 with Capsid Vertices and Components of the Capsid Vertex-Specific Complex. *J. Virol.* **2014**, *88*, 3815–3825. [CrossRef]
85. Takeshima, K.; Arii, J.; Maruzuru, Y.; Koyanagi, N.; Kato, A.; Kawaguchi, Y. Identification of the Capsid Binding Site in the Herpes Simplex Virus 1 Nuclear Egress Complex and Its Role in Viral Primary Envelopment and Replication. *J. Virol.* **2019**, *93*, e01290-19. [CrossRef]
86. McNab, A.R.; Desai, P.; Person, S.; Roof, L.L.; Thomsen, D.R.; Newcomb, W.W.; Brown, J.C.; Homa, F.L. The product of the herpes simplex virus type 1 UL25 gene is required for encapsidation but not for cleavage of replicated viral DNA. *J. Virol.* **1998**, *72*, 1060–1070. [CrossRef]
87. Trus, B.L.; Newcomb, W.W.; Cheng, N.; Cardone, G.; Marekov, L.; Homa, F.L.; Brown, J.C.; Steven, A.C. Allosteric Signaling and a Nuclear Exit Strategy: Binding of UL25/UL17 Heterodimers to DNA-Filled HSV-1 Capsids. *Mol. Cell* **2007**, *26*, 479–489. [CrossRef]
88. Toropova, K.; Huffman, J.B.; Homa, F.L.; Conway, J. The Herpes Simplex Virus 1 UL17 Protein Is the Second Constituent of the Capsid Vertex-Specific Component Required for DNA Packaging and Retention. *J. Virol.* **2011**, *85*, 7513–7522. [CrossRef]
89. Klupp, B.G.; Granzow, H.; Mettenleiter, T.C. Effect of the pseudorabies virus US3 protein on nuclear membrane localization of the UL34 protein and virus egress from the nucleus. *J. Gen. Virol.* **2001**, *82*, 2363–2371. [CrossRef] [PubMed]
90. Ryckman, B.J.; Roller, R.J. Herpes Simplex Virus Type 1 Primary Envelopment: UL34 Protein Modification and the US3-UL34 Catalytic Relationship. *J. Virol.* **2004**, *78*, 399–412. [CrossRef]
91. Roller, R.J.; Baines, J.D. Herpesvirus Nuclear Egress. *Adv. Anat. Embryol. Cell Biol.* **2017**, *223*, 143–169. [CrossRef] [PubMed]
92. Kuny, C.V.; Chinchilla, K.; Culbertson, M.R.; Kalejta, R.F. Cyclin-Dependent Kinase-Like Function Is Shared by the Beta- and Gamma- Subset of the Conserved Herpesvirus Protein Kinases. *PLoS Pathog.* **2010**, *6*, e1001092. [CrossRef] [PubMed]
93. Jacob, T.; van den Broeke, C.; Favoreel, H.W. Viral Serine/Threonine Protein Kinases. *J. Virol.* **2011**, *85*, 1158–1173. [CrossRef] [PubMed]
94. Kato, A.; Yamamoto, M.; Ohno, T.; Tanaka, M.; Sata, T.; Nishiyama, Y.; Kawaguchi, Y. Herpes Simplex Virus 1-Encoded Protein Kinase UL13 Phosphorylates Viral Us3 Protein Kinase and Regulates Nuclear Localization of Viral Envelopment Factors UL34 and UL31. *J. Virol.* **2006**, *80*, 1476–1486. [CrossRef] [PubMed]
95. Muradov, J.H.; Finnen, R.L.; Gulak, M.A.; Hay, T.J.M.; Banfield, B.W. pUL21 regulation of pUs3 kinase activity influences the nature of nuclear envelope deformation by the HSV-2 nuclear egress complex. *PLoS Pathog.* **2021**, *17*, e1009679. [CrossRef] [PubMed]
96. Wysocka, J.; Swigut, T.; Milne, T.; Dou, Y.; Zhang, X.; Burlingame, A.L.; Roeder, R.G.; Brivanlou, A.H.; Allis, C.D. WDR5 Associates with Histone H3 Methylated at K4 and Is Essential for H3 K4 Methylation and Vertebrate Development. *Cell* **2005**, *121*, 859–872. [CrossRef]
97. Wu, M.-Z.; Tsai, Y.-P.; Yang, M.-H.; Huang, C.-H.; Chang, S.-Y.; Chang, C.-C.; Teng, S.-C.; Wu, K.-J. Interplay between HDAC3 and WDR5 Is Essential for Hypoxia-Induced Epithelial-Mesenchymal Transition. *Mol. Cell* **2011**, *43*, 811–822. [CrossRef]

98. Wang, P.; Dreger, M.; Madrazo, E.; Williams, C.J.; Samaniego, R.; Hodson, N.W.; Monroy, F.; Baena, E.; Sánchez-Mateos, P.; Hurlstone, A.; et al. WDR5 modulates cell motility and morphology and controls nuclear changes induced by a 3D environment. *Proc. Natl. Acad. Sci. USA* **2018**, *115*, 8581–8586. [CrossRef]

99. Krause, M.; Yang, F.W.; Lindert, M.T.; Isermann, P.; Schepens, J.; Maas, R.J.A.; Venkataraman, C.; Lammerding, J.; Madzvamuse, A.; Hendriks, W.; et al. Cell migration through three-dimensional confining pores: Speed accelerations by deformation and recoil of the nucleus. *Philos. Trans. R. Soc. B Biol. Sci.* **2019**, *374*, 20180225. [CrossRef]

100. Mazumder, A.; Roopa, T.; Basu, A.; Mahadevan, L.; Shivashankar, G. Dynamics of Chromatin Decondensation Reveals the Structural Integrity of a Mechanically Prestressed Nucleus. *Biophys. J.* **2008**, *95*, 3028–3035. [CrossRef]

101. Nava, M.; Miroshnikova, Y.A.; Biggs, L.; Whitefield, D.B.; Metge, F.; Boucas, J.; Vihinen, H.; Jokitalo, E.; Li, X.; Arcos, J.M.G.; et al. Heterochromatin-Driven Nuclear Softening Protects the Genome against Mechanical Stress-Induced Damage. *Cell* **2020**, *181*, 800–817.e22. [CrossRef]

102. Bosse, J.B.; Hogue, I.B.; Feric, M.; Thiberge, S.Y.; Sodeik, B.; Brangwynne, C.P.; Enquist, L.W. Remodeling nuclear architecture allows efficient transport of herpesvirus capsids by diffusion. *Proc. Natl. Acad. Sci. USA* **2015**, *112*, E5725–E5733. [CrossRef]

103. Liu, S.Y.; Ikegami, K. Nuclear lamin phosphorylation: An emerging role in gene regulation and pathogenesis of laminopathies. *Nucleus* **2020**, *11*, 299–314. [CrossRef]

104. Machowska, M.; Piekarowicz, K.; Rzepecki, R. Regulation of lamin properties and functions: Does phosphorylation do it all? *Open Biol.* **2015**, *5*, 150094. [CrossRef]

105. Sharma, M.; Coen, N.M. Comparison of Effects of Inhibitors of Viral and Cellular Protein Kinases on Human Cytomegalovirus Disruption of Nuclear Lamina and Nuclear Egress. *J. Virol.* **2014**, *88*, 10982–10985. [CrossRef] [PubMed]

106. Forest, T.; Barnard, S.; Baines, J.D. Active intranuclear movement of herpesvirus capsids. *Nat. Cell Biol.* **2005**, *7*, 429–431. [CrossRef] [PubMed]

107. Simpson-Holley, M.; Colgrove, R.C.; Nalepa, G.; Harper, J.W.; Knipe, D.M. Identification and Functional Evaluation of Cellular and Viral Factors Involved in the Alteration of Nuclear Architecture during Herpes Simplex Virus 1 Infection. *J. Virol.* **2005**, *79*, 12840–12851. [CrossRef] [PubMed]

108. Feierbach, B.; Piccinotti, S.; Bisher, M.; Denk, W.; Enquist, L.W. Alpha-Herpesvirus Infection Induces the Formation of Nuclear Actin Filaments. *PLoS Pathog.* **2006**, *2*, e85. [CrossRef]

109. Bosse, J.B.; Virding, S.; Thiberge, S.; Scherer, J.; Wodrich, H.; Ruzsics, Z.; Koszinowski, U.H.; Enquist, L.W. Nuclear Herpesvirus Capsid Motility Is Not Dependent on F-Actin. *mBio* **2014**, *5*, e01909-14. [CrossRef]

110. Olmos, Y.; Hodgson, L.; Mantell, J.M.; Verkade, P.; Carlton, J.G. ESCRT-III controls nuclear envelope reformation. *Nat. Cell Biol.* **2015**, *522*, 236–239. [CrossRef]

111. Raab, M.; Gentili, M.; de Belly, H.; Thiam, H.-R.; Vargas, P.; Jimenez, A.J.; Lautenschlaeger, F.; Voituriez, R.; Lennon-Duménil, A.-M.; Manel, N.; et al. ESCRT III repairs nuclear envelope ruptures during cell migration to limit DNA damage and cell death. *Science* **2016**, *352*, 359–362. [CrossRef] [PubMed]

112. Penfield, L.; Shankar, R.; Szentgyörgyi, E.; Laffitte, A.; Mauro, M.S.; Audhya, A.; Müller-Reichert, T.; Bahmanyar, S. Regulated lipid synthesis and LEM2/CHMP7 jointly control nuclear envelope closure. *J. Cell Biol.* **2020**, *219*, e201908179. [CrossRef]

113. Tandon, R.; AuCoin, D.P.; Mocarski, E.S. Human Cytomegalovirus Exploits ESCRT Machinery in the Process of Virion Maturation. *J. Virol.* **2009**, *83*, 10797–10807. [CrossRef]

114. Streck, N.T.; Carmichael, J.C.; Buchkovich, N.J. Nonenvelopment Role for the ESCRT-III Complex during Human Cytomegalovirus Infection. *J. Virol.* **2018**, *92*, e02096-17. [CrossRef]

115. Crump, C.M.; Yates, C.; Minson, T. Herpes Simplex Virus Type 1 Cytoplasmic Envelopment Requires Functional Vps4. *J. Virol.* **2007**, *81*, 7380–7387. [CrossRef] [PubMed]

116. Farnsworth, A.; Wisner, T.W.; Webb, M.; Roller, R.; Cohen, G.; Eisenberg, R.; Johnson, D.C. Herpes simplex virus glycoproteins gB and gH function in fusion between the virion envelope and the outer nuclear membrane. *Proc. Natl. Acad. Sci. USA* **2007**, *104*, 10187–10192. [CrossRef] [PubMed]

117. Schulz, K.S.; Klupp, B.G.; Granzow, H.; Mettenleiter, T.C. Glycoproteins gB and gH Are Required for Syncytium Formation but Not for Herpesvirus-Induced Nuclear Envelope Breakdown. *J. Virol.* **2013**, *87*, 9733–9741. [CrossRef] [PubMed]

118. Dultz, E.; Ellenberg, J. Live imaging of single nuclear pores reveals unique assembly kinetics and mechanism in interphase. *J. Cell Biol.* **2010**, *191*, 15–22. [CrossRef]

119. Otsuka, S.; Ellenberg, J. Mechanisms of nuclear pore complex assembly—Two different ways of building one molecular machine. *FEBS Lett.* **2018**, *592*, 475–488. [CrossRef]

120. Hölper, J.E.; Klupp, B.G.; Luxton, G.W.G.; Franzke, K.; Mettenleiter, T.C. Function of Torsin AAA+ ATPases in Pseudorabies Virus Nuclear Egress. *Cells* **2020**, *9*, 738. [CrossRef]

121. Laudermilch, E.; Tsai, P.-L.; Graham, M.; Turner, E.; Zhao, C.; Schlieker, C. Dissecting Torsin/cofactor function at the nuclear envelope: A genetic study. *Mol. Biol. Cell* **2016**, *27*, 3964–3971. [CrossRef] [PubMed]

122. Rampello, A.J.; Laudermilch, E.; Vishnoi, N.; Prophet, S.M.; Shao, L.; Zhao, C.; Lusk, C.P.; Schlieker, C. Torsin ATPase deficiency leads to defects in nuclear pore biogenesis and sequestration of MLF2. *J. Cell Biol.* **2020**, *219*, e201910185. [CrossRef] [PubMed]

123. Sosa, B.A.; Rothballer, A.; Kutay, U.; Schwartz, T.U. LINC Complexes Form by Binding of Three KASH Peptides to Domain Interfaces of Trimeric SUN Proteins. *Cell* **2012**, *149*, 1035–1047. [CrossRef] [PubMed]
124. Gurusaran, M.; Davies, O.R. A molecular mechanism for LINC complex branching by structurally diverse SUN-KASH 6:6 assemblies. *eLife* **2021**, *10*, e60175. [CrossRef] [PubMed]

 *viruses*

Review

Conquering the Nuclear Envelope Barriers by EBV Lytic Replication

Chung-Pei Lee [1] and **Mei-Ru Chen** [2,*]

1   School of Nursing, National Taipei University of Nursing and Health Sciences, Taipei 112303, Taiwan; chungpei@ntunhs.edu.tw
2   Graduate Institute and Department of Microbiology, College of Medicine, National Taiwan University, Taipei 100233, Taiwan
*   Correspondence: mrc@ntu.edu.tw

Abstract: The nuclear envelope (NE) of eukaryotic cells has a highly structural architecture, comprising double lipid-bilayer membranes, nuclear pore complexes, and an underlying nuclear lamina network. The NE structure is held in place through the membrane-bound LINC (linker of nucleoskeleton and cytoskeleton) complex, spanning the inner and outer nuclear membranes. The NE functions as a barrier between the nucleus and cytoplasm and as a transverse scaffold for various cellular processes. Epstein–Barr virus (EBV) is a human pathogen that infects most of the world's population and is associated with several well-known malignancies. Within the nucleus, the replicated viral DNA is packaged into capsids, which subsequently egress from the nucleus into the cytoplasm for tegumentation and final envelopment. There is increasing evidence that viral lytic gene expression or replication contributes to the pathogenesis of EBV. Various EBV lytic proteins regulate and modulate the nuclear envelope structure in different ways, especially the viral BGLF4 kinase and the nuclear egress complex BFRF1/BFRF2. From the aspects of nuclear membrane structure, viral components, and fundamental nucleocytoplasmic transport controls, this review summarizes our findings and recently updated information on NE structure modification and NE-related cellular processes mediated by EBV.

Keywords: Epstein–Barr virus; BGLF4 kinase; nuclear egress; BFRF1; nuclear envelope modulation

Citation: Lee, C.-P.; Chen, M.-R. Conquering the Nuclear Envelope Barriers by EBV Lytic Replication. *Viruses* **2021**, *13*, 702. https://doi.org/10.3390/v13040702

Academic Editor: Donald M. Coen

Received: 10 March 2021
Accepted: 14 April 2021
Published: 18 April 2021

Publisher's Note: MDPI stays neutral with regard to jurisdictional claims in published maps and institutional affiliations.

1. The Structure and Function of the Nuclear Envelope

In higher eukaryotic cells, the nucleus is surrounded by an envelope composed of the inner and outer nuclear membranes, the perinuclear space between the membranes, and the underlying nuclear lamina network (Figure 1A). These three-layer structures are held in place by the nuclear pore complex (NPC), which functions as a transport gatekeeper for macromolecules trafficking in or out of the nucleus. The outer nuclear membrane is continuous with the rough endoplasmic reticulum. The nuclear lamina is a thin, electron-dense meshwork lining the nucleoplasmic face of the inner nuclear membrane (INM) [1,2] and provides firm structural support for the major components of the nuclear envelope (NE) [3,4]. The integrity of the envelope provides the structural support for nuclear morphology and functions as a platform for coordinating several cellular processes, including DNA repair, cell signaling, migration, and gene expression. The lamina is connected to the inner nuclear membrane through INM proteins (emerin and lamin B receptor), the chromatin proteins (histone H2A/H2B), and also the cytoskeleton-interacting LINC (linker of nucleoskeleton and cytoskeleton) complexes (reviewed in [5]). The physical linkages of LINC are instrumental in the plasticity of cellular organization and are required for processes such as nuclear migration and anchorage, meiotic chromosome movements, centrosome positioning, and the global organization of the cytoskeleton.

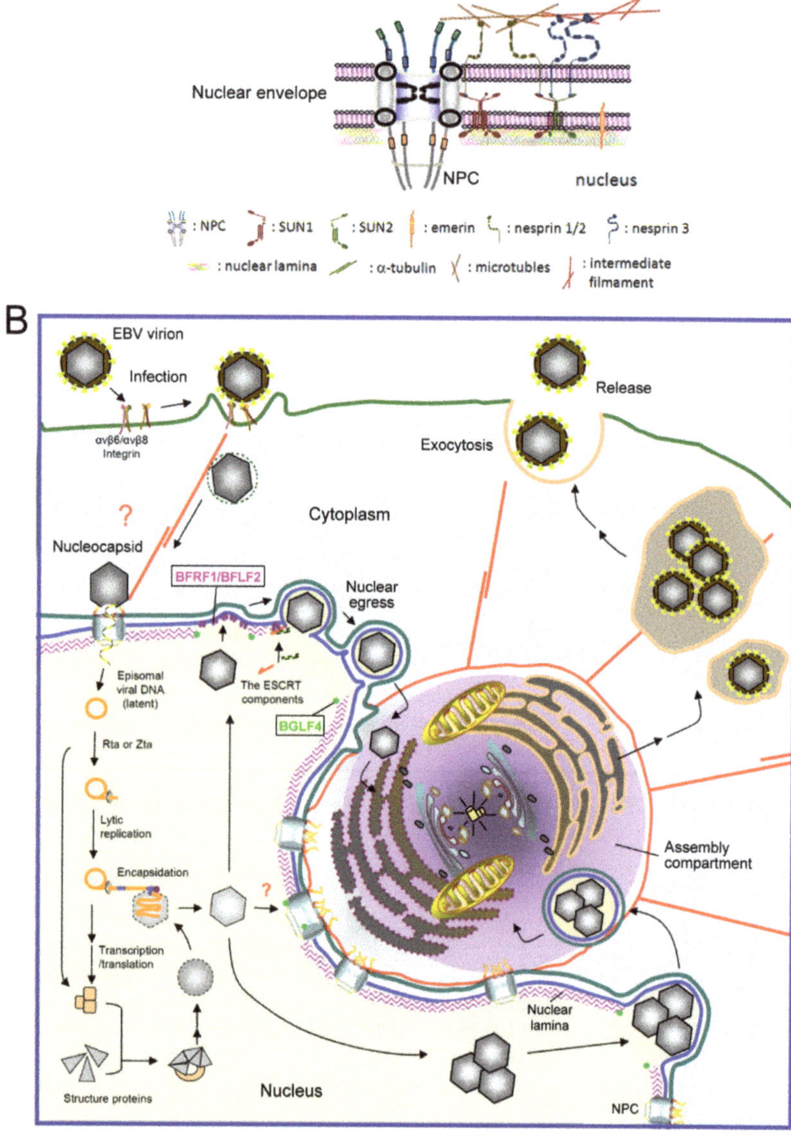

Figure 1. A diagram of EBV virion replication, production, and maturation. (**A**) An overview of the nuclear envelope (NE) structure with selected NE-associated proteins and nuclear pore complexes. The NE is composed of two lipid bilayers, inner and outer nuclear membranes (INM and ONM), and the nuclear lamina formed by lamin filaments. The nuclear pore complexes (NPCs) not only serve as gates for material transport, but also structurally hold the membranes and lamina together. The ONM, which is continuous with the ER, is characterized by cytoskeleton-associated nesprin proteins tethered by SUN1 and SUN2 in the INM. Emerin is an INM integrated protein that anchors on the nuclear lamina. (**B**) After EBV infects cells, the viral genome is injected into the nucleus through nuclear pore complexes. The linear genome is circularized into the episomal form through

the terminal repeats and maintained as an episomal form in the latently infected cells. With chemical stimulations or reactivation signals, the viral transactivator Zta or Rta activates replication-related viral proteins and lytic viral DNA replication. Along with the production and assembly of viral capsid proteins, the viral genome is encapsidated into the preformed procapsid to form the mature nucleocapsid. Simultaneously, the viral BGLF4 protein kinase mediates the phosphorylation and partial disassembly of nuclear membrane-underlying nuclear lamina, allowing the access of nucleocapsids to the nuclear membrane. In our observations, the nucleocapsids then traverse through the nuclear envelope with the coordination of the viral nuclear egress complex BFRF1/BFLF2, cellular ESCRT machinery, and Nedd4-like ubiquitin ligases. The cytoplasm-distributed nucleocapsids may subsequently transport into the juxtanuclear "viral assembly compartment". This specialized compartment, containing highly reorganized membrane structures and organelles, may provide a site for efficient tegumentation and secondary envelopment of the nucleocapsids. Finally, the mature virion is transported close to the cell margins and released from the infected cells by exocytosis. NPC, nuclear pore complex.

NE integrity is essential for cell homeostasis. Alteration of NE structure or components may impact the behavior and the phenotypes of cells in many ways. For decades, deleterious morphological changes of the NE in tumor cells have been recognized as an important parameter for the malignant criteria of tumors [5–7]. Many NE-associated diseases, collectively designated nuclear envelopathies or laminopathies, have been identified. Most of these disorders come from mutations in INM proteins or their interacting partners, diseases including muscular dystrophy, lipodystrophy, neuropathy, and progeria (premature aging) syndromes [8,9]. Diseases may be due to mutations or altered expression of alleles, the mechanical stability of the nucleus, or cell cycle regulation. However, the molecular mechanisms that give rise to these diseases and whether pathogen infection may modulate the nuclear structure and contribute to the pathogenesis remain poorly understood [10,11].

2. Nuclear Pore Complexes (NPC) Function as Gates and Guardians for Nuclear–Cytoplasmic Transport

The gatekeeper function of the NPC is important for preventing the abnormal distribution of cytoplasmic components or genome-toxic insults to the cellular chromatin. The NPC has a molecular mass of ~125 megaDaltons (MDa) in vertebrates and is composed of multiple copies of a set of 30 diverse proteins, termed nucleoporins (NUPs) [12]. In the middle of the nuclear pore, FG-NUPs, containing Phe-Gly (FG) repeats, play the most relevant role in controlling the nuclear transport receptors (NTR) that shuttle through the NPC. FG-NUPs and NTRs interact through hydrophobic contacts for the translocation step of nuclear transport through the NPC. The canonical transport of most macromolecules (>40 kDa) into the nucleus is controlled by nuclear localization signals (NLS) that are recognized by specific members of the importin receptor family. Proteins containing a bipartite motif or a stretch of basic amino acids bind to an importin alpha and beta heterodimeric receptor complex [13]. During transport, importin α binds to the NLS-bearing protein, and importin β mediates the binding of the transport complex to the FG-NUPs, which subsequently slide through the NPC [14,15]. The small GTPase Ran regulates the association and dissociation of the importin-cargo complex between GTP- and GDP-bound forms [16,17]. For the export of nuclear protein, XPO1 (Exportin-1/Chromosome Region Maintenance 1/CRM1) is the most known mediator. Nuclear export signals (NESs) are short leucine-rich sequences that can be found in many shuttling proteins. Cargo proteins containing NES are recognized by XPO1, which interacts with nucleoporins NUP214 and NUP88 in NPC to transport out of the nucleus in a RanGAP-dependent manner [18,19].

Nuclear transport is also an important step that controls specific signaling pathways. For example, activators of transcription, including IRF3 and NF-κB, are regularly expressed and sequestered in the cytoplasm. They are translocated into the nucleus through phosphorylation-mediated regulation and activate specific innate immune response genes upon stimulation [20,21]. Such regulation could occur in as little as 30 min to activate gene transcription in response to challenge by various pathogens.

3. Overview of EBV Lytic Replication

Epstein–Barr virus (EBV) is a human gamma-herpesvirus that infects most of the world's population. It has a 170–175 kb linear double-stranded DNA genome which encodes approximately 90 open reading frames. The genome is packaged within an icosahedral capsid, approximately 100–120 nm in diameter, surrounded by a proteinaceous tegument and a lipid bilayer envelope. EBV generally establishes two phases in its life cycle, latency and lytic replication [22]. EBV latent proteins help B cell proliferation, including LMP1 (latent membrane protein 1), which can mimic CD40 signaling to activate NF-κB signaling, and EBNA2 (EBV nuclear antigen 2), which can turn on Myc expression. EBV can transform primary B cells into lymphoblastoid cell lines in vitro and may cause post-transplantation lymphoproliferative disease (PTLD). Increasing evidence indicates that lytic phase gene expression also contributes to EBV oncogenesis [23].

EBV mainly infects naïve B lymphocytes and epithelial cells through different receptors, including CR2, HLA class II, and integrins [24]. B cell entry may occur via membrane fusion or endocytosis followed by fusion of the viral membrane with the membrane of the endocytic vesicle [25]. Ephrin receptor A2 was recently identified as the EBV receptor on epithelial cells [24,26,27]. The various steps of EBV replication are illustrated in Figure 1B. After the initial infection, mediated by sequential interactions of viral glycoproteins with host surface receptors, the nucleocapsid is internalized through endocytosis or membrane fusion [28–30]. It has been postulated that the nucleocapsid might undergo some structural changes and slide along the microtubule or actin cytoskeleton to the NPC boundary [31]. The linear viral genome is then injected into the nucleus [32]. However, how the EBV genome gets through the NPC and whether viral factors modify the NPC structure to facilitate transport of the viral genome remain unclear. The genome is then circularized and maintained by the EBV encoded nuclear antigen EBNA1 as an episomal form in the nucleus, remaining latent in the host [24].

Upon chemical or stress stimulation, or Ig cross-linking on the surface of B cells, EBV is reactivated and genome replication ensues. Viral lytic genes are expressed in a temporal and sequential order that is divided into three stages, immediate-early (IE), early (E), and late (L). Two immediate-early transactivators, Zta and Rta, are transcribed before viral protein synthesis and can turn on complete viral lytic replication [22]. Most of the early genes are required for viral DNA replication, such as the polymerase processivity factor BMRF1 and CDK1 (cyclin-dependent kinase 1)-like protein kinase BGLF4. The late viral products are mainly structural proteins that are expressed after viral DNA synthesis, such as viral capsid protein VCA and glycoprotein gp350/220 [33]. After DNA replication, the viral genome is packaged into the preformed capsid in the nucleus. With the coordination of cellular machinery and the viral egress complex BFRF1 and BFLF2, the intranuclear nucleocapsids subsequently move towards the inner leaflet of the nuclear membrane and bud from the NE [34,35]. In our unpublished observations, the transported nucleocapsids may obtain their tegument and secondary envelope at the "cytoplasmic assembly compartment", a specialized cell compartment containing highly reorganized membrane apparatus, cell organelles, and cytoskeletons. Finally, the mature virions containing the cis-Golgi derived membrane are transported and released from the cell through exocytosis [36] (Figure 1B).

4. The Nuclear Import of EBV Proteins

To initiate viral DNA replication, Zta and Rta, coordinately bind to the origin of lytic replication (oriLyt) and recruit essential lytic replication proteins to form a core replication complex [37,38]. The viral DNA replication machinery includes the DNA polymerase (BALF5), single-stranded DNA (ssDNA) binding protein (BALF2), DNA polymerase processivity factor (BMRF1), helicase (BBLF4), primase (BSLF1), primase accessory proteins (BBLF2/BBLF3), and uracil DNA glycosylase (BKRF3). Viral DNA replicates in a manner very similar to the mammalian DNA replication system [37–40]. For oriLyt binding and transactivator function, Zta contains a classical NLS for nuclear targeting. Interestingly, several EBV genes involved in viral DNA replication lack the canonical NLS signals in their coding region. For example, viral DNA polymerase (BALF5) and uracil DNA glycosylase (BKRF3) are transported into the nucleus through interaction with the DNA polymerase processivity factor (BMRF1) [39,41]. Interestingly, the three EBV primase/helicase complex components do not contain canonical NLS but can be imported into the nucleus through interaction with the origin binding protein Zta [42], suggesting EBV evolved with novel pathways for nuclear targeting. In addition, we found the EBV protein kinase BGLF4 enters the nucleus through the direct interaction of its carboxyl-terminal alpha-helical region with the FG-NUP proteins NUP60 and NUP153, independent of importins [43]. BGLF4 also modulates the structure and transport preference of the NPC to facilitate the nuclear import of several EBV lytic proteins, including all three primase/helicase complex proteins and the viral capsid protein BcLF1 [44]. We postulated that the novel nuclear targeting mechanism and BGLF4-mediated NPC preference may interfere with the cellular antiviral response and ensure dominance of the virus for efficient replication.

5. The Nuclear Export (Egress) of EBV Nucleocapsids

After genome replication, EBV DNA is packaged into the preassembled procapsid, which primarily consists of the major capsid protein, VCA, and is held together by 320 triplexes formed by two minor capsid proteins, BDLF1 and BORF1, and a small capsid protein, BFRF3, on the hexameric capsomers [45]. The nucleocapsid size is about 100–130 nm, too large to go through the nucleopore. The NE structure is a significant hindrance for the transport of the nucleocapsid into the cytoplasm. Our and others' studies demonstrated that at least three virus-encoded proteins, including BGLF4 protein kinase and nuclear egress complex BFRF1/BFLF2, modify the NE structure to facilitate the nucleocytoplasmic transport of nucleocapsids.

6. EBV BGLF4 Protein Kinase Functions through CDK Mimicry

The EBV BGLF4 protein kinase is a viral tegument protein essential for viral DNA replication and virion maturation [46–49]. In contrast to alpha-herpes viruses with Us3 and UL13 kinases, EBV encodes the unique BGLF4 kinase during the early through the late stage of virus replication. BGLF4 belongs to be the so-called conserved herpesvirus protein kinases (CHPKs), which recognize Ser/Thr-Proline motifs within their substrates [50]. Herpesviral CHPKs contribute to the initial infection by the virion, the regulation of expression of viral genes, the replication and encapsidation of the viral genome, and the egress and tegumentation of the nucleocapsid [48,50–53]. Since Ser/Thr-Proline motifs are also targeted by cellular cyclin-dependent kinases (CDKs) and other cell signaling-related kinases, CHPKs are believed to regulate the host cell environment in part through CDK mimicry. BGLF4 kinase phosphorylates multiple cellular substrates to regulate cellular environments for optimal virus replication and production through many processes, including molecules involved in cellular immunity, cell signaling, cell cycle, chromosome structure, nuclear envelope structure, and even cytoskeleton rearrangement [52–54], as summarized in Figure 2A. For example, the phosphorylation of BGLF4 on IRF3 and UXT is known to suppress the IRF3- and NFκB-mediated cellular suppression of virus replication [55,56]. BGLF4 regulates the cellular chromatin and chromosome structure by TIP60, condensin, and topoisomerase phosphorylation [51,57]. The expression of BGLF4 delays cell cycle progression at the S phase and results in retardation of cell growth [58]. Additionally, the presence of BGLF4 plays a crucial role in regulating virion production and maturation. The phosphorylation and reorganization of the nuclear lamina and NPC structure by BGLF4 facilitate the nuclear egress of nucleocapsids and viral late protein transportation [44,49]. A panel of viral proteins involved in virus gene regulation, nucleotide metabolism, and virion structure was also identified as BGLF4 substrates by proteomic approaches [51,59]. For instance, BGLF4 interacts with the viral transactivator Zta and regulates its transactivation activity [60]. EBV EBNA1, an essential protein for episomal EBV genome replication and maintenance during latency, is also a substrate and binding partner of BGLF4 [59]. Several reviews have summarized the functions of conserved herpesviral kinases [48,50,61,62]. Of note, a recent study using a quantitative proteomic approach indicates that CHPKs in gamma-herpesviruses have different interaction profiles from those of the α- and β-herpesviruses [53]. In contrast to all herpesviral CHPKs, which modulate cell transcription, replication and the cell cycle, EBV BGLF4 targets preferentially chromatin silencing and DNA replication-associated factors. It is suggested that, even though most CHPKs phosphorylate various cellular factors, the CHPKs of the gamma-herpesviruses have unique characteristics in altering the cellular environment during virus reactivation.

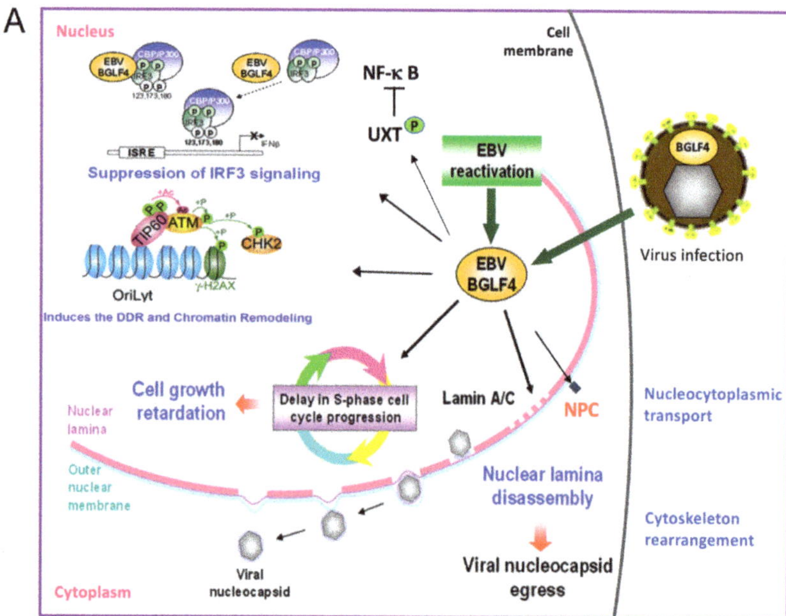

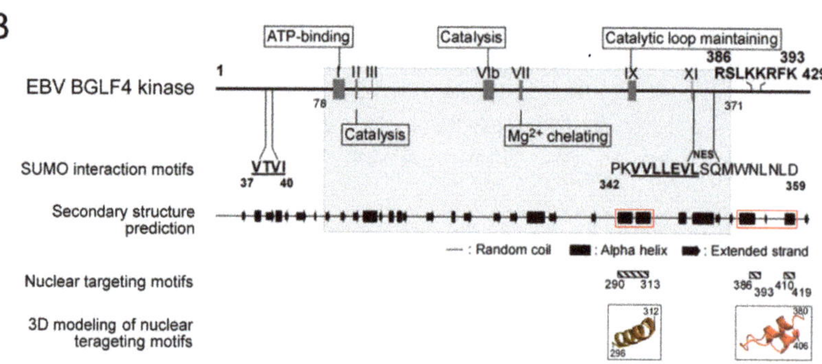

Figure 2. The regulatory functions and the conserved kinase motifs of EBV BGLF4 kinase. (**A**) BGLF4 is a virion-associated protein kinase. After virus infection or reactivation, the mainly nucleus-distributed BGLF4 regulates the cellular environment in many respects. The BGLF4-mediated phosphorylation of IRF3 and UXT is known to suppress the IRF3- and NFκB-mediated cellular

suppression of virus replication. BGLF4 regulates the cellular chromatin or chromosome structure by phosphorylating TIP60, condensin, and topoisomerase. The expression of BGLF4 delays cell cycle progression and results in cell growth retardation. These regulations can provide an optimal cellular environment for efficient virus replication. Additionally, BGLF4 plays a crucial role in regulating virion production and maturation. BGLF4 phosphorylates the nuclear lamin A/C and the components of the nuclear pore complex (NPC). The phosphorylation of these structural components attenuates the nuclear envelope barriers and facilitates the transport of viral late proteins and subsequently the nuclear egress of nucleocapsids. In our recent observation, reorganization of the cytoskeleton may also facilitate the maturation and production of the cytoplasm-located nucleocapsids. (**B**) Conserved kinase motifs of BGLF4 were identified by alignment with eukaryotic protein kinases. There are 11 conserved motifs defined in eukaryotic protein kinases [63]. Only motifs I, II, VIb, VII, IX, and XI can be identified in the kinase domain (gray region) of herpesviral kinases [64]. I, ATP binding; II and VIb, catalysis; VII, Mg^{2+} chelating; IX, catalytic loop maintaining. SUMO interaction motifs (SIMs) at a.a. 37–40 and 344–350 and the NES at a.a. 342–359 were identified in a recent study [65]. Amino acids 386–393 were predicted to comprise a putative NLS [66]. However, our study showed that the alpha-helical structure, rather than the positive charges of these amino acids, is critical for BGLF4 nuclear targeting [43]. The putative secondary structure of BGLF4 was analyzed by the GOR secondary structure prediction program (http://www.expasy.ch/tools/ (accessed on 23 May 2012)). Random coils, alpha helices, and extended strands are indicated as thin lines, boxes, and arrows, respectively. The red boxes indicate the BGLF4 unique regions crucial for nucleoporin interaction and nuclear targeting. The hatched boxes underneath the predicted secondary structure indicate the critical regions (a.a. 290–313, 386–393, and 410–419) for BGLF4 nuclear targeting. The critical helical regions for BGLF4 nuclear targeting were also predicted by Protein Structure Prediction Server (PS)^2 (version 2; http://ps2v2.life.nctu.edu.tw/ (accessed on 23 May 2012)) using casein kinase 2 alpha 1 polypeptide (PDB accession number 3bqcA) as the template and shown in enlarged 3D images for a.a. 296–312 and a.a. 380–406. This figure is adapted and reorganized from our previous study published in Journal of Virology 2012, 86:8072-85 [43].

We previously demonstrated that the expression of BGLF4 alone efficiently modulates chromatin structure and cytoskeleton arrangement, very similar to CDK1-induced promitotic events [57]. Additionally, we found that the ability to induce nuclear lamina disassembly is conserved in all CHPKs [49]. Correspondingly, nuclear retention of viral nucleocapsids was observed in cells infected with a CHPK-deficient recombinant virus or in kinase knockdown experiments [48,67–71]. These observations indicate the contribution of CHPKs to regulating virion maturation and release [49,51–53,72–75]. Besides, we found that BGLF4 targets the nucleus through an unconventional interaction with nucleoporins and is independent of cytosolic factors, unlike other CHPKs [43] (Figure 2B). Through this interaction, BGLF4 modulates the structure and preference of nuclear pore complexes to facilitate the nuclear import of EBV lytic proteins [44] (Figure 3). A recent study reported that the cellular CDK1-substrate SAMHD1 (sterile alpha motif and HD) is also regulated by BGLF4 and other CHPKs [76]. The phosphorylation of SAMHD1 leads to an increased cellular dNTP pool and therefore facilitates virus replication. Overall, these observations suggest that the CHPKs-mediated modulation of nuclear structure or the cellular environment is critical for overcoming the natural barrier or cellular limitation during herpesviral virion replication and maturation. For the nuclear egress process, BGLF4 plays an important role in phosphorylating nuclear lamin A/C and causes partial disassembly of the nuclear lamina. Therefore, the relaxed lamina structure facilitates the remodeling of nuclear membranes and subsequent recruitment of ESCRT components to transport nucleocapsids.

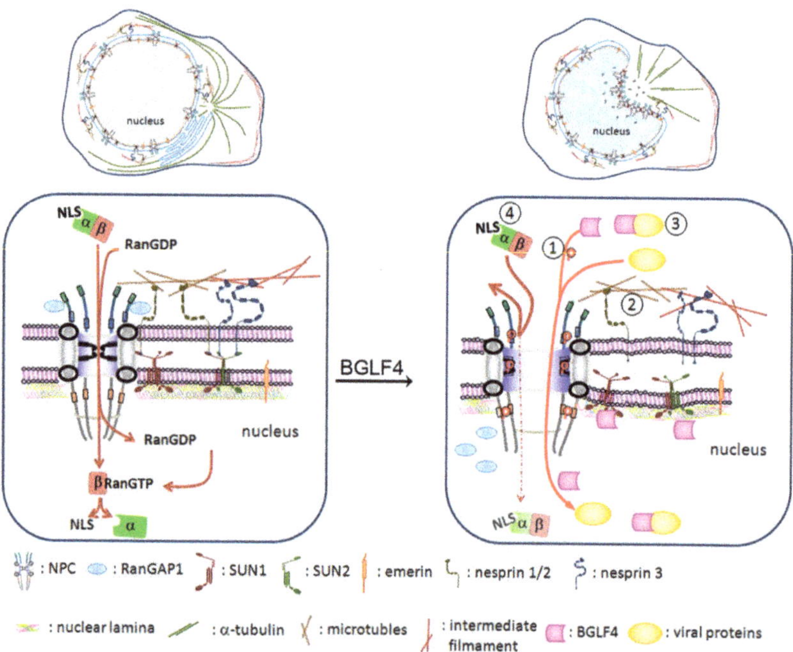

Figure 3. Hypothetical model of the EBV BGLF4-induced structural and functional changes of the nuclear pore complex. (Left part) Under physiological conditions, the nuclear import of NLS-containing proteins is mediated by an importin α-importin β complex and is dependent on RanGTP hydrolysis. (Right part) In the presence of BGLF4, BGLF4 is transported into the nucleus through direct interaction and phosphorylation of FG-NUPs, including NUP62 and NUP153, to dilate the nuclear pores. Simultaneously, BGLF4 also induces reorganization of microtubules and causes changes in the nuclear shape. The nuclear envelope becomes irregular, and the nuclear envelope-associated proteins, such as SUN1 and SUN2, are redistributed. Non-NLS-containing viral proteins can be transported into the nucleus at the same time through at least three different mechanisms: (1) dilated nuclear pores, (2) a microtubule reorganization-dependent mechanism, and (3) direct interaction with or phosphorylation by BGLF4. (4) At the same time, the nuclear import of canonical NLS-containing proteins is partly inhibited. This model is adapted from our previous paper published in Journal of Virology 2015, 89:1703-18 [44].

7. EBV Nuclear Egress Complex (NEC): BFRF1 and BFLF2 Proteins

The study regarding the nuclear egress of herpesviruses was pioneered by Roller and Baines' Labs that made the first and critical observations of how viruses encode these nuclear egress functions [77]. Based on the "envelopment–de-envelopment–re-envelopment" model, the large herpesviral nucleocapsids (100–130 nm) begin budding through the nuclear membrane via a transient envelopment. This is mediated first by the viral protein kinase and nuclear membrane-associated proteins at the inner nuclear membrane (INM), with the local disassembly of compact nuclear lamina for primary envelopment. The nucleocapsids with the primary envelope may then de-envelope by fusing with the outer nuclear membrane (ONM), be transported from the perinuclear space to rough endoplasmic reticulum cisternae or, alternatively, released from the disintegrated nuclear envelope [78,79]. After release from the NE-derived structures, the nucleocapsids subsequently become associated with viral tegument proteins and glycoproteins at cytoplasmic apparatuses for the final maturation of virions [34,80–82].

In herpesviruses, the conserved homologs of UL34 and UL31 are believed to be the major viral components that mediate the primary envelopment. The homologs of herpes simplex type 1 (HSV-1) UL34 are type II integral membrane proteins, which localize predominantly to the ONM, INM, and ER, whereas the UL31 homologs are nuclear matrix-associated phosphoproteins [83,84]. Transient overexpression of the UL34/UL31 homologs of HSV-1 or human cytomegalovirus (HCMV) induces subtle alterations of the nuclear lamina [85,86], suggesting the homologs of UL34 and UL31 regulate potentially the structure of the nuclear membrane. More recently, the NE modulation of the UL34/UL31 complexes, also called nuclear egress complexes (NECs), has received much attention among herpesviruses [35,81,87–90]. Using various super-resolution imaging techniques and biophysical analysis, the HCMV and HSV-1 heterodimeric core NECs were shown to assemble and form hexameric latticed structures [87,91]. NECs contribute to recruiting effectors and reorganizing the NE, allowing the docking and nuclear egress of viral nucleocapsids [81,87–93]. So far, the UL34 homologs are believed to be rate-limiting factors during the nuclear egress among the NEC components, such as pUL34 in HSV-1 [94], pUL50 in HCMV [95], and BFRF1 in EBV [84]. Intriguingly, even though herpesviral NECs all show a similar structural basis in their conserved regions, some unique activities are observed in individual herpesviruses.

The EBV gene products of BFRF1 and BFLF2, the positional homologs of UL34 and UL31 in HSV-1, have been shown to regulate the primary egress of nucleocapsids [96]. Coexpression of BFRF1 and BFLF2 induces modification of the nuclear membrane, which is frequently observed in cells with EBV replication [97,98]. Expression of BFRF1 in 293 cells induces the stacking of a multilayered nuclear membrane and vesicle-like structures [84,99]. Coexpression of BFRF1 and BFLF2 dramatically reorganizes the multilayered nuclear membrane and induces the dilated packaged cisternae and even concentric whorls [96]. Similar structural changes of nuclear envelope were also observed in cells expressing the homologous NEC proteins in other herpesviruses [100,101]. Consistently, the amounts of intracytoplasmic nucleocapsids are reduced in mutant EBV lacking BFRF1 or BFLF2 [84,102]. These observations indicate crucial roles of BFRF1 and BFLF2 in inducing the change of the nuclear membrane for efficient primary envelopment.

In contrast to BFRF1, knowledge regarding the function of EBV BFLF2 is limited. Deletion of the BFLF2 gene from an EBV bacmid induced the accumulation of empty capsids in the nucleus and reduced virion release, suggesting BFLF2 is required for correct DNA packaging and the nuclear egress of viral nucleocapsids [102,103]. When expressed alone, BFLF2 is distributed mainly in the nucleus. However, it colocalizes with BFRF1 at the nuclear rim and in NE-derived vesicles in coexpressing cells, suggesting that interaction between BFLF2 and BFRF1 is critical for their proper function [103]. Unlike its herpesviral homologs with a conserved nuclear localization signal (NLS), noncontinuous alkaline and histidine residues in BFLF2 function together as a noncanonical NLS for its nuclear targeting, in an importin β-dependent manner [103,104]. Interestingly, in a high-throughput yeast two-hybrid screening system, BFLF2 protein was highly connected with various human and EBV proteins, resembling a hub in the interactome network [105], suggesting that BFLF2 may function as a scaffold for nuclear cargo transport or cell signaling. Therefore, studying the regulatory function of BFLF2 may provide an insight into virus-host interactions and changes in the nuclear environment during EBV reactivation.

8. Cellular ESCRT Machinery and Ubiquitination Regulate the Nuclear Egress of EBV

The endosomal sorting complex required for transport (ESCRT) machinery is evolutionarily conserved and involved in catalyzing the scission of membrane necks in autophagy, cytokinesis, endosome sorting, biogenesis of multivesicular bodies (MVBs), NE reformation and repair, and the maturation and release of enveloped virions. In contrast to the cellular membrane-scission protein dynamin family, which cleaves membrane necks by constricting them from the outside, membrane scission mediated by the ESCRT machinery is from inside the neck [106,107]. The ESCRT components (also known as class E proteins) comprise five multiprotein complexes, ESCRT-0, -I, -II, -III, Vps4 (vacuolar protein sorting-4) ATPase, and several ESCRT-associated proteins [108,109]. ESCRT-0, -I, and -II are soluble complexes that shuttle between cytosolic and membrane-bound forms. These components coordinate together sequentially to curve the membrane and recruit ESCRT-III for scission of the membrane neck. ESCRT-III protein CHMPs (charged multivesicular body proteins) are soluble monomers that assemble on membranes to form tight filamentary spirals and are released from the membranes at the final stage, with other ESCRT proteins, by a transient ATP-dependent reaction of Vps4. In addition to the regular composition, cellular ESCRT-I protein TSG101 (tumor susceptibility gene 101) alternatively activates the spiral assembly of ESCRT-III through bridging by the ESCRT associated-protein, apoptosis linked gene-2 interacting protein X (Alix) [110].

Accumulating studies have reported that components of the ESCRT machinery are used by many enveloped viruses for budding and release from cells. For example, the dynamics of ESCRT protein recruitment in retroviruses were found to be extremely transient (~1–3 min) and sufficient for their functions on the cytoplasmic membrane for virus release [111]. In contrast to enveloped RNA viruses, the contribution of ESCRT machinery to the maturation of enveloped herpesvirus has emerged only in the past decade. For instance, the contribution of ESCRT to virion release and cytoplasmic re-envelopment of HSV-1 and HCMV has been characterized using dominant-negative (DN) inhibitors of ESCRT and siRNA strategies [112–115]. By immunofluorescence analysis, the cytoplasmic nucleocapsids and envelope components are associated with MVB and colocalized with endosomal markers in infected cells [116,117]. As observed by electron microscopy (EM), HHV-6 also induces MVB formation and cytoplasmic egress through an exosomal release pathway [118], suggesting that herpesviruses use the ESCRT machinery for their membrane-dependent maturation in the cytoplasm.

In EM images of EBV-infected cells, we observed that virus reactivation induces a vesicle-like ultrastructure [119,120]. Intriguingly, the expression of dominant-negative ESCRT proteins caused a blockade of virion release and retention of the viral NEC protein BFRF1 at the NE [99]. BFRF1 can interact with the adaptor protein Alix, recruit ESCRT components, and mediate vesicle formation from the nucleus-associated membrane. Remarkably, inhibition of the ESCRT machinery by dominant-negative components abolished BFRF1-induced vesicle formation, leading to the accumulation of viral DNA and capsid proteins in the nuclei of EBV-replicating cells. This observation suggests that the ESCRT machinery plays a crucial role during the nuclear egress of EBV. As revealed in Figure 4A, the membrane anchoring BFRF1 potentially oligomerizes and recruits Alix, Nedd4-like ubiquitin ligases and, sequentially, other ESCRT components to generate the vesicles from the nucleus-associated membrane after EBV reactivation. Cooperating with other viral factors, such as BFLF2, the BFRF1-mediated vesicles may further recruit and pack nucleocapsids for their subsequent nuclear egress. Even though the topological regulation of Alix in the process remains unclear, the EBV uses an alternative vesicular nucleocytoplasmic transport for its nuclear egress, other than the postulated envelopment–de-envelopment model proposed for alpha- and beta-herpes viruses. In addition to the ESCRT components, we further demonstrated that the membrane modulating functions of BFRF1 are regulated by ubiquitination [119]. Elimination of possible ubiquitination on BFRF1 hampers NE-derived vesicle formation and virus maturation. The Nedd4-like ubiquitin ligase Itch is preferentially associated with Alix and BFRF1, and required for BFRF1-induced vesicle formation. As BFRF1 enforces the dynamics of the NE and promotes the transfer of viral nuclear cargos, EBV BFRF1 may serve as a potent molecule to explore the vesicular nucleocytoplasmic transport pathway in higher eukaryotes [120]. Further investigation regarding how cellular Alix contributes topologically to this NE modulation may also provide a new understating of the dynamics and fundamental regulations of the NE.

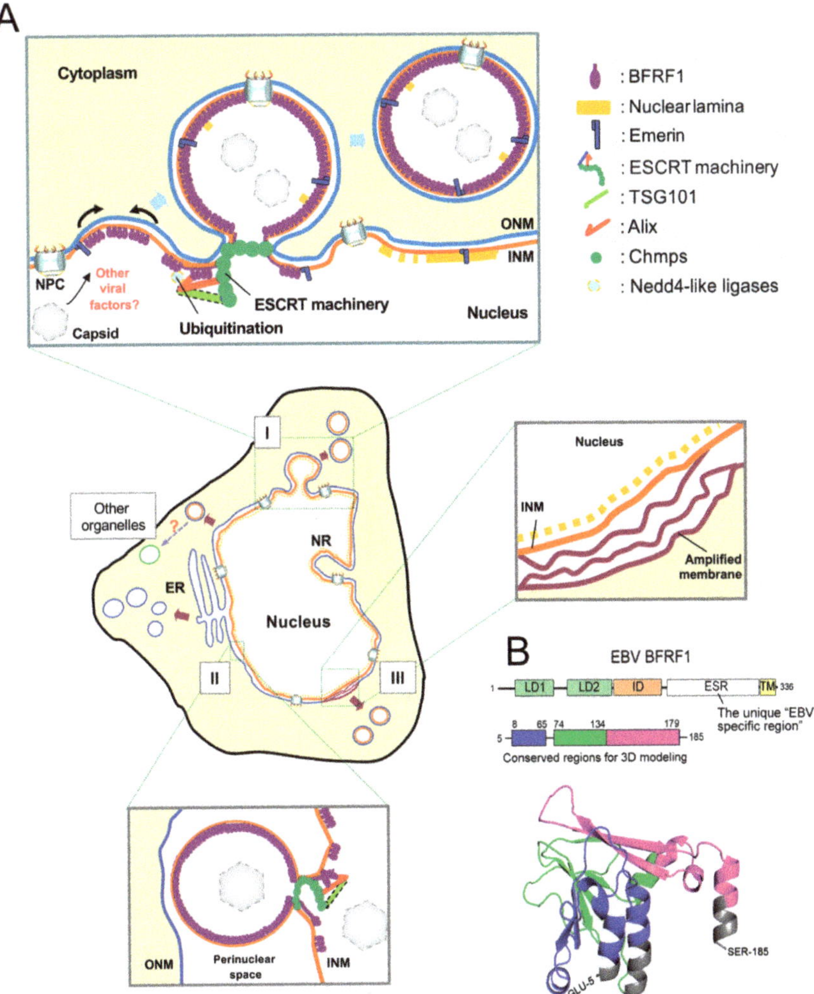

Figure 4. A hypothetical model of the interaction among BFRF1, cellular ESCRT components, and Nedd4-like ligases in vesicle formation and modulation of the nuclear membrane. (**A**) After EBV reactivation, membrane anchoring BFRF1 targets to the nucleus-associated membranes. BFRF1 potentially regulates the nuclear membrane through protein oligomerization. It then recruits Alix, Nedd4-like ubiquitin ligases, and, sequentially, other ESCRT components to generate vesicles derived from the double layer nuclear envelope (I), outer nuclear membrane (ONM), or from the inner nuclear membrane (INM) (II). These nuclear membrane-derived vesicles may contain some nucleoporins and other INM proteins, such as lamin A/C or emerin. Cooperating with other viral factors, such as BFLF2, the BFRF1-mediated vesicles may further recruit and pack some nucleocapsids for the subsequent nuclear egress. In addition, the expression of BFRF1 induces the formation of multilayered cisternal nuclear membranes (III), which may provide a more membranous structure for viral budding. ER, endoplasmic reticulum. NPC, nuclear pore complex. This model represents a refined version of BFRF1-ESCRT component interactions as analogously published in PLOS Pathogens 2012, e1002904 [99]. (**B**) Prediction of the 3D structural BFRF1, based on the HCMV 5DOB chain B. The EBV-specific region (ESR), consisting of amino acids 180–313, could not be predicted because of a lack of sequence similarity. The LD1 domain is shown in green, the LD2 domain is shown in blue, and the ID domain is shown in pink. The regions involved in BFRF1 interaction are indicated in the BFRF1 functional domains and in the 3D modeling. This figure is adapted and reorganized from a previous figure published in Journal of Virology 2020, 94: e01498-19 [103].

9. The Unique Sequences and Features of EBV NEC Proteins

Through protein alignments with herpesviral homologs, two putative L-like motifs (LDs, [62]YKFL[65], and [74]YPSSP[78]), but not conventional L motifs (PTAP, PPXY, and YXXL), an EBV-specific (ESR, a.a. 180–313), and a transmembrane region (TM, a.a. 314–336) were identified within the amino-terminus of EBV BFRF1 ([99] and unpublished data]) (Figure 4B). So far, the EBV-specific region is known to be critical for several BFRF1-mediated functions, such as vesicle formation, Itch-Alix complex, and BFLF2 interaction [99,103,119]. In contrast, EBV BFLF2 shares higher protein similarities with other homologs than does BFRF1 [103]. Intriguingly, the nuclear trafficking and critical residues responsible for regulating partner protein interaction and function are different from its UL31 homologs [103,104]. This suggests that EBV NEC proteins may have unique features different from those of herpesviral homologs. A recent study based on high-resolution protein crystal structures indicates that the γ-herpesviral EBV NEC shares some common folds with α- and β-herpesviruses; however, the primary amino acid sequences in herpesviral core NEC proteins are diverse. The structurally conserved regions are located mainly in the N-terminal regions of EBV BFRF1 (a.a. 2–192) and BFLF2 (a.a 78–110). Notably, both EBV BFRF1 and BFLF2 display several structural particularities [121]. Correspondingly, even though BFRF1 potentially mediates the phosphorylation and redistribution of nuclear emerin, similar to other herpesviral homologs [122], only EBV BFRF1 potently modulates nucleus-associated membranes and induces cytoplasmic vesicles independently of BFLF2 [99,123]. Altogether, these observations suggest that the core NEC of EBV accounts for its unique activities, in addition to the common function in regulating the nuclear structure and protein- or membrane-rearranging functions.

After nuclear egress, the nucleocapsids in the vesicles may be transported through the ER to Golgi pathways and obtain the final envelope with tegument proteins for morphogenesis [124]. Our current study also suggests the endoplasmic reticulum Golgi intermediate compartment (ERGIC), the origin of intracellular membrane genesis between ER and Golgi, may also be involved in the formation of viral cytoplasmic assembly compartment (manuscript in preparation)

10. Is the Vesicular Nucleocytoplasmic Transport Pathway Also Present in the Absence of Virus Infection?

As described in the first part of this review, the nuclear pore-mediated nucleocytoplasmic transport controls in the cell need to be tightly regulated. We may wonder whether EBV utilizes a pre-existed cellular pathway for its nuclear egress. Some studies in the past decade have suggested that an NPC-independent vesicular nucleocytoplasmic transport system is conserved in eukaryotes. For example, in response to nutrient starvation, *Saccharomyces* buds double-membrane vesicles (piecemeal microautophagy or late nucleophagy) from the NE into the cytoplasmic vacuole for selective degradation of non-essential nuclear components, such as portions of the NE [125,126]. Studies in *Drosophila* also indicate that large ribonucleoprotein particles are transported across the NE by a vesicle-mediated export for protein synthesis [127,128]. An unconventional export mechanism was also reported to mediate the selective degradation of nuclear lamin or emerin in the cytoplasm of mammalian cells [129,130]. These studies suggest that an NPC-independent vesicular nucleocytoplasmic transport system appears to be conserved among eukaryotes and may cooperate with cytoplasmic processes, such as autophagic proteolysis, for the hemostasis of nuclear components.

11. EBV Putatively Hijacks the Endogenous Transporting/Clearance Pathway for Its Maturation

Autophagy is an important mechanism for removing cytoplasmic protein aggregates and damaged organelles, especially when the cellular chaperone capacity or the ubiquitin-proteasome system is overwhelmed or ineffective. Several cellular factors, such as p62 (SQSTM1), ALFY (autophagy-linked FYVE protein), and yeast Atg8 homolog LC3s or

GABARAPs, have been shown to be crucial for the selective clearance of cytoplasmic aggresomes [131,132]. Starting with the binding of ubiquitinated substrates or cargos by the ubiquitin-binding receptor p62, the adaptor molecule ALFY binds sequentially to membrane-anchored GABARAP and induces autophagosome formation [133–135]. Autophagosomes may then fuse with endosomes to form amphisomes and, finally, fuse with lysosomes for autophagic proteolysis [34,136]. So far, autophagy dysregulation has been linked to several human diseases, such as autoimmune diseases, protein aggregation diseases, and cancers [137]. Herpesviruses, such as EBV and KSHV, also use various strategies to modulate the autophagy pathway for their replication and to overcome the immune response [137,138].

In EBV, the presence of BFRF1 was known to promote early autophagic steps, which is essential for EBV reactivation, and mediate the blockage of late autophagic steps [139,140]. We previously demonstrated that BFRF1 protein recruits cellular ESCRT components to modulate the NE for the nuclear egress of EBV [99]. Based on the characteristic of BFRF1 protein in driving membrane dynamics, we explored the possibility that elicitation of NE-derived vesicles by ectopic BFRF1 expression may improve the translocation and clearance of nucleus-located protein deposition [120]. We found BFRF1-induced vesicles fuse putatively with autophagic vacuoles in the cytoplasm. ALFY-mediated autophagic proteolysis subsequently leads to the clearance of nuclear aggregates (Figure 5). Therefore, it is suggested that viral nucleocapsids may be recognized as similar to the protein deposits in the cellular environment. EBV putatively employs or hijacks parts of the endogenous transporting/clearance pathway, originally responsible for protein aggregate catabolism, for nucleocapsid nucleus egress.

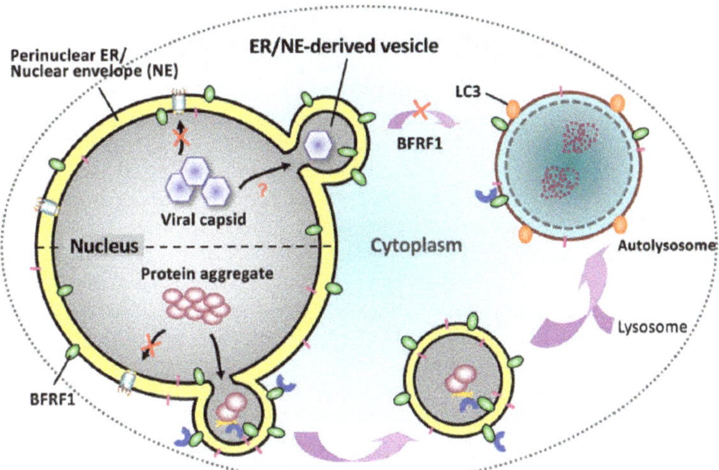

Figure 5. The hypothetical model for the cellular and viral regulation of EBV BFRF1-induced nuclear envelope (NE) modulation. The NE plays a pivotal role in cellular signaling and homeostasis. During EBV reactivation, the viral NE protein BFRF1 can modulate NE structure and form cytoplasmic vesicles by recruiting cellular ESCRT components and Itch ubiquitin ligase [99,119]. The ER/NE-derived vesicles provide nucleocytoplasmic transport for the nuclear egress of nucleocapsids (upper part). The presence of BFRF1 may mediate the blockage of late autophagic proteolysis for transported nucleocapsids in the cytoplasm [140]. Previously, we showed that BFRF1-induced vesicles potentially mediate the delivery and clearance of nuclear protein aggregates through autophagy proteolysis in the cytoplasm [120] (lower part). This information obtained may provide insights into the fundamental functions of the NE and knowledge of NE-related and protein aggregation diseases.

Correspondingly, some studies indicate that EBV uses autophagic machinery for transport and cytosolic maturation [139,141]. The presence of EBV BFRF1 is crucial for the early steps of autophagy and blocks the final proteolysis in the EBV life cycle [140]. These observations suggest that BFRF1-induced NE vesicles may provide the autophagic vehicles for cytoplasmic transport or the membrane source for cytoplasmic re-envelopment of EBV, and mediate virus escape from lysosomal proteolysis for efficient virion production. If that is the case, it will be worthwhile to investigate further how BFRF1 prevents the cytoplasmic fusion of lysosomes during virus maturation. On the other hand, recurrent EBV reactivation is known to enhance the genome instability and tumorigenicity of nasopharyngeal carcinoma [142]. If the presence of EBV BFRF1 improves the dynamics and trafficking of intracellular vesicles, it would be interesting to investigate whether repeated virus reactivation interferes with the regular transport pathway and contributes to the cytopathogenesis of cancer cells.

12. Coda: Is the Nucleocytoplasmic Pathway a New Target for Protein Aggregation Diseases?

Intracellular protein aggregates are a pathological hallmark of several neurodegenerative diseases that impact the global population today. Among the protein aggregation diseases, the polyglutamine (polyQ) expansion disorders, such as Huntington's disease, spinobular muscular atrophy, and the spinocerebellar ataxia 1, 3, and 7, involve cytotoxic, nonnative, aggregation-prone conformations because of the presence of long homopolymeric tracts of glutamine. So far, most strategies for eliminating protein aggregates have focused on reducing gene expression of the pathogenic proteins by RNA-targeting, reducing aggregate formation by pharmacologic or molecular chaperone treatment, or facilitating protein degradation by the ubiquitin–proteasome system [143,144]. Nevertheless, recent studies also indicated that autophagy or lysosome enhancer is a potential protein reduction strategy. For example, treatment of autophagy enhancer improves the clearance of proteolipid aggregates, reduces neuropathology, and prolongs survival of diseased mice [145]. This suggests that autophagy is a powerful system for clearing cell protein deposits. Since the unique EBV BFRF1 stimulates the protrusion of the NE-associated membrane for aggregated protein and links potentially to cytoplasmic autophagy, a similar vesicular nucleocytoplasmic transport across the intact membrane may serve as an ideal pathway for eliminating protein aggregates. Our idea was to test whether the protein deposits can be engulfed by the NE membrane, moved, and subjected to the cytoplasmic degradation system.

Using a system mimicking natural cell conditions in neurodegenerative diseases, we previously found that eliciting vesicular transport by ectopic BFRF1 expression reduces the accumulation of nuclear aggregate in neuroblastoma cells [120]. Thus, it suggests that activation of cellular component translocation by BFRF1 can be exploited to clear pathogenic protein aggregates from neuronal cells. It would be interesting and worthwhile to investigate further whether improving membrane dynamics using EBV BFRF1 can be adapted to an organismal level, such as transgenic animal systems, for the elimination of protein deposits. Nevertheless, the viral origin of BFRF1 may be a concern for its direct use in vivo. Further studies to characterize the safety of driving vesicular nucleocytoplasmic transport in vivo and explore the endogenous cellular factors or small molecule components capable of activating intrinsic vesicular transport are likely to be helpful for the therapy of protein aggregation diseases.

Author Contributions: Writing—Original Draft Preparation, C.-P.L. and M.-R.C.; Writing—Review and Editing, C.-P.L. and M.-R.C.; Visualization, C.-P.L. All authors have read and agreed to the published version of the manuscript.

Funding: This research was funded by the Ministry of Science and Technology, Taiwan, MOST106-2320-B-227-001-MY3 and MOST 107-2320-B-002-013-MY3, and by National Health Research Institute, Taiwan, NHRI-EX110-11013BI.

Institutional Review Board Statement: Not applicable.

Informed Consent Statement: Not applicable.

Acknowledgments: We are grateful to all previous students, postdoctoral fellows, and research assistants who participated in those studies at the National Taiwan University College of Medicine and National Taipei University of Nursing and Health Sciences, and technical supports from our collaborators. We also thank the staff of the imaging core at the First Core Labs, National Taiwan University College of Medicine, for technical assistance, and Tim J Harrison of University College London for the critical reading and editing of the manuscript.

Conflicts of Interest: All authors declare no conflict of interest.

References

1. Fawcett, D.W. On the occurrence of a fibrous lamina on the inner aspect of the nuclear envelope in certain cells of vertebrates. *Am. J. Anat.* **1966**, *119*, 129–145. [CrossRef] [PubMed]
2. Gruenbaum, Y.; Margalit, A.; Goldman, R.D.; Shumaker, D.K.; Wilson, K.L. The nuclear lamina comes of age. *Nat. Rev. Mol. Cell Biol.* **2005**, *6*, 21–31. [CrossRef] [PubMed]
3. McKeon, F.D.; Kirschner, M.W.; Caput, D. Homologies in both primary and secondary structure between nuclear envelope and intermediate filament proteins. *Nature* **1986**, *319*, 463–468. [CrossRef] [PubMed]
4. Moir, R.D.; Spann, T.P.; Goldman, R.D. The dynamic properties and possible functions of nuclear lamins. *Int. Rev. Cytol.* **1995**, *162B*, 141–182.
5. Ungricht, R.; Kutay, U. Mechanisms and functions of nuclear envelope remodelling. *Nat. Rev. Mol. Cell Biol.* **2017**, *18*, 229–245. [CrossRef]
6. Chow, K.H.; Factor, R.E.; Ullman, K.S. The nuclear envelope environment and its cancer connections. *Nat. Rev. Cancer* **2012**, *12*, 196–209. [CrossRef]
7. de Las Heras, J.I.; Batrakou, D.G.; Schirmer, E.C. Cancer biology and the nuclear envelope: A convoluted relationship. *Semin. Cancer Biol.* **2013**, *23*, 125–137. [CrossRef]
8. Romero-Bueno, R.; de la Cruz Ruiz, P.; Artal-Sanz, M.; Askjaer, P.; Dobrzynska, A. Nuclear Organization in stress and aging. *Cells* **2019**, *8*, 664. [CrossRef]
9. Stiekema, M.; van Zandvoort, M.; Ramaekers, F.C.S.; Broers, J.L.V. Structural and mechanical aberrations of the nuclear lamina in disease. *Cells* **2020**, *9*, 1884. [CrossRef]
10. Worman, H.J.; Fong, L.G.; Muchir, A.; Young, S.G. Laminopathies and the long strange trip from basic cell biology to therapy. *J. Clin. Investig.* **2009**, *119*, 1825–1836. [CrossRef]
11. Liu, S.Y.; Ikegami, K. Nuclear lamin phosphorylation: An emerging role in gene regulation and pathogenesis of laminopathies. *Nucleus* **2020**, *11*, 299–314. [CrossRef]
12. Lin, D.H.; Hoelz, A. The structure of the nuclear pore complex (An update). *Annu. Rev. Biochem.* **2019**, *88*, 725–783. [CrossRef]
13. Robbins, J.; Dilworth, S.M.; Laskey, R.A.; Dingwall, C. Two interdependent basic domains in nucleoplasmin nuclear targeting sequence: Identification of a class of bipartite nuclear targeting sequence. *Cell* **1991**, *64*, 615–623. [CrossRef]
14. Gorlich, D.; Henklein, P.; Laskey, R.A.; Hartmann, E. A 41 amino acid motif in importin-alpha confers binding to importin-beta and hence transit into the nucleus. *EMBO J.* **1996**, *15*, 1810–1817. [CrossRef]
15. Weis, K.; Ryder, U.; Lamond, A.I. The conserved amino-terminal domain of hSRP1 alpha is essential for nuclear protein import. *EMBO J.* **1996**, *15*, 1818–1825. [CrossRef]
16. Sorokin, A.V.; Kim, E.R.; Ovchinnikov, L.P. Nucleocytoplasmic transport of proteins. *Biochemistry* **2007**, *72*, 1439–1457. [CrossRef]
17. Wente, S.R.; Rout, M.P. The nuclear pore complex and nuclear transport. *Cold Spring Harb. Perspect. Biol.* **2010**, *2*, a000562. [CrossRef]
18. Fornerod, M.; Ohno, M.; Yoshida, M.; Mattaj, I.W. CRM1 is an export receptor for leucine-rich nuclear export signals. *Cell* **1997**, *90*, 1051–1060. [CrossRef]
19. Fung, H.Y.; Chook, Y.M. Atomic basis of CRM1-cargo recognition, release and inhibition. *Semin. Cancer Biol.* **2014**, *27*, 52–61. [CrossRef]
20. Fitzgerald, K.A.; McWhirter, S.M.; Faia, K.L.; Rowe, D.C.; Latz, E.; Golenbock, D.T.; Coyle, A.J.; Liao, S.M.; Maniatis, T. IKKepsilon and TBK1 are essential components of the IRF3 signaling pathway. *Nat. Immunol.* **2003**, *4*, 491–496. [CrossRef]
21. Hayden, M.S.; Ghosh, S. Signaling to NF-kappaB. *Genes Dev.* **2004**, *18*, 2195–2224. [CrossRef]
22. Longnecker, R.M.; Kieff, E.; Cohen, J.I. Epstein-Barr virus. In *Fields Virology*, 6th ed.; Lippincott Williams & Wilkins: Philadelphia, PA, USA, 2013; p. 1898.
23. Rosemarie, Q.; Sugden, B. Epstein-Barr virus: How its lytic phase contributes to oncogenesis. *Microorganisms* **2020**, *8*, 1824. [CrossRef]
24. Chen, J.; Longnecker, R. Epithelial cell infection by Epstein-Barr virus. *FEMS Microbiol. Rev.* **2019**, *43*, 674–683. [CrossRef]
25. Miller, N.; Hutt-Fletcher, L.M. Epstein-Barr virus enters B cells and epithelial cells by different routes. *J. Virol.* **1992**, *66*, 3409–3414. [CrossRef]

26. Chen, J.; Sathiyamoorthy, K.; Zhang, X.; Schaller, S.; Perez White, B.E.; Jardetzky, T.S.; Longnecker, R. Ephrin receptor A2 is a functional entry receptor for Epstein-Barr virus. *Nat. Microbiol.* **2018**, *3*, 172–180. [CrossRef]

27. Zhang, H.; Li, Y.; Wang, H.B.; Zhang, A.; Chen, M.L.; Fang, Z.X.; Dong, X.D.; Li, S.B.; Du, Y.; Xiong, D.; et al. Ephrin receptor A2 is an epithelial cell receptor for Epstein-Barr virus entry. *Nat. Microbiol.* **2018**, *3*, 1–8. [CrossRef]

28. Backovic, M.; Jardetzky, T.S.; Longnecker, R. Hydrophobic residues that form putative fusion loops of Epstein-Barr virus glycoprotein B are critical for fusion activity. *J. Virol.* **2007**, *81*, 9596–9600. [CrossRef]

29. Backovic, M.; Longnecker, R.; Jardetzky, T.S. Structure of a trimeric variant of the Epstein-Barr virus glycoprotein B. *Proc. Natl. Acad. Sci. USA* **2009**, *106*, 2880–2885. [CrossRef] [PubMed]

30. Grove, J.; Marsh, M. The cell biology of receptor-mediated virus entry. *J. Cell Biol.* **2011**, *195*, 1071–1082. [CrossRef] [PubMed]

31. Valencia, S.M.; Hutt-Fletcher, L.M. Important but differential roles for actin in trafficking of Epstein-Barr virus in B cells and epithelial cells. *J. Virol.* **2012**, *86*, 2–10. [CrossRef] [PubMed]

32. Chesnokova, L.S.; Jiang, R.; Hutt-Fletcher, L.M. Viral entry. In *Epstein Barr Virus Volume 2: One Herpes Virus: Many Diseases*; Münz, C., Ed.; Springer: Cham, Switzerland, 2015; pp. 221–235.

33. Lu, C.C.; Jeng, Y.Y.; Tsai, C.H.; Liu, M.Y.; Yeh, S.W.; Hsu, T.Y.; Chen, M.R. Genome-wide transcription program and expression of the Rta responsive gene of Epstein-Barr virus. *Virology* **2006**, *345*, 358–372. [CrossRef]

34. Lee, C.P.; Chen, M.R. Escape of herpesviruses from the nucleus. *Rev. Med. Virol.* **2010**, *20*, 214–230. [CrossRef]

35. Hellberg, T.; Passvogel, L.; Schulz, K.S.; Klupp, B.G.; Mettenleiter, T.C. Nuclear egress of herpesviruses: The prototypic vesicular nucleocytoplasmic transport. *Adv. Virus Res.* **2016**, *94*, 81–140.

36. Nanbo, A. Epstein-Barr Virus Exploits the Secretory Pathway to Release Virions. *Microorganisms* **2020**, *8*, 729. [CrossRef]

37. Fixman, E.D.; Hayward, G.S.; Hayward, S.D. Replication of Epstein-Barr virus oriLyt: Lack of a dedicated virally encoded origin-binding protein and dependence on Zta in cotransfection assays. *J. Virol.* **1995**, *69*, 2998–3006. [CrossRef]

38. El-Guindy, A.; Ghiassi-Nejad, M.; Golden, S.; Delecluse, H.J.; Miller, G. Essential role of Rta in lytic DNA replication of Epstein-Barr virus. *J. Virol.* **2013**, *87*, 208–223. [CrossRef]

39. Su, M.T.; Liu, I.H.; Wu, C.W.; Chang, S.M.; Tsai, C.H.; Yang, P.W.; Chuang, Y.C.; Lee, C.P.; Chen, M.R. Uracil DNA glycosylase BKRF3 contributes to Epstein-Barr virus DNA replication through physical interactions with proteins in viral DNA replication complex. *J. Virol.* **2014**, *88*, 8883–8899. [CrossRef]

40. Fixman, E.D.; Hayward, G.S.; Hayward, S.D. trans-acting requirements for replication of Epstein-Barr virus ori-Lyt. *J. Virol.* **1992**, *66*, 5030–5039. [CrossRef]

41. Kawashima, D.; Kanda, T.; Murata, T.; Saito, S.; Sugimoto, A.; Narita, Y.; Tsurumi, T. Nuclear transport of Epstein-Barr virus DNA polymerase is dependent on the BMRF1 polymerase processivity factor and molecular chaperone Hsp90. *J. Virol.* **2013**, *87*, 6482–6491. [CrossRef]

42. Gao, Z.; Krithivas, A.; Finan, J.E.; Semmes, O.J.; Zhou, S.; Wang, Y.; Hayward, S.D. The Epstein-Barr virus lytic transactivator Zta interacts with the helicase-primase replication proteins. *J. Virol.* **1998**, *72*, 8559–8567. [CrossRef]

43. Chang, C.W.; Lee, C.P.; Huang, Y.H.; Yang, P.W.; Wang, J.T.; Chen, M.R. Epstein-Barr virus protein kinase BGLF4 targets the nucleus through interaction with nucleoporins. *J. Virol.* **2012**, *86*, 8072–8085. [CrossRef]

44. Chang, C.W.; Lee, C.P.; Su, M.T.; Tsai, C.H.; Chen, M.R. BGLF4 kinase modulates the structure and transport preference of the nuclear pore complex to facilitate nuclear import of Epstein-Barr virus lytic proteins. *J. Virol.* **2015**, *89*, 1703–1718. [CrossRef]

45. Wang, W.H.; Kuo, C.W.; Chang, L.K.; Hung, C.C.; Chang, T.H.; Liu, S.T. Assembly of Epstein-Barr virus capsid in promyelocytic leukemia nuclear bodies. *J. Virol.* **2015**, *89*, 8922–8931. [CrossRef]

46. Wang, J.T.; Yang, P.W.; Lee, C.P.; Han, C.H.; Tsai, C.H.; Chen, M.R. Detection of Epstein-Barr virus BGLF4 protein kinase in virus replication compartments and virus particles. *J. Gen. Virol.* **2005**, *86 Pt 12*, 3215–3225. [CrossRef]

47. Asai, R.; Kato, A.; Kato, K.; Kanamori-Koyama, M.; Sugimoto, K.; Sairenji, T.; Nishiyama, Y.; Kawaguchi, Y. Epstein-Barr virus protein kinase BGLF4 is a virion tegument protein that dissociates from virions in a phosphorylation-dependent process and phosphorylates the viral immediate-early protein BZLF1. *J. Virol.* **2006**, *80*, 5125–5134. [CrossRef]

48. Gershburg, E.; Raffa, S.; Torrisi, M.R.; Pagano, J.S. Epstein-Barr virus-encoded protein kinase (BGLF4) is involved in production of infectious virus. *J. Virol.* **2007**, *81*, 5407–5412. [CrossRef]

49. Lee, C.P.; Huang, Y.H.; Lin, S.F.; Chang, Y.; Chang, Y.H.; Takada, K.; Chen, M.R. Epstein-Barr virus BGLF4 kinase induces disassembly of the nuclear lamina to facilitate virion production. *J. Virol.* **2008**, *82*, 11913–11926. [CrossRef]

50. Kawaguchi, Y.; Kato, K.; Tanaka, M.; Kanamori, M.; Nishiyama, Y.; Yamanashi, Y. Conserved protein kinases encoded by herpesviruses and cellular protein kinase cdc2 target the same phosphorylation site in eukaryotic elongation factor 1delta. *J. Virol.* **2003**, *77*, 2359–2368. [CrossRef]

51. Li, R.; Zhu, J.; Xie, Z.; Liao, G.; Liu, J.; Chen, M.R.; Hu, S.; Woodard, C.; Lin, J.; Taverna, S.D.; et al. Conserved herpesvirus kinases target the DNA damage response pathway and TIP60 histone acetyltransferase to promote virus replication. *Cell Host Microbe* **2011**, *10*, 390–400. [CrossRef]

52. Li, R.; Liao, G.; Nirujogi, R.S.; Pinto, S.M.; Shaw, P.G.; Huang, T.C.; Wan, J.; Qian, J.; Gowda, H.; Wu, X.; et al. Phosphoproteomic profiling reveals Epstein-Barr virus protein kinase integration of DNA damage response and mitotic signaling. *PLoS Pathog.* **2015**, *11*, e1005346. [CrossRef]

53. Bogdanow, B.; Schmidt, M.; Weisbach, H.; Gruska, I.; Vetter, B.; Imami, K.; Ostermann, E.; Brune, W.; Selbach, M.; Hagemeier, C.; et al. Cross-regulation of viral kinases with cyclin A secures shutoff of host DNA synthesis. *Nat. Commun.* **2020**, *11*, 4845. [CrossRef] [PubMed]

54. Wang, J.; Guo, W.; Long, C.; Zhou, H.; Wang, H.; Sun, X. The split Renilla luciferase complementation assay is useful for identifying the interaction of Epstein-Barr virus protein kinase BGLF4 and a heat shock protein Hsp90. *Acta Virol.* **2016**, *60*, 62–70. [CrossRef] [PubMed]

55. Wang, J.T.; Doong, S.L.; Teng, S.C.; Lee, C.P.; Tsai, C.H.; Chen, M.R. Epstein-Barr virus BGLF4 kinase suppresses the interferon regulatory factor 3 signaling pathway. *J. Virol.* **2009**, *83*, 1856–1869. [CrossRef] [PubMed]

56. Chang, L.S.; Wang, J.T.; Doong, S.L.; Lee, C.P.; Chang, C.W.; Tsai, C.H.; Yeh, S.W.; Hsieh, C.Y.; Chen, M.R. Epstein-Barr virus BGLF4 kinase downregulates NF-kappaB transactivation through phosphorylation of coactivator UXT. *J. Virol.* **2012**, *86*, 12176–12186. [CrossRef]

57. Lee, C.P.; Chen, J.Y.; Wang, J.T.; Kimura, K.; Takemoto, A.; Lu, C.C.; Chen, M.R. Epstein-Barr virus BGLF4 kinase induces premature chromosome condensation through activation of condensin and topoisomerase II. *J. Virol.* **2007**, *81*, 5166–5180. [CrossRef]

58. Chang, Y.H.; Lee, C.P.; Su, M.T.; Wang, J.T.; Chen, J.Y.; Lin, S.F.; Tsai, C.H.; Hsieh, M.J.; Takada, K.; Chen, M.R. Epstein-Barr virus BGLF4 kinase retards cellular S-phase progression and induces chromosomal abnormality. *PLoS ONE* **2012**, *7*, e39217. [CrossRef]

59. Zhu, J.; Liao, G.; Shan, L.; Zhang, J.; Chen, M.R.; Hayward, G.S.; Hayward, S.D.; Desai, P.; Zhu, H. Protein array identification of substrates of the Epstein-Barr virus protein kinase BGLF4. *J. Virol.* **2009**, *83*, 5219–5231. [CrossRef]

60. Asai, R.; Kato, A.; Kawaguchi, Y. Epstein-Barr virus protein kinase BGLF4 interacts with viral transactivator BZLF1 and regulates its transactivation activity. *J. Gen. Virol.* **2009**, *90 Pt 7*, 1575–1581. [CrossRef]

61. Prichard, M.N. Function of human cytomegalovirus UL97 kinase in viral infection and its inhibition by maribavir. *Rev. Med. Virol.* **2009**, *19*, 215–229. [CrossRef]

62. Steingruber, M.; Marschall, M. The cytomegalovirus protein kinase pUL97: Host interactions, regulatory mechanisms and antiviral drug targeting. *Microorganisms* **2020**, *8*, 515. [CrossRef]

63. Wagenaar, F.; Pol, J.M.; Peeters, B.; Gielkens, A.L.; de Wind, N.; Kimman, T.G. The US3-encoded protein kinase from pseudorabies virus affects egress of virions from the nucleus. *J. Gen. Virol.* **1995**, *76 Pt 7*, 1851–1859. [CrossRef]

64. Klupp, B.G.; Granzow, H.; Mettenleiter, T.C. Effect of the pseudorabies virus US3 protein on nuclear membrane localization of the UL34 protein and virus egress from the nucleus. *J. Gen. Virol.* **2001**, *82 Pt 10*, 2363–2371. [CrossRef]

65. Poon, A.P.; Benetti, L.; Roizman, B. U(S)3 and U(S)3.5 protein kinases of herpes simplex virus 1 differ with respect to their functions in blocking apoptosis and in virion maturation and egress. *J. Virol.* **2006**, *80*, 3752–3764. [CrossRef]

66. Murata, T.; Isomura, H.; Yamashita, Y.; Toyama, S.; Sato, Y.; Nakayama, S.; Kudoh, A.; Iwahori, S.; Kanda, T.; Tsurumi, T. Efficient production of infectious viruses requires enzymatic activity of Epstein-Barr virus protein kinase. *Virology* **2009**, *389*, 75–81. [CrossRef]

67. Sun, X.; Bristol, J.A.; Iwahori, S.; Hagemeier, S.R.; Meng, Q.; Barlow, E.A.; Fingeroth, J.D.; Tarakanova, V.L.; Kalejta, R.F.; Kenney, S.C. Hsp90 inhibitor 17-DMAG decreases expression of conserved herpesvirus protein kinases and reduces virus production in Epstein-Barr virus-infected cells. *J. Virol.* **2013**, *87*, 10126–10138. [CrossRef]

68. Marschall, M.; Marzi, A.; aus dem Siepen, P.; Jochmann, R.; Kalmer, M.; Auerochs, S.; Lischka, P.; Leis, M.; Stamminger, T. Cellular p32 recruits cytomegalovirus kinase pUL97 to redistribute the nuclear lamina. *J. Biol. Chem.* **2005**, *280*, 33357–33367. [CrossRef]

69. Kato, A.; Yamamoto, M.; Ohno, T.; Tanaka, M.; Sata, T.; Nishiyama, Y.; Kawaguchi, Y. Herpes simplex virus 1-encoded protein kinase UL13 phosphorylates viral Us3 protein kinase and regulates nuclear localization of viral envelopment factors UL34 and UL31. *J. Virol.* **2006**, *80*, 1476–1486. [CrossRef]

70. Mou, F.; Forest, T.; Baines, J.D. US3 of herpes simplex virus type 1 encodes a promiscuous protein kinase that phosphorylates and alters localization of lamin A/C in infected cells. *J. Virol.* **2007**, *81*, 6459–6470. [CrossRef]

71. Hamirally, S.; Kamil, J.P.; Ndassa-Colday, Y.M.; Lin, A.J.; Jahng, W.J.; Baek, M.C.; Noton, S.; Silva, L.A.; Simpson-Holley, M.; Knipe, D.M.; et al. Viral mimicry of Cdc2/cyclin-dependent kinase 1 mediates disruption of nuclear lamina during human cytomegalovirus nuclear egress. *PLoS Pathog.* **2009**, *5*, e1000275. [CrossRef]

72. Zhang, K.; Lv, D.W.; Li, R. Conserved herpesvirus protein kinases target SAMHD1 to facilitate virus replication. *Cell Rep.* **2019**, *28*, 449–459.e5. [CrossRef]

73. Roller, R.J.; Baines, J.D. Herpesvirus nuclear egress. *Adv. Anat. Embryol. Cell Biol.* **2017**, *223*, 143–169. [PubMed]

74. Leuzinger, H.; Ziegler, U.; Schraner, E.M.; Fraefel, C.; Glauser, D.L.; Heid, I.; Ackermann, M.; Mueller, M.; Wild, P. Herpes simplex virus 1 envelopment follows two diverse pathways. *J. Virol.* **2005**, *79*, 13047–13059. [CrossRef] [PubMed]

75. Klupp, B.G.; Granzow, H.; Mettenleiter, T.C. Nuclear envelope breakdown can substitute for primary envelopment-mediated nuclear egress of herpesviruses. *J. Virol.* **2011**, *85*, 8285–8292. [CrossRef] [PubMed]

76. Mettenleiter, T.C.; Klupp, B.G.; Granzow, H. Herpesvirus assembly: An update. *Virus Res.* **2009**, *143*, 222–234. [CrossRef]

77. Lye, M.F.; Wilkie, A.R.; Filman, D.J.; Hogle, J.M.; Coen, D.M. Getting to and through the inner nuclear membrane during herpesvirus nuclear egress. *Curr. Opin. Cell Biol.* **2017**, *46*, 9–16. [CrossRef]

78. Lv, Y.; Zhou, S.; Gao, S.; Deng, H. Remodeling of host membranes during herpesvirus assembly and egress. *Protein Cell* **2019**, *10*, 315–326. [CrossRef]

79. Klupp, B.G.; Granzow, H.; Mettenleiter, T.C. Primary envelopment of pseudorabies virus at the nuclear membrane requires the UL34 gene product. *J. Virol.* **2000**, *74*, 10063–10073. [CrossRef]

80. Farina, A.; Feederle, R.; Raffa, S.; Gonnella, R.; Santarelli, R.; Frati, L.; Angeloni, A.; Torrisi, M.R.; Faggioni, A.; Delecluse, H.J. BFRF1 of Epstein-Barr virus is essential for efficient primary viral envelopment and egress. *J. Virol.* **2005**, *79*, 3703–3712. [CrossRef]

81. Reynolds, A.E.; Liang, L.; Baines, J.D. Conformational changes in the nuclear lamina induced by herpes simplex virus type 1 require genes U(L)31 and U(L)34. *J. Virol.* **2004**, *78*, 5564–5575. [CrossRef]

82. Camozzi, D.; Pignatelli, S.; Valvo, C.; Lattanzi, G.; Capanni, C.; Dal Monte, P.; Landini, M.P. Remodelling of the nuclear lamina during human cytomegalovirus infection: Role of the viral proteins pUL50 and pUL53. *J. Gen. Virol.* **2008**, *89 Pt 3*, 731–740. [CrossRef]

83. Bigalke, J.M.; Heldwein, E.E. Structural basis of membrane budding by the nuclear egress complex of herpesviruses. *EMBO J.* **2015**, *34*, 2921–2936. [CrossRef]

84. Bigalke, J.M.; Heldwein, E.E. Have NEC coat, will travel: Structural basis of membrane budding during nuclear egress in herpesviruses. *Adv. Virus Res.* **2017**, *97*, 107–141.

85. Marschall, M.; Muller, Y.A.; Diewald, B.; Sticht, H.; Milbradt, J. The human cytomegalovirus nuclear egress complex unites multiple functions: Recruitment of effectors, nuclear envelope rearrangement, and docking to nuclear capsids. *Rev. Med. Virol.* **2017**, *27*, e1934. [CrossRef]

86. Marschall, M.; Hage, S.; Conrad, M.; Alkhashrom, S.; Kicuntod, J.; Schweininger, J.; Kriegel, M.; Losing, J.; Tillmanns, J.; Neipel, F.; et al. Nuclear egress complexes of HCMV and other herpesviruses: Solving the puzzle of sequence coevolution, conserved structures and subfamily-spanning binding properties. *Viruses* **2020**, *12*, 683. [CrossRef]

87. Walzer, S.A.; Egerer-Sieber, C.; Sticht, H.; Sevvana, M.; Hohl, K.; Milbradt, J.; Muller, Y.A.; Marschall, M. Crystal structure of the human cytomegalovirus pUL50-pUL53 core nuclear egress complex provides insight into a unique assembly scaffold for virus-host protein interactions. *J. Biol. Chem.* **2015**, *290*, 27452–27458. [CrossRef]

88. Hagen, C.; Dent, K.C.; Zeev-Ben-Mordehai, T.; Grange, M.; Bosse, J.B.; Whittle, C.; Klupp, B.G.; Siebert, C.A.; Vasishtan, D.; Bauerlein, F.J.; et al. Structural basis of vesicle formation at the inner nuclear membrane. *Cell* **2015**, *163*, 1692–1701. [CrossRef]

89. Draganova, E.B.; Zhang, J.; Zhou, Z.H.; Heldwein, E.E. Structural basis for capsid recruitment and coat formation during HSV-1 nuclear egress. *Elife* **2020**, *9*, e56627. [CrossRef]

90. Vu, A.; Poyzer, C.; Roller, R. Extragenic suppression of a mutation in herpes simplex virus 1 UL34 that affects lamina disruption and nuclear egress. *J. Virol.* **2016**, *90*, 10738–10751. [CrossRef]

91. Hage, S.; Sonntag, E.; Svrlanska, A.; Borst, E.M.; Stilp, A.C.; Horsch, D.; Muller, R.; Kropff, B.; Milbradt, J.; Stamminger, T.; et al. Phenotypical characterization of the nuclear egress of recombinant cytomegaloviruses reveals defective replication upon ORF-UL50 deletion but not pUL50 phosphosite mutation. *Viruses* **2021**, *13*, 165. [CrossRef]

92. Gonnella, R.; Farina, A.; Santarelli, R.; Raffa, S.; Feederle, R.; Bei, R.; Granato, M.; Modesti, A.; Frati, L.; Delecluse, H.J.; et al. Characterization and intracellular localization of the Epstein-Barr virus protein BFLF2: Interactions with BFRF1 and with the nuclear lamina. *J. Virol.* **2005**, *79*, 3713–3727. [CrossRef]

93. Epstein, M.A.; Achong, B.C. Morphology of the virus and virus-induced cytopathologic changes. In *The Epstein-Barr Virus*; Epstein, M.A., Achong, B.C., Eds.; Springer: New York, NY, USA, 1979; pp. 23–27.

94. Torrisi, M.R.; Cirone, M.; Pavan, A.; Zompetta, C.; Barile, G.; Frati, L.; Faggioni, A. Localization of Epstein-Barr virus envelope glycoproteins on the inner nuclear membrane of virus-producing cells. *J. Virol.* **1989**, *63*, 828–832. [CrossRef]

95. Lee, C.P.; Liu, P.T.; Kung, H.N.; Su, M.T.; Chua, H.H.; Chang, Y.H.; Chang, C.W.; Tsai, C.H.; Liu, F.T.; Chen, M.R. The ESCRT machinery is recruited by the viral BFRF1 protein to the nucleus-associated membrane for the maturation of Epstein-Barr Virus. *PLoS Pathog.* **2012**, *8*, e1002904. [CrossRef]

96. Klupp, B.G.; Granzow, H.; Fuchs, W.; Keil, G.M.; Finke, S.; Mettenleiter, T.C. Vesicle formation from the nuclear membrane is induced by coexpression of two conserved herpesvirus proteins. *Proc. Natl. Acad. Sci. USA* **2007**, *104*, 7241–7246. [CrossRef] [PubMed]

97. Luitweiler, E.M.; Henson, B.W.; Pryce, E.N.; Patel, V.; Coombs, G.; McCaffery, J.M.; Desai, P.J. Interactions of the Kaposi's Sarcoma-associated herpesvirus nuclear egress complex: ORF69 is a potent factor for remodeling cellular membranes. *J. Virol.* **2013**, *87*, 3915–3929. [CrossRef] [PubMed]

98. Granato, M.; Feederle, R.; Farina, A.; Gonnella, R.; Santarelli, R.; Hub, B.; Faggioni, A.; Delecluse, H.J. Deletion of Epstein-Barr virus BFLF2 leads to impaired viral DNA packaging and primary egress as well as to the production of defective viral particles. *J. Virol.* **2008**, *82*, 4042–4051. [CrossRef] [PubMed]

99. Dai, Y.C.; Liao, Y.T.; Juan, Y.T.; Cheng, Y.Y.; Su, M.T.; Su, Y.Z.; Liu, H.C.; Tsai, C.H.; Lee, C.P.; Chen, M.R. The novel nuclear targeting and BFRF1-interacting domains of BFLF2 are essential for efficient Epstein-Barr virus virion release. *J. Virol.* **2020**, *94*, e01498-19. [CrossRef] [PubMed]

100. Li, M.; Chen, T.; Zou, X.; Xu, Z.; Wang, Y.; Wang, P.; Ou, X.; Li, Y.; Chen, D.; Peng, T.; et al. Characterization of the Nucleocytoplasmic Transport Mechanisms of Epstein-Barr Virus BFLF2. *Cell Physiol. Biochem.* **2018**, *51*, 1500–1517. [CrossRef] [PubMed]

101. Gulbahce, N.; Yan, H.; Dricot, A.; Padi, M.; Byrdsong, D.; Franchi, R.; Lee, D.S.; Rozenblatt-Rosen, O.; Mar, J.C.; Calderwood, M.A.; et al. Viral perturbations of host networks reflect disease etiology. *PLoS Comput. Biol.* **2012**, *8*, e1002531. [CrossRef]

102. Gatta, A.T.; Carlton, J.G. The ESCRT-machinery: Closing holes and expanding roles. *Curr. Opin. Cell Biol.* **2019**, *59*, 121–132. [CrossRef]
103. Vietri, M.; Radulovic, M.; Stenmark, H. The many functions of ESCRTs. *Nat. Rev. Mol. Cell Biol.* **2020**, *21*, 25–42. [CrossRef]
104. Hurley, J.H.; Ren, X. The circuitry of cargo flux in the ESCRT pathway. *J. Cell Biol.* **2009**, *185*, 185–187. [CrossRef]
105. Raiborg, C.; Stenmark, H. The ESCRT machinery in endosomal sorting of ubiquitylated membrane proteins. *Nature* **2009**, *458*, 445–452. [CrossRef]
106. Strack, B.; Calistri, A.; Craig, S.; Popova, E.; Gottlinger, H.G. AIP1/ALIX is a binding partner for HIV-1 p6 and EIAV p9 functioning in virus budding. *Cell* **2003**, *114*, 689–699. [CrossRef]
107. Jouvenet, N.; Zhadina, M.; Bieniasz, P.D.; Simon, S.M. Dynamics of ESCRT protein recruitment during retroviral assembly. *Nat. Cell Biol.* **2011**, *13*, 394–401. [CrossRef]
108. Crump, C.M.; Yates, C.; Minson, T. Herpes simplex virus type 1 cytoplasmic envelopment requires functional Vps4. *J. Virol.* **2007**, *81*, 7380–7387. [CrossRef]
109. Fraile-Ramos, A.; Pelchen-Matthews, A.; Risco, C.; Rejas, M.T.; Emery, V.C.; Hassan-Walker, A.F.; Esteban, M.; Marsh, M. The ESCRT machinery is not required for human cytomegalovirus envelopment. *Cell Microbiol.* **2007**, *9*, 2955–2967. [CrossRef]
110. Pawliczek, T.; Crump, C.M. Herpes simplex virus type 1 production requires a functional ESCRT-III complex but is independent of TSG101 and ALIX expression. *J. Virol.* **2009**, *83*, 11254–11264. [CrossRef]
111. Tandon, R.; AuCoin, D.P.; Mocarski, E.S. Human cytomegalovirus exploits ESCRT machinery in the process of virion maturation. *J. Virol.* **2009**, *83*, 10797–10807. [CrossRef]
112. Calistri, A.; Sette, P.; Salata, C.; Cancellotti, E.; Forghieri, C.; Comin, A.; Gottlinger, H.; Campadelli-Fiume, G.; Palu, G.; Parolin, C. Intracellular trafficking and maturation of herpes simplex virus type 1 gB and virus egress require functional biogenesis of multivesicular bodies. *J. Virol.* **2007**, *81*, 11468–11478. [CrossRef]
113. Das, S.; Pellett, P.E. Spatial relationships between markers for secretory and endosomal machinery in human cytomegalovirus-infected cells versus those in uninfected cells. *J. Virol.* **2011**, *85*, 5864–5879. [CrossRef]
114. Mori, Y.; Koike, M.; Moriishi, E.; Kawabata, A.; Tang, H.; Oyaizu, H.; Uchiyama, Y.; Yamanishi, K. Human herpesvirus-6 induces MVB formation, and virus egress occurs by an exosomal release pathway. *Traffic* **2008**, *9*, 1728–1742. [CrossRef]
115. Lee, C.P.; Liu, G.T.; Kung, H.N.; Liu, P.T.; Liao, Y.T.; Chow, L.P.; Chang, L.S.; Chang, Y.H.; Chang, C.W.; Shu, W.C.; et al. The Ubiquitin Ligase Itch and Ubiquitination Regulate BFRF1-Mediated Nuclear Envelope Modification for Epstein-Barr Virus Maturation. *J. Virol.* **2016**, *90*, 8994–9007. [CrossRef]
116. Liu, G.T.; Kung, H.N.; Chen, C.K.; Huang, C.; Wang, Y.L.; Yu, C.P.; Lee, C.P. Improving nuclear envelope dynamics by EBV BFRF1 facilitates intranuclear component clearance through autophagy. *FASEB J.* **2018**, *32*, 3968–3983. [CrossRef]
117. Muller, Y.A.; Hage, S.; Alkhashrom, S.; Hollriegl, T.; Weigert, S.; Dolles, S.; Hof, K.; Walzer, S.A.; Egerer-Sieber, C.; Conrad, M.; et al. High-resolution crystal structures of two prototypical beta- and gamma-herpesviral nuclear egress complexes unravel the determinants of subfamily specificity. *J. Biol. Chem.* **2020**, *295*, 3189–3201. [CrossRef]
118. Yadav, S.; Libotte, F.; Buono, E.; Valia, S.; Farina, G.A.; Faggioni, A.; Farina, A. EBV early lytic protein BFRF1 alters emerin distribution and post-translational modification. *Virus Res.* **2017**, *232*, 113–122. [CrossRef]
119. Hage, S.; Sonntag, E.; Borst, E.M.; Tannig, P.; Seyler, L.; Bauerle, T.; Bailer, S.M.; Lee, C.P.; Muller, R.; Wangen, C.; et al. Patterns of autologous and nonautologous interactions between core Nuclear Egress Complex (NEC) Proteins of alpha-, beta- and gamma-Herpesviruses. *Viruses* **2020**, *12*, 303. [CrossRef]
120. Nanbo, A.; Noda, T.; Ohba, Y. Epstein-Barr virus acquires its final envelope on intracellular compartments with golgi markers. *Front. Microbiol.* **2018**, *9*, 454. [CrossRef]
121. Mijaljica, D.; Prescott, M.; Devenish, R.J. A late form of nucleophagy in Saccharomyces cerevisiae. *PLoS ONE* **2012**, *7*, e40013. [CrossRef]
122. Mijaljica, D.; Prescott, M.; Devenish, R.J. The intricacy of nuclear membrane dynamics during nucleophagy. *Nucleus* **2010**, *1*, 213–223. [CrossRef]
123. Speese, S.D.; Ashley, J.; Jokhi, V.; Nunnari, J.; Barria, R.; Li, Y.; Ataman, B.; Koon, A.; Chang, Y.T.; Li, Q.; et al. Nuclear envelope budding enables large ribonucleoprotein particle export during synaptic Wnt signaling. *Cell* **2012**, *149*, 832–846. [CrossRef]
124. Jokhi, V.; Ashley, J.; Nunnari, J.; Noma, A.; Ito, N.; Wakabayashi-Ito, N.; Moore, M.J.; Budnik, V. Torsin mediates primary envelopment of large ribonucleoprotein granules at the nuclear envelope. *Cell Rep.* **2013**, *3*, 988–995. [CrossRef] [PubMed]
125. Park, Y.E.; Hayashi, Y.K.; Bonne, G.; Arimura, T.; Noguchi, S.; Nonaka, I.; Nishino, I. Autophagic degradation of nuclear components in mammalian cells. *Autophagy* **2009**, *5*, 795–804. [CrossRef] [PubMed]
126. Dou, Z.; Xu, C.; Donahue, G.; Shimi, T.; Pan, J.A.; Zhu, J.; Ivanov, A.; Capell, B.C.; Drake, A.M.; Shah, P.P.; et al. Autophagy mediates degradation of nuclear lamina. *Nature* **2015**, *527*, 105–109. [CrossRef] [PubMed]
127. Garcia-Mata, R.; Gao, Y.S.; Sztul, E. Hassles with taking out the garbage: Aggravating aggresomes. *Traffic* **2002**, *3*, 388–396. [CrossRef]
128. Hyttinen, J.M.; Amadio, M.; Viiri, J.; Pascale, A.; Salminen, A.; Kaarniranta, K. Clearance of misfolded and aggregated proteins by aggrephagy and implications for aggregation diseases. *Ageing Res. Rev.* **2014**, *18*, 16–28. [CrossRef]
129. Bjorkoy, G.; Lamark, T.; Pankiv, S.; Overvatn, A.; Brech, A.; Johansen, T. Monitoring autophagic degradation of p62/SQSTM1. *Methods Enzymol.* **2009**, *452*, 181–197.
130. Isakson, P.; Holland, P.; Simonsen, A. The role of ALFY in selective autophagy. *Cell Death Differ.* **2013**, *20*, 12–20. [CrossRef]

131. Lystad, A.H.; Ichimura, Y.; Takagi, K.; Yang, Y.; Pankiv, S.; Kanegae, Y.; Kageyama, S.; Suzuki, M.; Saito, I.; Mizushima, T.; et al. Structural determinants in GABARAP required for the selective binding and recruitment of ALFY to LC3B-positive structures. *EMBO Rep.* **2014**, *15*, 557–565. [CrossRef]

132. Nixon, R.A. The role of autophagy in neurodegenerative disease. *Nat. Med.* **2013**, *19*, 983–997. [CrossRef]

133. Romeo, M.A.; Santarelli, R.; Gilardini Montani, M.S.; Gonnella, R.; Benedetti, R.; Faggioni, A.; Cirone, M. Viral infection and autophagy dysregulation: The case of HHV-6, EBV and KSHV. *Cells* **2020**, *9*, 2624. [CrossRef]

134. Silva, L.M.; Jung, J.U. Modulation of the autophagy pathway by human tumor viruses. *Semin. Cancer Biol.* **2013**, *23*, 323–328. [CrossRef]

135. Granato, M.; Santarelli, R.; Farina, A.; Gonnella, R.; Lotti, L.V.; Faggioni, A.; Cirone, M. Epstein-Barr virus blocks the autophagic flux and appropriates the autophagic machinery to enhance viral replication. *J. Virol.* **2014**, *88*, 12715–12726. [CrossRef]

136. Gonnella, R.; Dimarco, M.; Farina, G.A.; Santarelli, R.; Valia, S.; Faggioni, A.; Angeloni, A.; Cirone, M.; Farina, A. BFRF1 protein is involved in EBV-mediated autophagy manipulation. *Microbes Infect.* **2020**, *22*, 585–591. [CrossRef]

137. Nowag, H.; Guhl, B.; Thriene, K.; Romao, S.; Ziegler, U.; Dengjel, J.; Munz, C. Macroautophagy proteins assist Epstein Barr virus production and get incorporated into the virus particles. *EBioMedicine* **2014**, *1*, 116–125. [CrossRef]

138. Fang, C.Y.; Huang, S.Y.; Wu, C.C.; Hsu, H.Y.; Chou, S.P.; Tsai, C.H.; Chang, Y.; Takada, K.; Chen, J.Y. The synergistic effect of chemical carcinogens enhances Epstein-Barr virus reactivation and tumor progression of nasopharyngeal carcinoma cells. *PLoS ONE* **2012**, *7*, e44810. [CrossRef]

139. Magana, J.J.; Velazquez-Perez, L.; Cisneros, B. Spinocerebellar ataxia type 2: Clinical presentation, molecular mechanisms, and therapeutic perspectives. *Mol. Neurobiol.* **2013**, *47*, 90–104. [CrossRef]

140. Bourdenx, M.; Daniel, J.; Genin, E.; Soria, F.N.; Blanchard-Desce, M.; Bezard, E.; Dehay, B. Nanoparticles restore lysosomal acidification defects: Implications for Parkinson and other lysosomal-related diseases. *Autophagy* **2016**, *12*, 472–483. [CrossRef]

141. Palmieri, M.; Pal, R.; Nelvagal, H.R.; Lotfi, P.; Stinnett, G.R.; Seymour, M.L.; Chaudhury, A.; Bajaj, L.; Bondar, V.V.; Bremner, L.; et al. Corrigendum: mTORC1-independent TFEB activation via Akt inhibition promotes cellular clearance in neurodegenerative storage diseases. *Nat. Commun.* **2017**, *8*, 15793. [CrossRef]

142. Hanks, S.K.; Quinn, A.M. Protein kinase catalytic domain sequence database: Identification of conserved features of primary structure and classification of family members. *Methods Enzymol.* **1991**, *200*, 38–62.

143. Leader, D.P. Viral protein kinases and protein phosphatases. *Pharmacol. Ther.* **1993**, *59*, 343–389. [CrossRef]

144. Li, R.; Wang, L.; Liao, G.; Guzzo, C.M.; Matunis, M.J.; Zhu, H.; Hayward, S.D. SUMO binding by the Epstein-Barr virus protein kinase BGLF4 is crucial for BGLF4 function. *J. Virol.* **2012**, *86*, 5412–5421. [CrossRef] [PubMed]

145. Gershburg, E.; Marschall, M.; Hong, K.; Pagano, J.S. Expression and localization of the Epstein-Barr virus-encoded protein kinase. *J. Virol.* **2004**, *78*, 12140–12146. [CrossRef] [PubMed]

Article

Role of Vesicle-Associated Membrane Protein-Associated Proteins (VAP) A and VAPB in Nuclear Egress of the Alphaherpesvirus Pseudorabies Virus

Anna D. Dorsch [1], Julia E. Hölper [1], Kati Franzke [2], Luca M. Zaeck [1], Thomas C. Mettenleiter [1] and Barbara G. Klupp [1,*]

[1] Institute of Molecular Virology and Cell Biology, Friedrich-Loeffler-Institut, 17493 Greifswald, Insel Riems, Germany; anna_dorsch_187@web.de (A.D.D.); julia.hoelper@fli.de (J.E.H.); luca.zaeck@fli.de (L.M.Z.); thomasc.mettenleiter@fli.de (T.C.M.)

[2] Institute of Infectology, Friedrich-Loeffler-Institut, 17493 Greifswald, Insel Riems, Germany; kati.franzke@fli.de

* Correspondence: barbara.klupp@fli.de

Abstract: The molecular mechanism affecting translocation of newly synthesized herpesvirus nucleocapsids from the nucleus into the cytoplasm is still not fully understood. The viral nuclear egress complex (NEC) mediates budding at and scission from the inner nuclear membrane, but the NEC is not sufficient for efficient fusion of the primary virion envelope with the outer nuclear membrane. Since no other viral protein was found to be essential for this process, it was suggested that a cellular machinery is recruited by viral proteins. However, knowledge on fusion mechanisms involving the nuclear membranes is rare. Recently, vesicle-associated membrane protein-associated protein B (VAPB) was shown to play a role in nuclear egress of herpes simplex virus 1 (HSV-1). To test this for the related alphaherpesvirus pseudorabies virus (PrV), we mutated genes encoding VAPB and VAPA by CRISPR/Cas9-based genome editing in our standard rabbit kidney cells (RK13), either individually or in combination. Single as well as double knockout cells were tested for virus propagation and for defects in nuclear egress. However, no deficiency in virus replication nor any effect on nuclear egress was obvious suggesting that VAPB and VAPA do not play a significant role in this process during PrV infection in RK13 cells.

Keywords: herpesvirus; pseudorabies virus; PrV; nuclear egress; vesicle-associated membrane protein associated protein; VAPA; VAPB; CRISPR/Cas9 genome editing

Citation: Dorsch, A.D.; Hölper, J.E.; Franzke, K.; Zaeck, L.M.; Mettenleiter, T.C.; Klupp, B.G. Role of Vesicle-Associated Membrane Protein-Associated Proteins (VAP) A and VAPB in Nuclear Egress of the Alphaherpesvirus Pseudorabies Virus. *Viruses* **2021**, *13*, 1117. https://doi.org/10.3390/v13061117

Academic Editor: Donald M. Coen

Received: 15 April 2021
Accepted: 7 June 2021
Published: 10 June 2021

Publisher's Note: MDPI stays neutral with regard to jurisdictional claims in published maps and institutional affiliations.

1. Introduction

Besides being important pathogens, viruses are also renowned as pioneers in cell biology. Numerous basic principles in molecular cell biology have been identified using viruses or viral proteins [1]. Among these pathways is the vesicular transport of (viral) glycoproteins from the endoplasmic reticulum (ER) through the Golgi towards the plasma membrane [2]. This vesicle-mediated transport pathway, which encompasses a series of membranous compartments in the eukaryotic cell, has been intensively studied and is already understood in detail. Surprisingly, the nuclear envelope seems to be excluded from this rather ubiquitous vesicular transport, and it was long believed that all exchange of material between the nucleus and the cytoplasm occurs exclusively through the nuclear pores [3].

Herpesviruses are complex viruses and their double-stranded DNA genome encodes for at least 70 different viral proteins, including enzymes involved in nucleotide metabolism and DNA replication [4]. For virus replication and assembly, herpesviruses not only use the cytoplasm but also the host cell nucleus. In the nucleus, viral transcription and DNA replication take place. The newly synthesized viral DNA is incorporated in the nucleus into a preformed capsid, which is assembled autocatalytically from capsid subcomplexes. The mature nucleocapsid, however, is far too bulky to pass through the nuclear pores. Therefore, herpesviruses are compelled to use other means to overcome the nuclear envelope barrier to access the site of final virion maturation [5]. Although expansion of nuclear pores has been suggested to occur during herpesvirus infection [6,7], the common pathway involves a vesicle-mediated process, which is designated as envelopment–de-envelopment pathway [8,9]. Nucleocapsids bud at the inner nuclear membrane (INM), thereby acquiring an envelope, which results in a primary virion located in the perinuclear space (PNS). Subsequently, in a still largely enigmatic process, the primary envelope fuses with the outer nuclear membrane (ONM) to finally release the nucleocapsid into the cytoplasm (reviewed in [10–12]). Once in the cytoplasm, tegument proteins are added and the final virion envelope is acquired from membranes of trans-Golgi network-derived vesicles or endocytic membranes [8,13]. Our research on the nuclear egress of pseudorabies virus (PrV) in rabbit kidney cells (RK13) was one of the first to describe this process on a molecular basis and added important biochemical and structural data supporting the envelopment–de-envelopment pathway, which was long disputed for herpes simplex virus 1 (HSV-1) (reviewed in [8]).

The nuclear egress complex (NEC) mediates the first step in this vesicle-mediated translocation (reviewed in [10–12]). Two conserved viral proteins designated as pUL34 and pUL31 in the alphaherpesviruses HSV-1 and PrV form the NEC. pUL34 is a type II, tail-anchored membrane protein, which is autonomously targeted to the nuclear envelope whereas pUL31 is channeled into the nucleus through nuclear pores using the cellular import machinery [14–17]. Both proteins interact at the nucleoplasmic site of the INM to form the dimeric NEC. Oligomerization of NEC dimers into hexons and further into honeycomb-like lattices is thought to induce membrane bending and finally scission of vesicles into the PNS (reviewed in [18]). This process can be induced in eukaryotic cells by co-expressing only the two NEC components [19,20], indicating that no other viral protein is necessary. Vesicle formation from synthetic membranes can also be mediated by co-expression of only pUL31 and pUL34 demonstrating that also no other cellular protein is required for this vesiculation process [21,22]. However, fusion of these vesicles is only very rarely observed, indicating that other proteins are involved in an efficient de-envelopment.

A viral protein already known to participate in this fusion process is the alphaherpesvirus specific protein kinase pUS3. In the absence of pUS3 or its kinase function, primary enveloped virions accumulate in the PNS within large herniations of the INM [23–27]. Enzymatically active pUS3, however, is not essential for alphaherpesvirus replication and progeny virus titers are only slightly reduced in its absence, indicating that phosphorylation of viral and/or cellular proteins may modulate efficiency of nuclear egress. Since no other viral proteins were identified to play a major role, we speculated that a cellular fusion machinery acting at the nuclear envelope is hijacked by herpesviruses [10]. Unfortunately, knowledge on fusion proteins or fusion machineries active at or within the nuclear membranes is still limited.

SUN2, a component of the linker of the nucleoskeleton and cytoskeleton (LINC) complex regulating the spacing between the nuclear membranes [28], was shown to modulate fusion of the primary virion envelope with the ONM. Overexpression of a dominant-negative SUN2 resulted in accumulations of primary virions in a dilated PNS, indicating that SUN2 in the LINC complex keeps the membranes at a certain distance to allow for efficient membrane fusion [29]. In line with this, an accumulation of primary virions was found in cells deficient for the AAA + ATPases Torsin A and B with a possible impact on scission of the primary enveloped particles from the INM [30]. Torsin A is reported to be involved in LINC complex assembly and/or disassembly [31,32], again pointing to a role for the LINC complex in nuclear egress.

Knockdown of CD98 heavy chain and its binding partner ß1 integrin resulted in accumulations of primary HSV-1 virions similar to those found in the absence of pUS3 [33] or in simultaneous absence of glycoproteins (g)B and gH [34], which are part of the core fusion machinery required for virus entry (reviewed in [35]). CD98 heavy chain was shown to interact with gB, gH, pUL31 and pUS3 and it was proposed that this interaction drives membrane fusion between the primary envelope and the ONM [33]. For PrV, however, gB and gH play no role in nuclear egress and virion morphogenesis occurs independently of these glycoproteins. In addition, neither of these glycoproteins could be detected in primary virions or in the INM further arguing against a functional role in de-envelopment [36]. This evidence renders a similar function for CD98 in nuclear egress of PrV nucleocapsids unlikely, but this remains to be tested experimentally.

The endosomal sorting complex required for transport (ESCRT) III has also been described to be involved in nuclear egress of HSV-1. In infected cells, ESCRT-III is recruited to the INM and scission of primary enveloped particles is impaired in ESCRT-III depleted cells [37,38], whereas another study showed no defect in nuclear egress by using dominant negative versions of several ESCRT components [39]. However, none of the abovementioned cellular proteins is essential for nuclear egress of herpesvirus nucleocapsids.

In a recent study, the cellular vesicle fusion protein vesicle-associated membrane protein (VAMP)-associated protein B (VAPB) was shown to play a role in nuclear egress of HSV-1 [40]. Mass spectrometric analysis of membranous vesicles isolated from infected HeLa cells identified several cellular proteins known to be involved in vesicular trafficking in the cytoplasm. Of those, VAPB, which showed the highest enrichment score, was further analyzed [40]. Using immune labeling, VAPB was found to co-localize with the NEC component pUL34 in the nuclear membrane and knockdown of VAPB by siRNA resulted in an around two-$\log_{10}$ drop in virus titers accompanied by an accumulation of viral particles in the PNS, strongly suggesting a role in nuclear egress [40]. Moreover, expression of a mutant VAPB, which is found in patients suffering from amyotrophic lateral sclerosis, resulted in a dilation of the PNS with membrane vesicles accumulating in the PNS pointing to a function in nuclear envelope maintenance [41].

VAPB and VAPA are members of the highly conserved VAP family and are ubiquitously expressed in eukaryotic cells of various tissues and organs (reviewed in [42]). The less well characterized VAPC is a differently spliced isoform of VAPB sharing 70 N-terminal amino acids followed by additional 29 unique amino acids [42]. VAPs play a role in diverse cellular functions, such as regulation of lipid transport and homeostasis, and membrane trafficking. VAPs are mainly localized in the ER, where they interact with many different proteins to tether the ER to other intracellular organelles at defined membrane contact sites. A possible role of VAPs in membrane fusion is based on their interaction with SNARE (soluble N-ethylmaleimide-sensitive factor attachment protein receptor) proteins, but the structural and functional role of this interaction remains to be established (reviewed in [42]). VAPA and VAPB are type II membrane proteins with their C-terminal end anchored in the membrane. Thus, a major part of the protein extends into the cytosol when anchored in the ER or into the nucleoplasm when they reach the INM [40]. In mammalian cells, homo- and heterodimeric complexes of VAPA and VAPB have been described [43,44] pointing to an at least partial redundancy.

In addition to HSV-1, members of the VAP-family have already been described to play an important role for replication of different viruses. For example, VAPA interacts with the nonstructural protein 1 (NS1) of human norovirus and replication efficiency of murine norovirus was significantly decreased in VAPA or VAPB deficient cells [45]. It was also shown that NS5A and NS5B of hepatitis C virus interact with VAPB and that this interaction is essential for virus replication most likely by providing a scaffold in the appropriate membrane compartment [42,46].

Here, we analyzed the role of VAPB and VAPA for replication of PrV. Although HSV-1 and PrV belong to the alphaherpesvirus subfamily, they differ in several aspects of the viral life cycle. Regarding nuclear egress, HSV-1 seems to add a number of tegument proteins already in the nucleus, e.g., pUL47 and pUL51. These proteins could not be detected on nuclear capsids of PrV (reviewed in [47]) [48], indicating that in addition to the conserved alphaherpesviral nuclear egress proteins pUL34, pUL31 and pUS3, these proteins might recruit a different set of cellular proteins to modulate nuclear egress.

To test for the involvement of VAPs in nuclear egress of PrV, we targeted the corresponding genes by CRISPR/Cas9-based genome editing in RK13 cells. RK13 cells have been our standard PrV host cells for more than 20 years, revealing basic principles of herpesvirus morphogenesis and release (reviewed in, e.g., [8,10,49]). RK13 cell clones with deletions in *VAPB*, *VAPA* or both coding genes were tested for efficient replication of PrV and for impairment of nuclear egress. However, in contrast to HSV-1 in HeLa cells, neither single knockout (KO) of VAPB, VAPA nor VAPA/B double KO (DKO) had a significant effect on PrV replication or on nuclear egress in RK13 cells.

2. Materials and Methods

2.1. Cells and Virus

RK13 (CCLV-Rie 0109), HeLa (CCLV-Rie 0082) and Vero (CCLV-Rie 0228) cells were grown in minimum essential medium supplemented with 10 % fetal calf serum at 37 °C and 5 % CO_2 in a humid atmosphere. PrV strain Kaplan (PrV-Ka; [50]) was propagated in RK13 cells.

2.2. CRISPR/Cas9-Based Gene Editing

The transcript ID for the corresponding genes in the host species *Oryctolagus cuniculus* (OryCun2.0) was identified using the genome browser database e!Ensembl (www.ensembl.org) [51]. Entering the transcript ID into the online program ChopChop (chopchop.cbu.uib.no [52]), specific gRNAs targeting exon 2 in the corresponding genes were designed. To generate knockout cell lines, single guide RNA (sgRNAs) expressing plasmids were constructed. Three or four sgRNAs (each 20 nucleotides) were selected per gene and oligonucleotides were ordered in forward and reverse orientation (Eurofins, Ebersberg, Germany) with overhangs compatible with the BbsI restriction enzyme site. In addition, primers to amplify the target gene regions were synthesized (Table 1).

Table 1. Primer and oligonucleotide sequences. Compatible 5′ overhangs for restriction enzyme BbsI used for cloning are underlined.

Name	Sequence (5′–3′)
VAPA_ctrl-seq_Fwd	TCGGGTTTAGATTTCTGCAGTT
VAPA_ctrl-seq_Rev	GCGTAATTTCATACACTGGCAA
VAPB_ctrl-seq_Fwd	ATCCTAACTGCTGCTAACTGGC
VAPB_ctrl-seq_Rev	CTCCAATTCTGAAATCCAGTCC
VAPA gRNA #1_Fwd	<u>CAC</u>CAAACAGTCACAGTCGACCCT
VAPA gRNA #1_Rev	<u>AAAC</u>AGGGTCGACTGTGACTGTTT
VAPA gRNA #2_Fwd	<u>CAC</u>CGGCCTCACACAGTACCGGCG
VAPA gRNA #2_Rev	<u>AAAC</u>CGCCGGTACTGTGTGAGGCC
VAPA gRNA #3_Fwd	<u>CAC</u>CAACAGTGGAATTATTGACCC
VAPA gRNA #3_Rev	<u>AAAC</u>GGGTCAATAATTCCACTGTT
VAPA gRNA #4_Fwd	<u>CAC</u>CTCTTAAATTGCGAAATCCAT
VAPA gRNA #4_Rev	<u>AAAC</u>ATGGATTTCGCAATTTAAGA
VAPB gRNA #1_Fwd	<u>CAC</u>CAACAGCGGGATCATTGACGC
VAPB gRNA #1_Rev	<u>AAAC</u>GCGTCAATGATCCCGCTGTT
VAPB gRNA #2_Fwd	<u>CAC</u>CGGGGCCTCCATTAACGTGTC
VAPB gRNA #2_Rev	<u>AAAC</u>GACACGTTAATGGAGGCCCC
VAPB gRNA #3_Fwd	<u>CAC</u>CGTGCTTTAAAGTGAAGACGA
VAPB gRNA #3_Rev	<u>AAAC</u>TCGTCTTCACTTTAAAGCAC
HU6-F	ATAATTTCTTGGGTAGTTTGCAG

The corresponding forward and reverse oligonucleotides were hybridized, phosphorylated and cloned into BbsI-cleaved vector pX330-NeoR [53]. Correct synthesis and cloning were verified by sequencing using primer HU6-F (Table 1). All three or four sgRNA expressing plasmids were co-transfected into RK13 cells by calcium phosphate co-precipitation [54] using equal amounts of each plasmid. For VAPA/B DKO cells, all seven plasmids were transfected simultaneously.

From each transfection assay, eleven G418-resistant single cell colonies were selected and screened for an effect on virus propagation. For this, cells were seeded in 24 well dishes and infected with PrV-Ka at a multiplicity of infection (MOI) of 5. Cells were harvested after 24 h and progeny titers were determined on RK13 cells. Since titers derived from the different single cell clones showed no clear difference, two or three clones each were randomly selected. The targeted gene region was amplified, and the PCR product was sequenced. For this, genomic DNA was isolated by lysing the cells in TEN/Sarkosyl buffer (20 mM TrisHCl pH 7.4, 1 mM EDTA, 150 mM NaCl, 3% N-lauroylsarcosinate) at 65 °C for 20 min. Then, RNA and proteins were digested by treatment with RNase (Roche Diagnostics, Mannheim, Germany) for 30 min at 37 °C followed by incubation with Pronase (Serva, Heidelberg, Germany) for 2 h at 45 °C. DNA was extracted by phenol/chloroform, precipitated with ethanol, and resuspended in Tris-HCl, pH 7.4. Gene specific primers (Table 1) were used to amplify the target gene region using either Phusion® High Fidelity DNA Polymerase (ThermoFisherScientific, Darmstadt, Germany) or the KAPA HiFi HotStart Ready Mix PCR Kit (Roche Diagnostics) as recommended by the manufacturers. PCR products were first sequenced directly using the corresponding forward or reverse primers and the BigDye Terminator Cycle Sequencing Kit (ThermoFisherScientific). PCR products, whose sequence differed from the parental RK13 sequence, were cloned into the vector pBluescript SK + (Stratagene, Frankfurt am Main, Germany). As a standard, for each selected cell clone, plasmid inserts of 10 randomly picked bacterial colonies were sequenced using the vector specific T7 primer. Sequences were analyzed with Geneious Prime software (version 2019.2.3). For the in-depth analyses, one clone each, which preferably carried an out-of-frame mutation in the targeted exon on both alleles was used.

2.3. Western Blot Analysis

Lysates of parental, single and double knockout RK13 as well as of Vero and HeLa cells were separated in sodium dodecyl sulfate (SDS)-10% polyacrylamide gels. Proteins were transferred to nitrocellulose membranes and blocked with 6% skimmed milk. Parallel blots were incubated with the polyclonal rabbit sera specific for VAPA (1:2.500; Invitrogen # PA5-22188; raised against a synthetic peptide corresponding to a region within amino acids 29 and 122 of human VAPA ID#Q9P0L0; ThermoFisherScientific), VAPB (1:5.000; Invitrogen # PA5-53023; corresponding to amino acids 105-216 of human VAPB ID#O95292; ThermoFisherScientific), or with a monoclonal α-tubulin antibody (1:10.000; Sigma, Taufkirchen, Germany). Bound antibodies were detected after incubation with peroxidase-conjugated anti-rabbit or anti-mouse antibodies (Dianova, Hamburg, Germany) using the Clarity Western ECL substrate (Bio-Rad Laboratories, Feldkirchen, Germany). Signals were recorded with a VersaDoc 4000 MP imager (Bio-Rad Laboratories) using the Quantity One software (version 4.6.9).

2.4. Actin Staining

To visualize a putative reorganization of actin in cells lacking VAPA/B as reported previously for HeLa cells [55], RK13 as well as the single and double KO cells were fixed with 4% paraformaldehyde for 20 min followed by permeabilization with 0.1% Triton-X-100. Actin was labelled with phalloidin-Alexa Fluor 488 (dilution 1:50; ThermoFisherScientific) and nuclei were counterstained with DAPI. Samples were imaged with a confocal laser scanning microscope (Leica DMI 6000 TCS SP5, 63× oil-immersion objective, NA = 1.4; Leica, Wetzlar, Germany). Image processing and visualization were done with Fiji (v1.53) [56].

2.5. Growth Properties

Cell clones carrying mutations in the targeted genes were tested for efficient PrV propagation. To this end RK13-VAPA KO, RK13-VAPB KO, RK13-VAPA/B DKO and parental RK13 cells were infected with PrV-Ka using either a high MOI of 5 or a low MOI of 0.05. For this, precooled cells and diluted virus suspensions were incubated on ice for 1 h followed by addition of prewarmed medium to allow for synchronous infection. Non-penetrated virus was inactivated by low pH treatment after 1 h at 37 °C [57]. Cells and supernatant were harvested 24 h and 48 h post infection and titers were determined on RK13 cells. Mean values of five independent assays were calculated and plotted with the corresponding standard deviation. No statistically significant difference in virus titers was found using a two-way ANOVA with Dunnett's multiple comparison test performed with GraphPad Prism (version 9.0.0.121).

2.6. Ultrastructural Analysis

RK13, RK13-VAPA KO, RK13-VAPB KO and RK13-VAPA/B DKO cells were infected with PrV-Ka at an MOI of 1 and processed for electron microscopy 14 h post infection as described recently [29].

3. Results

3.1. Generation of RK13-VAPB KO, RK13-VAPA KO and RK13-VAPA/B DKO Cell Lines

CRISPR/Cas9-based genome editing is a highly efficient tool to target any gene within a given host genome for which sequence information is available. We used the online tool ChopChop (chopchop.cbu.uib.no [52]) to select either four or three sgRNAs targeting the second exon of VAPA or VAPB with the lowest probability for off-site targets. Oligonucleotides were cloned into the plasmid pX330-NeoR, which encodes the Cas9 nuclease and a resistance for neomycin/G418 [53]. RK13 cells were co-transfected with the four or three sgRNA-expressing plasmids or with all seven plasmids simultaneously to generate single and double knockout cells, respectively. Transfected cells were selected in medium containing 0.5 mg/mL G418 and resistant single cell colonies were analyzed. Based on the data shown for HSV-1 [40], we expected that at least RK13-VAPB KO would be easily identifiable by a drop in progeny virus titers compared to parental RK13 cells. For each transfection assay, 11 cell clones were first screened for efficient virus propagation. Since differences in progeny virus titers were only marginal (data not shown), two or three randomly selected cell clones from the originally tested 11 single cell clones were further analyzed. For this, the CRISPR/Cas9-targeted region was amplified by PCR with genomic DNA as template and sequenced. One cell clone each preferentially with an out-of-frame mutation and/or a large deletion in the coding region of the targeted gene were further tested.

As shown in Figure 1, the selected clone RK13-VAPA KO carried a deletion of 41 nucleotides (nt) in exon 2 of *VAPA*, while in exon 2 of RK13-VAPB KO a 50 nt and an additional 9 nt stretch were deleted. In the cloned PCR products from genomic DNA of both cell lines, only one type of mutation was found indicating that the mutation is biallelic. Sequencing of the RK13-VAPA/B DKO cell line revealed a huge deletion comprising 379 nt in *VAPB* thereby removing the complete exon 2 including parts of the intron regions, but only a short deletion of either 15 nt or only one nucleotide in exon 2 of *VAPA* (Figure 1).

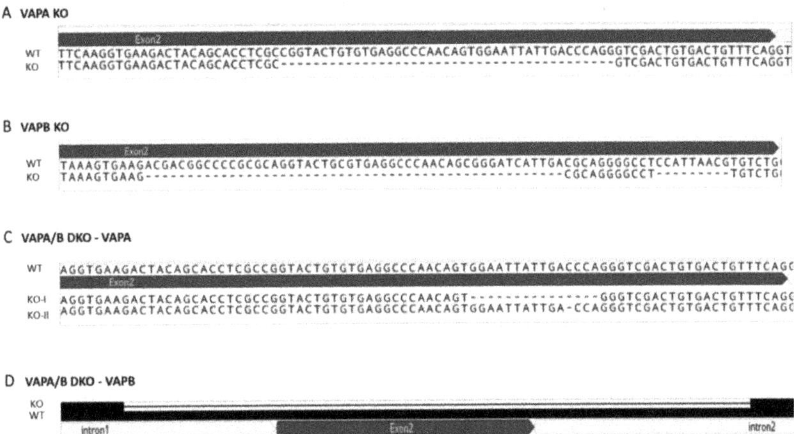

Figure 1. CRISPR/Cas9 genome editing of VAPA- and VAPB-encoding genes in RK13 cells. Shown are the wildtype sequences (WT) and the deletions uncovered in the respective genes of the selected knockout cells (KO). In RK13-VAPA KO cells, 41 nt were deleted in the *VAPA* ORF (**A**), while 50 nt and additional 9 nt were removed in the VAPB coding region in RK13-VAPB KO (**B**). In both single KO cell lines, the alleles comprise identical changes. *VAPA* in RK13-VAPA/B double knockout (DKO) contained deletions of 15 nt (KO-I) or 1 nt (KO-II) (**C**), while the complete exon 2 in *VAPB* is missing (**D**). Due to the large deletion identical regions are indicated by black bars and the deleted sequence is represented by a thin line.

3.2. Characterization of RK13-VAPB KO, RK13-VAPA KO and RK13-VAPA/B DKO

The selected single and double KO cell lines all carried deletions in the targeted gene region. To test for protein expression, immunoblotting was performed. Reactivity of the VAPA- and VAPB-specific antisera with the rabbit homologs in RK13 cells was compared to Vero and HeLa cells. As presented in Figure 2A, proteins of comparable electrophoretic mobility as in Vero and HeLa cells could be detected in RK13 cells, highlighting the high conservation of VAPs (reviewed in [42]). It is difficult to speculate on the relative expression levels since the sera might detect the homologs with different affinities. However, the comparable α-tubulin-specific signal indicates that a similar amount of cell lysate was applied.

To test whether the corresponding proteins are missing in the knockout cells, lysates of RK13, RK13-VAPA KO, RK13-VAPB KO and RK13-VAPA/B DKO cells were separated on SDS-10 %-polyacrylamide gels and transferred to nitrocellulose. Parallel blots were probed with polyclonal rabbit antisera specific for VAPA or VAPB. A signal corresponding to VAPA, as detected in Vero, HeLa and the parental RK13 cells (Figure 2A), is missing in RK13-VAPA KO and RK13-VAPA/B DKO cells. The 33 kDa VAPB band is absent in lysates of RK13-VAPB KO and RK13-VAPA/B DKO cells (Figure 2B), indicating that the gene specific knockout was successful in eliminating protein expression in the selected cell clones. The faint signal visible in immunoblots with the VAPA-specific serum might be due to a weak cross-reactivity with VAPB, while the upper bands represent unspecific signals since they are identical in all cell lines. As loading control, both blots were re-probed with anti-α-tubulin.

Previously it was shown that actin organization is drastically perturbed in HeLa VAPA/B DKO cells [55]. To investigate whether VAPs function in a similar way in RK13 cells, we analyzed the actin skeleton organization using confocal laser-scanning microscopy. For this, we fixed and stained the parental RK13 as well as the single and double KO cells with fluorophore-conjugated phalloidin. As evident in Figure 3, stress fibers as present in RK13 and the single KO cells, are less dominant in the RK13-VAPA/B DKO cells and comparable to data shown for HeLa cells [55], supporting a functional double knockout of VAPA and VAPB. The single KO cells showed no obvious perturbation of actin, which is in line with the notion that both proteins feature at least some functional redundancy [42].

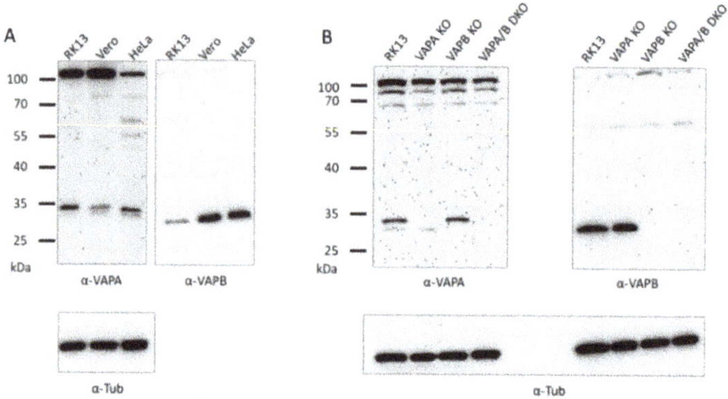

Figure 2. Immunoblots showing VAPA and VAPB in RK13, Vero and HeLa cells and the absence of VAPA and/or VAPB in the corresponding RK13 KO and DKO cell lines. Proteins in lysates of RK13, Vero and HeLa cells (**A**) or RK13 and the single as well as the double VAP KO cell lines (**B**) were separated on SDS-10 % polyacrylamide gels, and parallel blots were incubated with polyclonal sera against VAPA or VAPB. Masses of marker proteins are given on the left in kDa. As loading control, parallel (**A**) or the same blots were (re-)probed with anti-α-tubulin.

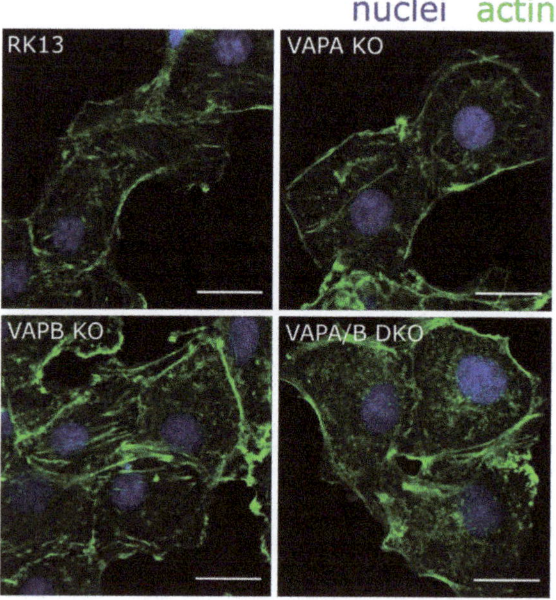

Figure 3. Actin organization is perturbed in RK13-VAPA/B DKO cells. RK13, RK13-VAPA KO, RK13-VAPB KO and RK13-VAPA/B DKO cells were fixed with 4% paraformaldehyde and cytoskeletal actin was stained with fluorophore-conjugated phalloidin (green). A loss of actin stress fibers and an accumulation of actin comets is obvious in RK13-VAPA/B DKO as reported for HeLa cells lacking VAPA/B [55]. Nuclei were counterstained with DAPI (blue). Shown are representative images. Scale bar = 20 μm.

3.3. Absence of VAPA, VAPB or Both Has No Major Impact on PrV Replication

To analyze the influence of the introduced modifications on replication of PrV-Ka, cells were infected either with an MOI of 5 or 0.05. Cells and supernatant were harvested 24 h or 48 h post infection and progeny viral titers were determined on RK13 cells (Figure 4). Only titers of PrV-Ka derived from RK13-VAPB KO cells were marginally lower at 24 h post infection, which, however, was not statistically significant. In addition, parallel infection of cells expressing neither VAPA nor VAPB showed no titer reduction, indicating that the minor decrease is most likely not due to the VAPB deficiency.

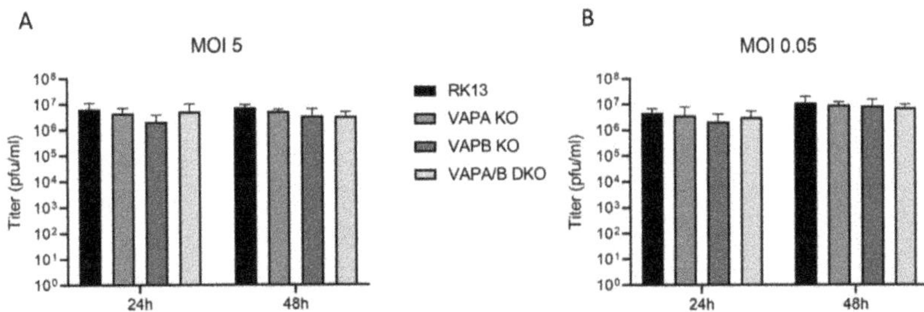

Figure 4. Growth properties of PrV-Ka derived from parental RK13, VAP single and double KO cells. RK13, RK13-VAPA KO, RK13-VAPB KO and RK13-VAPA/B DKO were infected with PrV-Ka at an MOI of 5 (**A**) or 0.05 (**B**) and harvested after 24 or 48 h. Shown are mean values of progeny virus titers in plaque forming units per milliliter (pfu/ml) of five replicates with the corresponding standard deviation.

3.4. Absence of VAPA or VAPB or Both Has No Significant Impact on PrV Nuclear Egress or Final Virion Maturation

The growth analyses indicated only a minor effect on titers for PrV-Ka propagated on RK13-VAPB KO cells. To investigate this in more detail and with special focus on nuclear egress, parental RK13, RK13-VAPA KO, RK13-VAPB KO and RK13-VAPA/B DKO were infected with PrV-Ka at an MOI of 1 and processed for ultrastructural analysis 14 h post infection (Figure 5). In the representative images, all stages of virion morphogenesis were observed in cells with defects in VAPA (Figure 5B), VAPB (Figure 5C) or VAPA and VAPB expression (Figure 5D), comparable to the parental RK13 cells (Figure 5A). Mature capsids in the nucleus, single primary enveloped virions in the PNS, enveloped virus particles in cytoplasmic transport vesicles and mature virions on the cell surface are easily discernible. In contrast to the study for HSV-1 after siRNA-based knockdown of VAPB in HeLa cells [40], no accumulations of capsids in the nucleus or primary virions in the PNS were obvious, indicating that neither VAPA nor VAPB individually or in combination play an important role in nuclear egress or virion morphogenesis of PrV in RK13 cells.

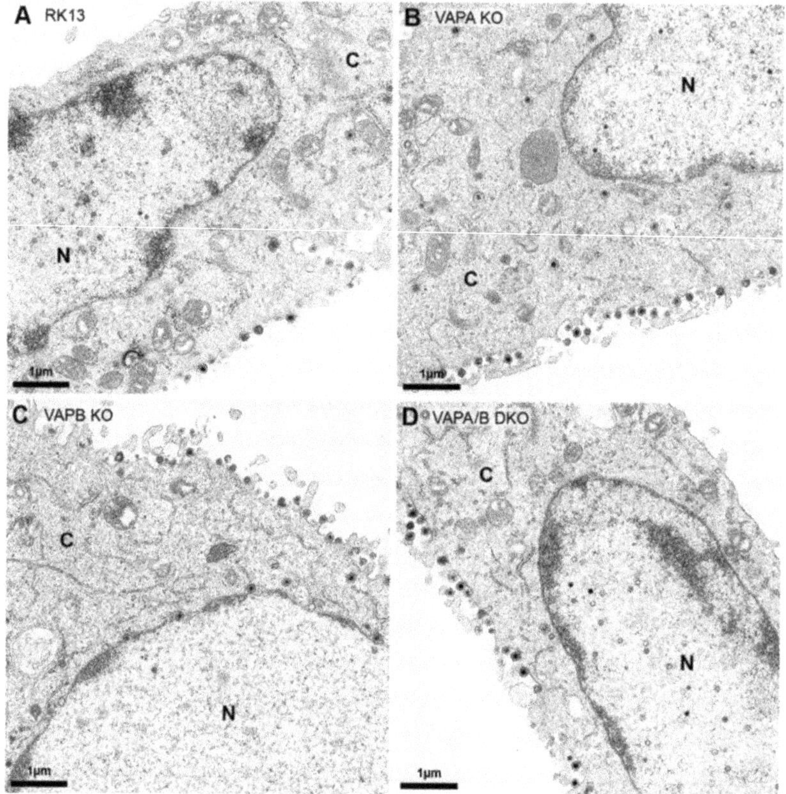

Figure 5. Ultrastructural analysis of PrV-Ka infected RK13, VAP KO and DKO cells. (**A**) RK13, (**B**) RK13-VAPA KO, (**C**) RK13-VAPB KO and (**D**) RK13-VAPA/B DKO cells were infected with PrV-Ka (MOI 1) and processed for electron microscopy 14 h post infection. Shown are presentative images. N: nucleus, C: cytoplasm.

4. Discussion

The molecular mechanism of nuclear egress of herpesvirus nucleocapsids is still not fully understood. The first step in this process, the formation of primary enveloped virions by budding through the INM into the PNS, is well studied. Using a multimodal imaging approach, the NEC coat could be visualized in situ [58] and the elucidation of the crystal structures for the NEC heterodimers from several herpesviruses provided detailed insights into this characteristic vesicle formation and scission machinery at and from the INM [59–62]. In contrast, the second step encompassing fusion of the primary virion envelope with the ONM or, after escape into the lumen of the ER, with the ER membrane remains elusive. The NEC itself seems unable to mediate efficient fusion and vesicles accumulate in the PNS when both NEC components are co-expressed [19–22]. Recently, it was proposed that vesicle fusion proteins redirected from the ER to the nuclear envelope might facilitate herpesvirus nuclear egress [40]. Cell fractionation coupled with mass spectrometry found especially VAPB enriched in nuclear membrane fractions of HSV-1 infected cells. Additionally, siRNA-mediated knockdown of VAPB expression in HeLa cells resulted in approximately two-$\log_{10}$ reduced viral titers concomitant with an accumulation of virus particles in the nucleus and in the perinuclear cleft, supporting a functional role of VAPB in nuclear egress of HSV-1 [40].

These data sparked our interest to investigate the role of VAPB and VAPA in nuclear egress of PrV. VAPA and VAPB, which are highly conserved in eukaryotes, share 63 % amino acid sequence identity and are supposed to serve similar cellular functions forming not only homo- but also heterodimeric complexes [42]. We used CRISPR/Cas9 gene editing technology as provided with the vector pX330-NeoR [53] to impair VAPA and VAPB expression. We targeted the second exon of each gene by three or four different sgRNAs. The sgRNA expressing plasmids were co-transfected into RK13 cells to enhance the efficiency of the knockout and to increase the chance to introduce larger deletions, specifically aiming for out-of-frame mutations to ensure absence of protein expression. Using this approach we already generated Torsin A and Torsin B single as well as double knockout cells [30].

Preliminary testing of several single cell clones showed no drastic impact on progeny viral titers, already indicating a difference to the effect on HSV-1 after siRNA-mediated VAPB knockdown in HeLa cells [40]. Sequencing of the targeted gene region in the selected CRISPR/Cas9-mutated KO cell clones showed large biallelic out-of-frame deletions in the *VAPA* gene (41 nt) in RK13-VAPA KO and in the gene encoding VAPB (50 and 9 nt) in RK13-VAPB KO cells. In the RK13-VAPA/B DKO cells, the complete exon 2 of the *VAPB* gene, including part of the surrounding intron sequences was removed, while the *VAPA* gene contained only a single nucleotide deletion in one allele and an in-frame deletion of 15 nt in the other allele. The 5-codon in-frame deletion might still result in a truncated but functional VAPA, but immunoblot analysis using the VAPA specific serum showed no residual protein expression.

The VAPA- and VAPB-specific antisera were raised against the human proteins, but homologs are well conserved between species [42]. The VAPA-specific antiserum was generated against amino acids 29–122 of human VAPA (#Q9P0L0), which are identical in the predicted rabbit protein sequence (data not shown). The sequence of the immunogen used for generation of anti-VAPB corresponds to amino acids 105–217 of the rabbit homologue (accession no. XP_008250457) and shares 83% identical residues. Both sera specifically detected 34 and 33 kDa proteins in the parental RK13 lysates comparable to the reactivity in Vero and HeLa cells, but corresponding signals were absent in the single and double KO cells, pointing to a successful protein knockout. Loss of actin stress fibers was reported for a HeLa VAPA/B DKO cell line [55] and a comparable perturbation was found in RK13-VAPA/B DKO further supporting a successful functional knockout. The single VAP KO cells had no obvious difference in F-actin staining, indicating that loss of function can be compensated by the remaining protein.

In contrast to HSV-1 [40], growth analyses with wildtype strain PrV-Ka showed no obvious impairment in RK13 cells deficient for VAPB expression. Progeny virus titers derived from the single and VAPA/B double KO cells were comparable to parental RK13 cells independent of the MOI used and the time point analyzed. Only titers derived from RK13-VAPB KO at 24 h after infection showed a marginal, approximately, 3-fold decrease in virus progeny. Infection of RK13-VAPA/B DKO cells, which carry a large deletion in the *VAPB* gene region resulting in a complete removal of exon 2, did not show a corresponding titer reduction, arguing against a specific effect of VAPB on infectious virus production. In-depth structural analysis of infected single and double KO cells supported these results. No impairment of nuclear egress as described for HSV-1 in HeLa cells [40] was observed. In contrast, all stages of virion maturation including numerous released virions on the cell surface could be detected in agreement with the results of the single and multistep growth assays. In our preliminary studies, we tested eleven single cell clones of each of the single and double KO for an effect on progeny viral titers in parallel. However, titers were comparable for all randomly picked clones arguing against selection for second-site mutations, which may compensate a potential VAPA/B deficit.

Despite the close phylogenetic relationship between HSV-1 and PrV, previous research uncovered multiple differences between these two alphaherpesviruses. Although it is reasonable to speculate that the molecular machinery executing de-envelopment at the

ONM is as conserved as is the budding and scission apparatus at the INM, various viral and host proteins seem to modulate both stages (reviewed in [47]). In growth analyses of HSV-1 on the single and double KO RK13 cells, we were unable to observe the reported two-$\log_{10}$ reduction at 48 h post infection after siRNA-mediated knockdown of VAPB in HeLa cells [40]. No titer reduction was evident for HSV-1 derived from RK13-VAPA/B DKO cells (data not shown). The reason for the different results between HSV-1 and PrV, which were generated in different cell lines and by different technical approaches, siRNA-mediated knock-down versus CRISPR/Cas9 knockout, and for HSV-1 in HeLa cells compared to RK13 cells remains to be tested. However, while VAPs might play a role in nuclear egress, they seem not to be essentially involved in the process, indicating that the key players have not been targeted yet.

Despite the prominent role of VAPs in organelle tethering, lipid homeostasis and transport, VAP-deficient cells are viable [55,63]. It has been reported that the inter-organelle membrane contact sites, in which VAPA and VAPB play a prominent role, often rely on several independent tethers to secure the essential functions in the cell [64]. In line with this, in addition to VAPA and VAPB, a third ER receptor, MOSPD2, has been identified recently by a proteomic approach [63]. It needs to be investigated whether this protein plays a role singly or in combination with VAPA and VAPB in herpesvirus nuclear egress.

In future studies it will be interesting to also investigate the influence of a dominant negative version of VAPB. A proline to serine substitution (P56S) in VAPB, which was identified in an autosomal dominant form of amyotrophic lateral sclerosis, was shown to result in a nuclear envelope defect [41]. Blocking the normal function of VAPs by overexpressing the defective isoform might result in more drastic effects as single and double knockouts since several other cellular proteins might serve similar functions.

Author Contributions: Conceptualization, B.G.K., J.E.H. and T.C.M.; methodology, A.D.D., J.E.H., L.M.Z. and K.F.; software, A.D.D., J.E.H. and L.M.Z.; validation, J.E.H. and B.G.K.; investigation, A.D.D., K.F., L.M.Z. and B.G.K.; data curation, A.D.D., J.E.H. and B.G.K.; writing—original draft preparation, B.G.K.; writing—review and editing, A.D.D., B.G.K., J.E.H., L.M.Z. and T.C.M.; visualization, A.D.D., K.F., L.M.Z. and B.G.K.; supervision, J.E.H. and B.G.K.; project administration, T.C.M. and B.G.K.; funding acquisition, B.G.K. and T.C.M. All authors have read and agreed to the published version of the manuscript.

Funding: This research was funded by the Deutsche Forschungsgemeinschaft (DFG), grant number ME 854/12-2.

Data Availability Statement: The data presented in this study are available on request from the corresponding author.

Acknowledgments: We thank Karla Günther and Petra Meyer for technical support and Mandy Jörn for the help with electron microscopic images. The pX330-NeoR vector was kindly provided by Walter Fuchs. We also thank the Collection of Cell Lines in Veterinary Medicine (CCLV) at Friedrich-Loeffler-Institut for the assistance in supplying cell lines and media.

References

1. Fields, B.K.D. *Fields Virology*; Wolters Kluwer Health: Philadelphia, PA, USA, 2007; Volume 1.
2. Rothman, J.E.; Bursztyn-Pettegrew, H.; Fine, R.E. Transport of the membrane glycoprotein of vesicular stomatitis virus to the cell surface in two stages by clathrin-coated vesicles. *J. Cell Biol.* **1980**, *86*, 162–171. [CrossRef] [PubMed]
3. Tran, E.J.; Wente, S.R. Dynamic nuclear pore complexes: Life on the edge. *Cell* **2006**, *125*, 1041–1053. [CrossRef] [PubMed]
4. McGeoch, D.J.; Rixon, F.J.; Davison, A.J. Topics in herpesvirus genomics and evolution. *Virus Res.* **2006**, *117*, 90–104. [CrossRef] [PubMed]
5. Mettenleiter, T.C. Breaching the Barrier-The Nuclear Envelope in Virus Infection. *J. Mol. Biol* **2016**, *428*, 1949–1961. [CrossRef]
6. Wild, P.; Senn, C.; Manera, C.L.; Sutter, E.; Schraner, E.M.; Tobler, K.; Ackermann, M.; Ziegler, U.; Lucas, M.S.; Kaech, A. Exploring the nuclear envelope of herpes simplex virus 1-infected cells by high-resolution microscopy. *J. Virol.* **2009**, *83*, 408–419. [CrossRef]
7. Leuzinger, H.; Ziegler, U.; Schraner, E.M.; Fraefel, C.; Glauser, D.L.; Heid, I.; Ackermann, M.; Mueller, M.; Wild, P. Herpes simplex virus 1 envelopment follows two diverse pathways. *J. Virol.* **2005**, *79*, 13047–13059. [CrossRef]
8. Mettenleiter, T.C. Herpesvirus assembly and egress. *J. Virol.* **2002**, *76*, 1537–1547. [CrossRef]

9. Skepper, J.N.; Whiteley, A.; Browne, H.; Minson, A. Herpes simplex virus nucleocapsids mature to progeny virions by an envelopment –> deenvelopment –> reenvelopment pathway. *J. Virol.* **2001**, *75*, 5697–5702. [CrossRef]

10. Mettenleiter, T.C.; Muller, F.; Granzow, H.; Klupp, B.G. The way out: What we know and do not know about herpesvirus nuclear egress. *Cell. Microbiol.* **2013**, *15*, 170–178. [CrossRef]

11. Johnson, D.C.; Baines, J.D. Herpesviruses remodel host membranes for virus egress. *Nat. Rev. Microbiol.* **2011**, *9*, 382–394. [CrossRef]

12. Mettenleiter, T.C.; Klupp, B.G.; Granzow, H. Herpesvirus assembly: An update. *Virus Res.* **2009**, *143*, 222–234. [CrossRef]

13. Hollinshead, M.; Johns, H.L.; Sayers, C.L.; Gonzalez-Lopez, C.; Smith, G.L.; Elliott, G. Endocytic tubules regulated by Rab GTPases 5 and 11 are used for envelopment of herpes simplex virus. *EMBO J.* **2012**, *31*, 4204–4220. [CrossRef]

14. Klupp, B.G.; Granzow, H.; Mettenleiter, T.C. Primary envelopment of pseudorabies virus at the nuclear membrane requires the UL34 gene product. *J. Virol.* **2000**, *74*, 10063–10073. [CrossRef]

15. Passvogel, L.; Klupp, B.G.; Granzow, H.; Fuchs, W.; Mettenleiter, T.C. Functional characterization of nuclear trafficking signals in pseudorabies virus pUL31. *J. Virol.* **2015**, *89*, 2002–2012. [CrossRef]

16. Funk, C.; Ott, M.; Raschbichler, V.; Nagel, C.H.; Binz, A.; Sodeik, B.; Bauerfeind, R.; Bailer, S.M. The Herpes Simplex Virus Protein pUL31 Escorts Nucleocapsids to Sites of Nuclear Egress, a Process Coordinated by Its N-Terminal Domain. *PLoS Pathog.* **2015**, *11*, e1004957. [CrossRef]

17. Schmeiser, C.; Borst, E.; Sticht, H.; Marschall, M.; Milbradt, J. The cytomegalovirus egress proteins pUL50 and pUL53 are translocated to the nuclear envelope through two distinct modes of nuclear import. *J. Gen. Virol.* **2013**, *94*, 2056–2069. [CrossRef]

18. Bigalke, J.M.; Heldwein, E.E. Nuclear Exodus: Herpesviruses Lead the Way. *Annu. Rev. Virol.* **2016**, *3*, 387–409. [CrossRef]

19. Desai, P.J.; Pryce, E.N.; Henson, B.W.; Luitweiler, E.M.; Cothran, J. Reconstitution of the Kaposi's sarcoma-associated herpesvirus nuclear egress complex and formation of nuclear membrane vesicles by coexpression of ORF67 and ORF69 gene products. *J. Virol.* **2012**, *86*, 594–598. [CrossRef]

20. Klupp, B.G.; Granzow, H.; Fuchs, W.; Keil, G.M.; Finke, S.; Mettenleiter, T.C. Vesicle formation from the nuclear membrane is induced by coexpression of two conserved herpesvirus proteins. *Proc. Natl. Acad. Sci. USA* **2007**, *104*, 7241–7246. [CrossRef]

21. Lorenz, M.; Vollmer, B.; Unsay, J.D.; Klupp, B.G.; Garcia-Saez, A.J.; Mettenleiter, T.C.; Antonin, W. A single herpesvirus protein can mediate vesicle formation in the nuclear envelope. *J. Biol. Chem.* **2015**, *290*, 6962–6974. [CrossRef]

22. Bigalke, J.M.; Heuser, T.; Nicastro, D.; Heldwein, E.E. Membrane deformation and scission by the HSV-1 nuclear egress complex. *Nat. Commun.* **2014**, *5*, 4131. [CrossRef]

23. Sehl, J.; Portner, S.; Klupp, B.G.; Granzow, H.; Franzke, K.; Teifke, J.P.; Mettenleiter, T.C. Roles of the different isoforms of the pseudorabies virus protein kinase pUS3 in nuclear egress. *J. Virol.* **2020**. [CrossRef]

24. Mou, F.; Wills, E.; Baines, J.D. Phosphorylation of the U(L)31 protein of herpes simplex virus 1 by the U(S)3-encoded kinase regulates localization of the nuclear envelopment complex and egress of nucleocapsids. *J. Virol.* **2009**, *83*, 5181–5191. [CrossRef]

25. Reynolds, A.E.; Wills, E.G.; Roller, R.J.; Ryckman, B.J.; Baines, J.D. Ultrastructural localization of the herpes simplex virus type 1 UL31, UL34, and US3 proteins suggests specific roles in primary envelopment and egress of nucleocapsids. *J. Virol.* **2002**, *76*, 8939–8952. [CrossRef]

26. Klupp, B.G.; Granzow, H.; Mettenleiter, T.C. Effect of the pseudorabies virus US3 protein on nuclear membrane localization of the UL34 protein and virus egress from the nucleus. *J. Gen. Virol.* **2001**, *82*, 2363–2371. [CrossRef]

27. Wagenaar, F.; Pol, J.M.; Peeters, B.; Gielkens, A.L.; de Wind, N.; Kimman, T.G. The US3-encoded protein kinase from pseudorabies virus affects egress of virions from the nucleus. *J. Gen. Virol.* **1995**, *76 Pt 7*, 1851–1859. [CrossRef]

28. Rothballer, A.; Schwartz, T.U.; Kutay, U. LINCing complex functions at the nuclear envelope: What the molecular architecture of the LINC complex can reveal about its function. *Nucleus* **2013**, *4*, 29–36. [CrossRef]

29. Klupp, B.G.; Hellberg, T.; Granzow, H.; Franzke, K.; Dominguez Gonzalez, B.; Goodchild, R.E.; Mettenleiter, T.C. Integrity of the Linker of Nucleoskeleton and Cytoskeleton Is Required for Efficient Herpesvirus Nuclear Egress. *J. Virol.* **2017**, *91*. [CrossRef]

30. Holper, J.E.; Klupp, B.G.; Luxton, G.W.G.; Franzke, K.; Mettenleiter, T.C. Function of Torsin AAA+ ATPases in Pseudorabies Virus Nuclear Egress. *Cells* **2020**, *9*, 738. [CrossRef]

31. Atai, N.A.; Ryan, S.D.; Kothary, R.; Breakefield, X.O.; Nery, F.C. Untethering the nuclear envelope and cytoskeleton: Biologically distinct dystonias arising from a common cellular dysfunction. *Int. J. Cell Biol.* **2012**, *2012*, 634214. [CrossRef]

32. Nery, F.C.; Zeng, J.; Niland, B.P.; Hewett, J.; Farley, J.; Irimia, D.; Li, Y.; Wiche, G.; Sonnenberg, A.; Breakefield, X.O. TorsinA binds the KASH domain of nesprins and participates in linkage between nuclear envelope and cytoskeleton. *J. Cell Sci.* **2008**, *121*, 3476–3486. [CrossRef] [PubMed]

33. Hirohata, Y.; Arii, J.; Liu, Z.; Shindo, K.; Oyama, M.; Kozuka-Hata, H.; Sagara, H.; Kato, A.; Kawaguchi, Y. Herpes Simplex Virus 1 Recruits CD98 Heavy Chain and beta1 Integrin to the Nuclear Membrane for Viral De-Envelopment. *J. Virol.* **2015**, *89*, 7799–7812. [CrossRef] [PubMed]

34. Farnsworth, A.; Wisner, T.W.; Webb, M.; Roller, R.; Cohen, G.; Eisenberg, R.; Johnson, D.C. Herpes simplex virus glycoproteins gB and gH function in fusion between the virion envelope and the outer nuclear membrane. *Proc. Natl. Acad. Sci. USA* **2007**, *104*, 10187–10192. [CrossRef] [PubMed]

35. Vallbracht, M.; Backovic, M.; Klupp, B.G.; Rey, F.A.; Mettenleiter, T.C. Common characteristics and unique features: A comparison of the fusion machinery of the alphaherpesviruses Pseudorabies virus and Herpes simplex virus. *Adv. Virus Res.* **2019**, *104*, 225–281. [CrossRef]

36. Klupp, B.; Altenschmidt, J.; Granzow, H.; Fuchs, W.; Mettenleiter, T.C. Glycoproteins required for entry are not necessary for egress of pseudorabies virus. *J. Virol.* **2008**, *82*, 6299–6309. [CrossRef]

37. Lee, C.P.; Liu, P.T.; Kung, H.N.; Su, M.T.; Chua, H.H.; Chang, Y.H.; Chang, C.W.; Tsai, C.H.; Liu, F.T.; Chen, M.R. The ESCRT machinery is recruited by the viral BFRF1 protein to the nucleus-associated membrane for the maturation of Epstein-Barr Virus. *PLoS Pathog.* **2012**, *8*, e1002904. [CrossRef]

38. Arii, J.; Watanabe, M.; Maeda, F.; Tokai-Nishizumi, N.; Chihara, T.; Miura, M.; Maruzuru, Y.; Koyanagi, N.; Kato, A.; Kawaguchi, Y. ESCRT-III mediates budding across the inner nuclear membrane and regulates its integrity. *Nat. Commun.* **2018**, *9*, 3379. [CrossRef]

39. Pawliczek, T.; Crump, C.M. Herpes simplex virus type 1 production requires a functional ESCRT-III complex but is independent of TSG101 and ALIX expression. *J. Virol.* **2009**, *83*, 11254–11264. [CrossRef]

40. Saiz-Ros, N.; Czapiewski, R.; Epifano, I.; Stevenson, A.; Swanson, S.K.; Dixon, C.R.; Zamora, D.B.; McElwee, M.; Vijayakrishnan, S.; Richardson, C.A.; et al. Host Vesicle Fusion Protein VAPB Contributes to the Nuclear Egress Stage of Herpes Simplex Virus Type-1 (HSV-1) Replication. *Cells* **2019**, *8*, 120. [CrossRef]

41. Tran, D.; Chalhoub, A.; Schooley, A.; Zhang, W.; Ngsee, J.K. A mutation in VAPB that causes amyotrophic lateral sclerosis also causes a nuclear envelope defect. *J. Cell Sci.* **2012**, *125*, 2831–2836. [CrossRef]

42. Lev, S.; Ben Halevy, D.; Peretti, D.; Dahan, N. The VAP protein family: From cellular functions to motor neuron disease. *Trends Cell Biol.* **2008**, *18*, 282–290. [CrossRef]

43. Nishimura, Y.; Hayashi, M.; Inada, H.; Tanaka, T. Molecular cloning and characterization of mammalian homologues of vesicle-associated membrane protein-associated (VAMP-associated) proteins. *Biochem. Biophys. Res. Commun.* **1999**, *254*, 21–26. [CrossRef]

44. Teuling, E.; Ahmed, S.; Haasdijk, E.; Demmers, J.; Steinmetz, M.O.; Akhmanova, A.; Jaarsma, D.; Hoogenraad, C.C. Motor neuron disease-associated mutant vesicle-associated membrane protein-associated protein (VAP) B recruits wild-type VAPs into endoplasmic reticulum-derived tubular aggregates. *J. Neurosci.* **2007**, *27*, 9801–9815. [CrossRef]

45. McCune, B.T.; Tang, W.; Lu, J.; Eaglesham, J.B.; Thorne, L.; Mayer, A.E.; Condiff, E.; Nice, T.J.; Goodfellow, I.; Krezel, A.M.; et al. Noroviruses Co-opt the Function of Host Proteins VAPA and VAPB for Replication via a Phenylalanine-Phenylalanine-Acidic-Tract-Motif Mimic in Nonstructural Viral Protein NS1/2. *mBio* **2017**, *8*. [CrossRef]

46. Hamamoto, I.; Nishimura, Y.; Okamoto, T.; Aizaki, H.; Liu, M.; Mori, Y.; Abe, T.; Suzuki, T.; Lai, M.M.; Miyamura, T.; et al. Human VAP-B is involved in hepatitis C virus replication through interaction with NS5A and NS5B. *J. Virol.* **2005**, *79*, 13473–13482. [CrossRef]

47. Arii, J. Host and Viral Factors Involved in Nuclear Egress of Herpes Simplex Virus 1. *Viruses* **2021**, *13*, 754. [CrossRef]

48. Kopp, M.; Klupp, B.G.; Granzow, H.; Fuchs, W.; Mettenleiter, T.C. Identification and characterization of the pseudorabies virus tegument proteins UL46 and UL47: Role for UL47 in virion morphogenesis in the cytoplasm. *J. Virol.* **2002**, *76*, 8820–8833. [CrossRef]

49. Mettenleiter, T.C. Intriguing interplay between viral proteins during herpesvirus assembly or: The herpesvirus assembly puzzle. *Vet. Microbiol.* **2006**, *113*, 163–169. [CrossRef]

50. Kaplan, A.S.; Vatter, A.E. A comparison of herpes simplex and pseudorabies viruses. *Virology* **1959**, *7*, 394–407. [CrossRef]

51. Yates, A.D.; Achuthan, P.; Akanni, W.; Allen, J.; Allen, J.; Alvarez-Jarreta, J.; Amode, M.R.; Armean, I.M.; Azov, A.G.; Bennett, R.; et al. Ensembl 2020. *Nucleic Acids Res.* **2020**, *48*, D682–D688. [CrossRef]

52. Labun, K.; Montague, T.G.; Krause, M.; Torres Cleuren, Y.N.; Tjeldnes, H.; Valen, E. CHOPCHOP v3: Expanding the CRISPR web toolbox beyond genome editing. *Nucleic Acids Res.* **2019**, *47*, W171–W174. [CrossRef] [PubMed]

53. Hubner, A.; Petersen, B.; Keil, G.M.; Niemann, H.; Mettenleiter, T.C.; Fuchs, W. Efficient inhibition of African swine fever virus replication by CRISPR/Cas9 targeting of the viral p30 gene (CP204L). *Sci. Rep.* **2018**, *8*, 1449. [CrossRef] [PubMed]

54. Graham, F.L.; van der Eb, A.J. A new technique for the assay of infectivity of human adenovirus 5 DNA. *Virology* **1973**, *52*, 456–467. [CrossRef]

55. Dong, R.; Saheki, Y.; Swarup, S.; Lucast, L.; Harper, J.W.; De Camilli, P. Endosome-ER Contacts Control Actin Nucleation and Retromer Function through VAP-Dependent Regulation of PI4P. *Cell* **2016**, *166*, 408–423. [CrossRef] [PubMed]

56. Schindelin, J.; Arganda-Carreras, I.; Frise, E.; Kaynig, V.; Longair, M.; Pietzsch, T.; Preibisch, S.; Rueden, C.; Saalfeld, S.; Schmid, B.; et al. Fiji: An open-source platform for biological-image analysis. *Nat. Methods* **2012**, *9*, 676–682. [CrossRef] [PubMed]

57. Mettenleiter, T.C. Glycoprotein gIII deletion mutants of pseudorabies virus are impaired in virus entry. *Virology* **1989**, *171*, 623–625. [CrossRef]

58. Hagen, C.; Dent, K.C.; Zeev-Ben-Mordehai, T.; Grange, M.; Bosse, J.B.; Whittle, C.; Klupp, B.G.; Siebert, C.A.; Vasishtan, D.; Bauerlein, F.J.; et al. Structural Basis of Vesicle Formation at the Inner Nuclear Membrane. *Cell* **2015**, *163*, 1692–1701. [CrossRef] [PubMed]

59. Walzer, S.A.; Egerer-Sieber, C.; Sticht, H.; Sevvana, M.; Hohl, K.; Milbradt, J.; Muller, Y.A.; Marschall, M. Crystal Structure of the Human Cytomegalovirus pUL50-pUL53 Core Nuclear Egress Complex Provides Insight into a Unique Assembly Scaffold for Virus-Host Protein Interactions. *J. Biol. Chem.* **2015**, *290*, 27452–27458. [CrossRef]

60. Lye, M.F.; Sharma, M.; El Omari, K.; Filman, D.J.; Schuermann, J.P.; Hogle, J.M.; Coen, D.M. Unexpected features and mechanism of heterodimer formation of a herpesvirus nuclear egress complex. *EMBO J.* **2015**, *34*, 2937–2952. [CrossRef]

61. Zeev-Ben-Mordehai, T.; Weberruss, M.; Lorenz, M.; Cheleski, J.; Hellberg, T.; Whittle, C.; El Omari, K.; Vasishtan, D.; Dent, K.C.; Harlos, K.; et al. Crystal Structure of the Herpesvirus Nuclear Egress Complex Provides Insights into Inner Nuclear Membrane Remodeling. *Cell Rep.* **2015**, *13*, 2645–2652. [CrossRef]
62. Bigalke, J.M.; Heldwein, E.E. Structural basis of membrane budding by the nuclear egress complex of herpesviruses. *EMBO J.* **2015**, *34*, 2921–2936. [CrossRef]
63. Di Mattia, T.; Wilhelm, L.P.; Ikhlef, S.; Wendling, C.; Spehner, D.; Nomine, Y.; Giordano, F.; Mathelin, C.; Drin, G.; Tomasetto, C.; et al. Identification of MOSPD2, a novel scaffold for endoplasmic reticulum membrane contact sites. *EMBO Rep.* **2018**, *19*. [CrossRef] [PubMed]
64. Eisenberg-Bord, M.; Shai, N.; Schuldiner, M.; Bohnert, M. A Tether Is a Tether Is a Tether: Tethering at Membrane Contact Sites. *Dev. Cell* **2016**, *39*, 395–409. [CrossRef]

 *viruses*

Article

The HSV1 Tail-Anchored Membrane Protein pUL34 Contains a Basic Motif That Supports Active Transport to the Inner Nuclear Membrane Prior to Formation of the Nuclear Egress Complex

Christina Funk [1], Débora Marques da Silveira e Santos [1,2], Melanie Ott [3], Verena Raschbichler [3] and Susanne M. Bailer [1,2,3,*]

1   Fraunhofer Institute for Interfacial Engineering and Biotechnology IGB, 70569 Stuttgart, Germany; Christina.Funk@igb.fraunhofer.de (C.F.); deboramarquesss@hotmail.com (D.M.d.S.eS.)

2   Institute for Interfacial Engineering and Plasma Technology IGVP, University of Stuttgart, 70174 Stuttgart, Germany

3   Max von Pettenkofer-Institute, Ludwig-Maximilians-University Munich, 80539 Munich, Germany; melanieott@gmx.de (M.O.); verenarasch@gmail.com (V.R.)

*   Correspondence: Susanne.Bailer@igvp.uni-stuttgart.de; Tel.: +49-711-970-4180

Abstract: Herpes simplex virus type 1 nucleocapsids are released from the host nucleus by a budding process through the nuclear envelope called nuclear egress. Two viral proteins, the integral membrane proteins pUL34 and pUL31, form the nuclear egress complex at the inner nuclear membrane, which is critical for this process. The nuclear import of both proteins ensues separately from each other: pUL31 is actively imported through the central pore channel, while pUL34 is transported along the peripheral pore membrane. With this study, we identified a functional bipartite NLS between residues 178 and 194 of pUL34. pUL34 lacking its NLS is mislocalized to the TGN but retargeted to the ER upon insertion of the authentic NLS or a mimic NLS, independent of the insertion site. If co-expressed with pUL31, either of the pUL34-NLS variants is efficiently, although not completely, targeted to the nuclear rim where co-localization with pUL31 and membrane budding seem to occur, comparable to the wild-type. The viral mutant HSV1(17⁺)Lox-UL34-NLS mt is modestly attenuated but viable and associated with localization of pUL34-NLS mt to both the nuclear periphery and cytoplasm. We propose that targeting of pUL34 to the INM is facilitated by, but not dependent on, the presence of an NLS, thereby supporting NEC formation and viral replication.

Keywords: HSV1; pUL34; pUL31; nuclear egress; NEC; targeting integral membrane proteins; TA membrane proteins; inner nuclear membrane (INM); nuclear import; importins

Citation: Funk, C.; Marques da Silveira e Santos, D.; Ott, M.; Raschbichler, V.; Bailer, S.M. The HSV1 Tail-Anchored Membrane Protein pUL34 Contains a Basic Motif That Supports Active Transport to the Inner Nuclear Membrane Prior to Formation of the Nuclear Egress Complex. *Viruses* **2021**, *13*, 1544. https://doi.org/10.3390/v13081544

Academic Editor: Donald M. Coen

Received: 24 June 2021
Accepted: 3 August 2021
Published: 5 August 2021

Publisher's Note: MDPI stays neutral with regard to jurisdictional claims in published maps and institutional affiliations.

1. Introduction

Herpesviral propagation is an intricate process initiated in the infected nucleus. Capsids formed in the nucleoplasm and packaged with one copy of the viral DNA are too large for export through the nuclear pore. To overcome the nuclear envelope barrier, herpesviruses have established a conserved membrane-budding process called nuclear egress. Primary envelopment occurs by association of nucleocapsids with the inner nuclear membrane (INM) and subsequent membrane budding resulting in enveloped nucleocapsids that transit the perinuclear space. Fusion of primary envelope with the outer nuclear membrane (ONM) allows for the de-envelopment of the nucleocapsids, further maturation in the cytoplasm and re-envelopment at the cytoplasmic membranes, resulting in infectious virions [1–5].

Two essential viral proteins, pUL34 and pUL31, with orthologs in all herpesviruses, form the nuclear egress complex (NEC), which is essential for this process [6–9]. Herpes simplex virus type 1 (HSV1) pUL34 represents an integral membrane protein featured by a cyto-/nucleoplasmically exposed N-terminal domain and a C-terminal transmembrane domain (TM) called the tail-anchor (TA) domain [10,11]. The nuclear phosphoprotein pUL31 consists of a variable N-terminal part enriched in basic residues ([12] and references therein) and a larger conserved C-terminal domain. Interaction of both NEC partners is mediated by an N-terminal domain of pUL34 (aa 137–181) together with the first of four conserved regions CR1 of pUL31 (for review see [6]). Both in vivo and in vitro, recombinantly expressed pUL31 and pUL34 are sufficient to induce the membrane-budding process, which results in vesicles with a highly ordered coat on the inside [13–18]. The NEC thus represents the minimal virus-encoded membrane-budding machinery that is sufficient to drive membrane budding and scission of vesicles during nuclear egress. Crystal structures of dimeric NECs of various orthologs revealed an elongated heterodimer that is organized and stabilized by two interfaces between pUL34 and pUL31 [15,19–22]. Formation of the NEC at the INM requires that both pUL31 and pUL34 are transported into the nucleus. While pUL31 harbors a bipartite nuclear localization sequence (NLS) that is redundant in the viral context and enters the nucleus by transport through the central nuclear pore [12], nuclear import of pUL34 is poorly understood.

The transport of integral membrane proteins to the INM is essential for various nuclear processes and thus crucial for the entire life cycle of eukaryotes. Numerous membrane proteins of the INM were identified that are involved in inheritable diseases (for review see [23]). Despite its importance, the targeting of integral membrane proteins to the INM is one of the least understood cellular transport processes. Essentially, two models are debated for transport of integral membrane proteins to the INM (as reviewed by [24]). First, there is the diffusion-retention model, where integral membrane proteins are inserted into the endoplasmic reticulum (ER) membrane and laterally diffuse to the INM, where retention occurs by INM residents; secondly, there is the transport factor-mediated model, where integral membrane proteins are actively transported to the INM using soluble transport factors of the importin α/β family ([25] and references therein). TA membrane proteins like pUL34 represent a particular kind of integral membrane proteins [10,11]. Due to their C-terminal transmembrane domain, TA membrane proteins are released from the ribosome to expose this domain for post-translational membrane insertion. For this reason, TA membrane proteins may be soluble for a short period of time, potentially providing them access to transport through the central pore channel and subsequent membrane insertion at the INM.

In order to gain further insight into transport of TA membrane proteins to the INM, we made use of the HSV1 infection as a functional reporter system. We identified a basic motif between residues 178 and 194 of pUL34 resembling a bipartite NLS. We demonstrate that this NLS is functional and important for the targeting of pUL34 to the nuclear rim, and most likely to the INM, and that it plays a role in viral replication. We propose that the pUL34-NLS targets pUL34 to the POM for efficient translocation to the INM prior to herpesviral NEC formation and capsid nuclear egress.

2. Materials and Methods

2.1. Cells, Viruses and General Methods

HeLa (ATCC-No. CCL-2), Hep2 (ATCC-No. CCl-23) and Vero cells (ATCC-No. CCL-81) were cultured in Dulbecco's modified Eagle's medium containing 10% FCS. The BAC-DNA pHSV1(17$^+$)Lox (kindly provided by B. Sodeik) [26–28] and virus thereof (HSV1(17$^+$)Lox) were used for all experiments. Plasmid transfection was done using Effectene Transfection Reagent (Qiagen GmbH, Hilden, Germany) according to manufacturer's instructions. In contrast, BAC transfection was done using Lipofectamine 2000 (Thermo Fisher Scientific Inc., Waltham, MA, USA). HSV1 propagation, plaque formation, virus titration and kinetics [11,29], as well as the Y2H method [30], were described previously.

2.2. Plasmids

Plasmids were generated by classical restriction cloning, Gateway recombination cloning (Thermo Fisher Scientific Inc., Waltham, MA, USA) or Gibson assembly cloning (New England Biolabs Inc., Ipswich, UK) according to the manufacturer's protocols. Single base-pair exchanges were introduced using the QuikChange site-directed mutagenesis kit (Agilent Technologies, Santa Clara, CA, USA) and verified by sequencing. The primers used to generate plasmids are listed in Table 1.

Table 1. Primer sequences used for plasmid cloning, site-directed mutagenesis and BAC mutagenesis.

No.	Name	Sequence	Use
1	UL34_vec_fw	GCAGACTGGAAGCTGGCACCAGGCCCTGCGG	Gibson
2	UL34_vec_rv	TGCTTCCCGGGCGGGAGGGCCCTTGGGTTAAC	Gibson
3	MBP_fw	TCCCGCCCGGGAAGCAAAATCGAAGAAGGTAAA	Gibson
4	MBP_rev	CCTGGTGCCAGCTTCCAGTCTGCGCGTCTTTCAGG	Gibson
5	UL34-NLS-mt_fw	CGCCGAGCAGGCTATTACCCGTAACAACAACACCCGGCGGTCCCGGGG	mutagenesis
6	UL34-NLS-mt_rev	CCCCGGGACCGCCGGGTGTTGTTGTTACGGGTAATAGCCTGCTCGGCG	mutagenesis
7	NLS-UL34-NLS-mt_vector_fw	GCGGGACTGGGCAAGCCCTACAC	Gibson
8	NLS-UL34-NLS-mt_vector_rev	CCTCGAGGGATCCCCGGGTACCGAG	Gibson
9	NLS-UL34-NLS-mt_fw	CCGGGGATCCCTCGAGGAGACGGATCCTGTGC	Gibson
10	NLS-UL34-NLS-mt_rev1	AGTCCCGCGGATCCTCGGCGGCGACGGGTAATAG	Gibson
11	NLS-UL34-NLS-mt_rev2	TGTAGGGCTTGCCCAGTCCCGCGGATCCTC	Gibson
12	UL34-NLS-mt-NLS_fw	CCGAGGCCGGGCTGCGGCGGCGCCGAACGGGTTTC	mutagenesis
13	UL34-NLS-mt-NLS_rev	CGGAAACCCGTTCGGCGCCGCCGCAGCCCGGCCTC	mutagenesis
14	SV40NLS-UL34-NLS-mt_fw1	GCGGTCCCGAATTCGAGCTCGGTAC	Gibson
15	SV40NLS-UL34-NLS-mt_rev1	CCTACCTTTCTCTTCTTTTTTGGCCTCGAGGGATCCC	Gibson
16	SV40NLS-UL34-NLS-mt_fw2	GCCGAGACCGCGGTCCCGAATTCGAG	Gibson
17	SV40NLS-UL34-NLS-mt_rev2	CTTGCCCAGTCCCGCGGATCCTACCTTTCTCTTCTTTTTTG	Gibson
18	EYFP-UL34-NLS_fw	TCAGATCCGCTAGCGCTACCGGTCGCCACCATGGTGAGCAAGGGCGAG	restriction cloning
19	EYFP-UL34-NLS_rev1	CTGCTCGGCGGCGCGGCACAGAATCCGTCTCGTAGCTCGAGATCTGAG	restriction cloning
20	EYFP-UL34-NLS_rev2	TTTGGATCCTAAGGTTCGGCGGCGACGGGTAATAGCCTGCTCGGCGGCGC	restriction cloning
21	UL34/gk_fw	GAACCCTTTGGTGGGTTTACGCGGGCACGCACGCTCCCATCGCGGGCGCCCCTGTTGACAATTAATCATCGGCA	BAC
22	UL34/gk_rev	GCGAAGGCGTCCGGAACGCACTGGCGATTAGGGCGGCGGTGCGTCCTTTTGCCAGTGTTACAACCAATTAACC	BAC
23	UL34_fw	GAACCCTTTGGTGGGTTTACGCGGGCACGCACGCTCCCATCGCGGGCGCCATGGCGGGACTGGGCAAGCCCTAC	BAC
24	UL34_rev	GCGAAGGCGTCCGGAACGCACTGGCGATTAGGGCGGCGGTGCGTCCTTTTTTATAGGCGCGCGCCAGCACCAAC	BAC

Constructs encoding pUL34 or mutants thereof were cloned into Gateway-compatible pCR3-C-HA, pCR3-C-MBP, pCDNA6.2-N-YFP, pCDNA6.2-C-YFP or pGBKT7 vectors. Truncation mutants of pUL34 as well as plasmids used for Y2H were generated using Gateway cloning as described previously [31]. Plasmid pCR3-N-Myc-UL31 as well as pEXPR-IBA5-UL34 expressing Strep-pUL34 were described previously [11]. Oligonucleotides used to generate pUL34-MBP-TM and different Strep-pUL34 mutants are listed in Table 1. The plasmid encoding EYFP-UL34-NLS was cloned based on EYFP-Nuc (Takara Bio Inc, Shiga, Japan), which was also used as a control. Plasmids pQE60-KPNA5, pQE60-KPNB1 and pQE32-RanQ69L were kindly provided by D. Görlich and U. Kutay, and pGPS-galK-kan was provided by Z. Ruzsics.

2.3. Immunofluorescence Microscopy

HeLa, Hep2 or Vero cells were grown on coverslips, transfected or infected and fixed with 2% formaldehyde/PBS for 15 min at room temperature (RT). Cells were permeabilized using 0.5% Triton X-100 (5 min, 4 °C). The binding of antibodies to the HSV1 Fc receptor-like proteins gE/gI was blocked with human IgG of seronegative individuals/PBS for at least 3 h at room temperature. Following incubation with primary antibodies (15 min, room temperature) and washing with PBS, the binding of anti-mouse, anti-rabbit or anti-chicken secondary fluorescently labelled antibodies (Thermo Fisher Scientific Inc., Waltham, MA, USA) was allowed (15 min, RT). Mouse monoclonal antibodies anti-HA (Thermo Fisher Scientific Inc., Waltham, MA, USA), anti-ICP0 (Santa Cruz Biotechnology, Dallas, TX, USA), anti-Lamin A/C (Santa Cruz Biotechnology, Dallas, TX, USA), anti-Myc (clone 9E10, kindly provided by J. von Einem) and anti-Strep tag II (IBA Lifesciences GmbH, Göttingen, Germany), as well as polyclonal rabbit antibodies anti-MBP (New England Biolabs Inc., Ipswich, MA, USA), anti-TGN46 (Bio-Rad, Hercules, CA, USA) and anti-chicken anti-pUL34 (kindly provided by R. Roller) (further blocking step using BlokHen® solution (Aves Labs, Davis, CA, USA) for 30 min), were employed. Confocal laser scanning microscopes LSM710 (Zeiss, Oberkochen, Germany) and TCS SP5 (Leica, Mannheim, Germany) were used for sample examinations, and Zen-Lite (Zeiss, Oberkochen, Germany) and Adobe Photoshop (Adobe, San José, CA, USA) were used for processing of microscope recordings.

2.4. BAC Mutagenesis

The HSV1 UL34 mutants were generated using pHSV1(17$^+$)Lox and a modified *galK* positive counterselection scheme essentially as described [11,32]. First, the UL34 coding region was replaced by a *galK*-kan cassette (Table 2), which was previously amplified using the pGPS-galK-kan plasmid (kindly provided by Z. Ruzsics) and primers equipped with 50 bp homologies flanking the UL34 locus (Table 1). In a second step, the *galK*-kan cassette was substituted with a UL34 region encoding UL34-NLS mt or wild-type UL34 (Table 2). Virus reconstitution was performed through transfection of Vero cells with the respective BAC-DNA using Lipofectamin 2000 (Thermo Fisher Scientific Inc., Waltham, MA, USA).

Table 2. BAC mutagenesis.

No.	BAC	Inserted Fragment	Recipient BAC
1	pHSV1(17$^+$)Lox		
2	pHSV1(17$^+$)Lox ΔUL34/gK	*galK*-kan in UL34 locus (bp 1-825); PCR with oligos no. 17 and 18 and pGPS-*galK*-kan as template	No. 1
3	pHSV1(17$^+$)Lox-UL34 wt rescue	UL34 wt; PCR with oligos no. 19 and 20 and pDONR207-UL34$_{1\text{-}275}$ as template	No. 2
4	pHSV1(17$^+$)Lox-UL34-NLS mt	UL34-NLS mt; PCR with no. 19 and 20 and pDONR207-UL34$_{1\text{-}275}$-NLS mt as template	No. 2

2.5. Formation of Nuclear Transport Complexes

All proteins used to analyze the formation of nuclear transport complexes were recombinantly expressed in suitable *E. coli* strains (JM101 (pQE60-KPNB1, pQE32-RanQ69L), BL21(DE3) (pET9d-GST-UL34ΔTM), M15/pREP4 (pQE60-KPNA5)). Protein expression was induced by addition of 1 mM IPTG (culture with OD$_{600}$ 0.6) for 4 h at RT. Bacterial lysates were generated in universal buffer (20 mM HEPES, 100 mM potassium acetate, 2 mM magnesium acetate, 0,1% Tween 20, 10% glycerin; pH 7.5), in the presence of protease inhibitors by mechanical disruption and subsequently cleared by centrifugation. Recombinantly expressed His- and GST-tagged proteins were affinity purified using Ni-NTA agarose (Qiagen GmbH, Hilden, Germany) and GSH sepharose (GE Healthcare, Chicago, IL, USA), respectively. Protein purification was performed in a universal buffer with the buffer additionally containing 20 mM of Imidazol in the case of His-tagged proteins. To allow for complex formation, the GST-tagged protein of interest was again bound to GSH sepharose followed by the addition of the His-tagged, putative binding partners. Immobilized proteins were released from the sepharose by incubation with 4x Lämmli buffer (15 min, RT) and analyzed by SDS-PAGE and Western blotting using goat anti-GST (Sigma-Aldrich, St. Louis, MO, USA) and mouse anti-His antibodies (Qiagen GmbH, Hilden, Germany), followed by POX-conjugated secondary reagents.

3. Results

3.1. Membrane Insertion of pUL34 Occurs Prior to Its Nuclear Import

Previous data suggested that herpes simplex virus 1 (HSV1) pUL34 and pUL31 enter the nucleus separate of each other [12]. Full-length pUL34 carrying an N-terminal MBP tag, thereby exposing a cyto-/nucleoplasmic domain of about 75 kDa, was detected at ER-like structures and at the nuclear periphery irrespective of the presence of its NEC partner Myc-pUL31, the latter of which accumulated in the nucleus [12]. To corroborate and extend these findings, the cyto-/nucleoplasmic domain of pUL34 was N-terminally fused to YFP, resulting in a smaller cyto-/nucleoplasmic mass of about 60 kDa (Figure 1A). Again YFP-pUL34 was detected at ER-like structures and at the nuclear periphery (Figure 1B). In the presence of Myc-pUL31, the subcellular distribution of YFP-pUL34 resembled that of MBP-pUL34 (Figure 1B; [12]).

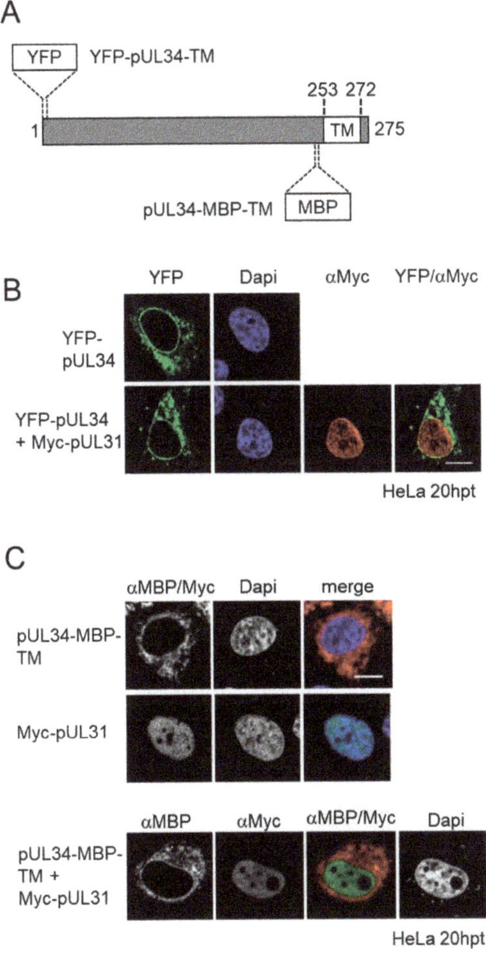

Figure 1. Membrane insertion of pUL34 occurs prior to its nuclear import. (**A**) Schematic depiction of the TA membrane protein pUL34 with the C-terminal transmembrane domain (TM) and either the N-terminal YFP tag (YFP-pUL34-TM) or a pre-TM-inserted MBP-tag (pUL34-MBP-TM); (**B,C**) HeLa cells were transfected with plasmids encoding YFP-pUL34-TM, pUL34-MBP-TM or Myc-pUL31 either alone ((**B,C**) top panels) or in combinations ((**B,C**), bottom panels). pUL31 localization was determined by using anti-Myc antibodies, while pUL34-MBP-TM was detected by anti-MBP antibodies followed by secondary reagents. YFP-tagged pUL34 was visualized directly using confocal microscopy. Nuclei were visualized by Dapi. The scale bar corresponds to 10 μm.

To exclude the interference of N-terminal tagging of pUL34 with its structural integrity potentially having an influence on its interaction with pUL31, an MBP-tag was inserted between residues 235 and 236 prior to the pUL34-TM domain, representing an unstructured region (Figure 1A,C). The transient expression of pUL34-MBP-TM, again exposing a cyto-/nucleoplasmic domain of about 75 kDa, resulted in a similar ER-like distribution as observed with YFP- and MBP-pUL34. In presence of Myc-pUL31, the subcellular distribution of pUL34-MBP-TM was unaltered, and both proteins located to separate subcellular entities. Together, these data indicate that pUL34 and pUL31 do not interact in the cytoplasm and that the membrane-anchored pUL34 needs to access a nuclear transport mode different from pUL31.

3.2. The NEC Protein pUL34 Contains a Functional Bipartite Nuclear Localization Sequence

The finding that pUL34 reaches the nucleus independent of pUL31 suggested it contains its own import mechanism. Using bioinformatic tools, a basic motif was identified within the TA membrane protein pUL34, composed of $RR(X)_{11}RRRR$, resembling a classical bipartite NLS (http://www.expasy.org/; Figure 2A). To be classified as a functional NLS, a given sequence has to fulfil several criteria, e.g., to trigger the nuclear import of an unrelated, cytoplasmic protein, and to mediate physical interaction with transport factors of the importin α/β family [33].

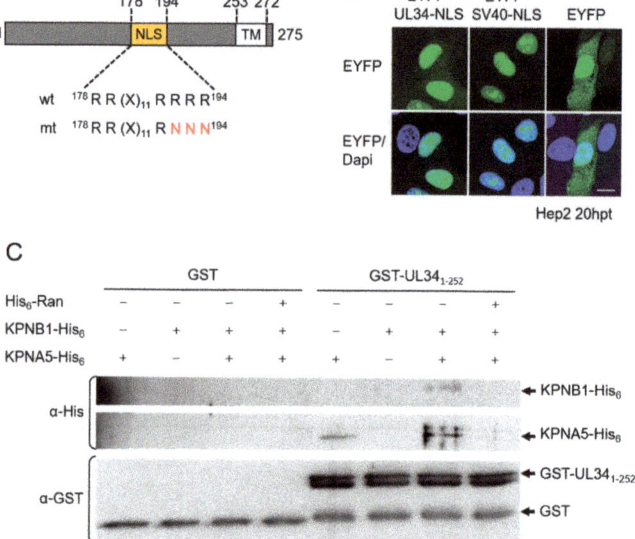

Figure 2. The NEC protein pUL34 contains a functional bipartite NLS. (**A**) Schematic depiction of the TA membrane protein pUL34 with the bioinformatically predicted bipartite NLS (residues 178–194) and the exchanged residues in case of the NLS mutation (red). (**B**) To determine whether the UL34-NLS can mediate the nuclear import of a cytoplasmic protein, Hep2 cells were transfected with plasmids encoding EYFP-UL34-NLS and EYFP-SV40-NLS as the positive control and EYFP alone as the negative control. Then, 20 h post transfection (hpt), EYFP was visualized directly and nuclei were detected by Dapi. Scale bar corresponds to 10 µm. (**C**) Formation of a nuclear import complex in vitro was started with binding of GST-pUL34$_{1\text{-}252}$ or GST alone as the negative control to GSH sepharose followed by incubation with KPNB1-His$_6$, KPNA5-His$_6$ and His$_6$-RanGTP, either alone or in combinations. Eluates were analyzed by SDS-PAGE followed by Western blotting using mouse anti-His or goat anti-GST antibodies, followed by POX-conjugated secondary reagents.

EYFP-pUL34-NLS, where pUL34 residues 178–194 (Figure 2A) were fused to EYFP, strongly accumulated in the nucleus (and nucleolus) comparable to EYFP-SV40-NLS containing the classical SV40-NLS, while EYFP lacking an NLS was distributed between the cytoplasm and nucleus (Figure 2B). Thus, residues 178–194 of pUL34 represent a potent NLS. The yeast two-hybrid (Y2H) assay revealed that pUL34$_{1\text{-}252}$ redundantly interacted with α importins KPNA2, 4 and 5. pUL34$_{1\text{-}203}$, which still contained the complete NLS, also interacted with all tested importins, while pUL34$_{1\text{-}180}$ contained only the first half of the bipartite NLS, and pUL34$_{1\text{-}168}$, where the complete NLS was deleted, showed no interaction. Finally, pUL34$_{1-252}$-NLS mt, where three positively charged residues within one half of the predicted bipartite NLS were replaced by alanines (Figure 2A), also lost the ability to bind α importins. pUL34$_{1-252}$ was able to form a trimeric complex with α importin KPNA5 and β importin KPNB1 (Figure 2C), and complex formation was inhibited in the presence of RanGTP, demonstrating the functionality of the pUL34-NLS (Figure 2C).

To analyze whether the identified classical NLS was active within the pUL34 context, the soluble cyto-/nucleoplasmic domain of pUL34 of about 27 kDa was C-terminally tagged and expressed as wild-type or NLS mutant. Indirect immunofluorescence analysis revealed that both pUL34$_{1-252}$-HA and pUL34$_{1-252}$-NLS mt-HA were exclusively located in the nucleoplasm, reflecting the ability of small proteins to enter the nucleus by diffusion [34]. Interestingly, in either case, soluble pUL34 showed an almost exclusive nuclear localization pointing to an undefined nuclear retention mechanism. pUL34$_{1-252}$-MB, P where the HA-tag was replaced by MBP resulting in a protein of about 75 kDa, efficiently accumulated in the nucleus, while a residual protein was found in the cytoplasm. In contrast, pUL34$_{1-252}$-NLS mt-MBP was exclusively detected in the cytoplasm and thus unable to enter the nucleus. Thus, the NLS of pUL34$_{1-252}$ is able to mediate active transport in its protein context. In summary, we conclude that residues 178–194 of the NEC protein pUL34 compose a functional bipartite NLS.

3.3. Targeting of pUL34 to the Nuclear Rim Combines Active Transport and Retention by Its NEC Partner pUL31

To determine the relevance of the pUL34-NLS for the targeting of the full-length protein to the nuclear rim, the subcellular distribution of Strep-pUL34 was compared to the NLS mutant Strep-pUL34-NLS mt in the absence and presence of its NEC partner pUL31. In the absence of pUL31, wild-type Strep-pUL34 was localized to the ER and the nuclear periphery (Figure 3B). In contrast, the pUL34-NLS mt was primarily found at Golgi-like structures (Figure 3B). Upon co-expression with dsRed-pUL31, that alone was located to the nucleoplasm (Figure 3A), and wild-type pUL34 was efficiently recruited to the nuclear rim (Figure 3C), consistent with previous results [12,35,36]. If co-expressed with dsRed-pUL31, pUL34-NLS mt is recruited to the nuclear periphery with some residual localization of pUL34-NLS mt to Golgi-like structures. In either case, however, intense co-localizations of pUL34 and pUL31 resembling vesicular structures were observed at the nuclear periphery, consistent with NEC formation and budding activity (Figure 3C).

The NLS (residues 178–194) is located in close vicinity to the pUL31 interaction domain of pUL34. To determine whether mutagenesis of the NLS potentially affects pUL34 and pUL31 interaction, the Y2H system was applied. The bait plasmids expressing pUL34$_{1-252}$, pUL34$_{1-252}$-NLS mt or pUL34$_{1-180}$ were co-transformed with a prey plasmid expressing pUL31 or a negative control. As shown in Figure 3D, either version of pUL34, whether containing an intact or a truncated NLS, was able to interact with pUL31 consistent with the pUL34 domain (residues 1–180), which lacks the sufficient NLS for pUL31 interaction [7,9]. We thus conclude that the interaction of pUL34 and pUL31, and thus NEC formation, is independent of the pUL34 region where the NLS is located and the NLS itself. Together, these data show that the NLS contributes to the efficient and directed targeting of pUL34 to the nucleus, while it is not required for retention by its NEC partner pUL31.

3.4. The pUL34-NLS Functions Independently of Its Position and Can Be Replaced by an SV40-NLS

The position of the NLS within pUL34 could be relevant for the NLS function. To analyze the impact of the NLS position on the subcellular localization of pUL34, the pUL34-NLS mt was further engineered by fusion of the authentic NLS to its N-terminal end, resulting in NLS-pUL34-NLS mt (Figure 4A). In addition, pUL34-NLS mt-NLS was generated, where an NLS mimicking the authentic pUL34-NLS was inserted C-terminally to the mutated one (Figure 4A). These mutants were transiently expressed and analyzed regarding their subcellular distribution. In contrast to pUL34-NLS mt, which perfectly co-localized with the TGN marker TGN46, both pUL34-NLS mt versions containing an additional NLS (NLS-pUL34-NLS mt and pUL34-NLS mt-NLS) were located at the nuclear envelope and the ER comparable to wild-type pUL34 (Figure 4B), indicating that an additional authentic or unrelated NLS can restore its localization to the ER. If co-expressed with dsRed-pUL31, either of the pUL34-NLS mutants was recruited to the nuclear rim. Strikingly, intense vesicular co-localizations of pUL34 with pUL31 in the nuclear periphery were observed in all pUL34-NLS variants, indicating that neither mutagenesis of the authentic NLS nor insertion of compensatory NLSs irrespective of their site seemed to abrogate NEC formation and budding (Figure 4C). In summary, the pUL34-NLS is required for efficient localization of pUL34 to the ER and the nuclear periphery, a function that is independent of the position and context of the NLS.

The authentic bipartite NLS of pUL34 is largely composed of arginines. To compare the authentic NLS of pUL34 to that of SV40-NLS, a classical monopartite NLS enriched in lysins, the SV40-NLS-pUL34-NLS mt was generated, transiently expressed and analyzed for its subcellular distribution (Figure 4B,C). Upon isolated expression, SV40-NLS-pUL34-NLS mt was found at the nuclear envelope as well as at cytoplasmic ER-like structures similar to wild-type pUL34 and in contrast to pUL34-NLS mt (Figure 5B). Thus, a classical SV40-NLS can replace the authentic NLS of pUL34. In the presence of dsRed-pUL31, SV40-NLS-pUL34-NLS mt was recruited to the nuclear rim (Figure 4C), as seen for all pUL34-NLS mutants (Figure 4C). To conclude, the authentic bipartite pUL34-NLS composed of arginines can be substituted by a monopartite classical NLS without altering the subcellular distribution of pUL34 in the presence or absence of pUL31. Furthermore, intense vesicular co-localizations of pUL34 with pUL31 were observed both in the absence and presence of the pUL34-NLS and potential phosphosites therein [37,38], consistent with budding activity and a specific role of this motif in the transport and targeting of pUL34 distinct from the NEC formation and function.

3.5. Viral Replication Is Modestly Compromised in Absence of the pUL34-NLS

To determine the importance of the pUL34-NLS for viral replication, the UL34 locus of pHSV1(17$^+$)Lox was replaced by the *galK*-kan selection cassette, which, in turn, was replaced either by wild-type UL34, resulting in pHSV1(17$^+$)Lox-UL34 wt rescue (Lox), or by UL34-NLS mt, resulting in pHSV1(17$^+$)Lox-UL34-NLS mt (Lox-UL34-NLS mt; Figure 2A, Figure 5A). Viral propagation of the mutant virus compared to the wild-type virus was investigated based on plaque formation as well as growth kinetics. Lox as well as Lox-UL34-NLS mt revealed comparable plaque sizes observed three days after the transfection of Vero cells with the respective BAC-DNA (Figure 5B). A detailed analysis of growth kinetics revealed that Lox-UL34-NLS mt showed 36–48 hpi when the virus yield plateaued, and about half a log difference in virus yield compared to Lox was observed, indicating a modest impairment in virus replication (Figure 5C).

To analyze the subcellular distribution of pUL34-NLS mt during infection, Vero cells were infected with Lox or Lox-UL34-NLS mt at an MOI of 1 and analyzed at 12 hpi by indirect immunofluorescence using a combination of anti-Lamin A/C with anti-pUL34 antibodies (Figure 5D). In Lox-infected cells, pUL34 perfectly co-localized with Lamin A/C in the nuclear periphery. In contrast, in Lox-UL34-NLS mt-infected cells, pUL34-NLS mt revealed two different localization patterns, either at the nuclear periphery comparable to the wild-type pUL34, or at both the nuclear periphery and ER-like structures, indicating an incomplete recruitment to the nuclear rim during infection.

Taken together, the mutation of pUL34-NLS results in a viable although attenuated virus associated with an incomplete or delayed transport of pUL34 to the nuclear periphery, potentially causing a delay in capsid nuclear egress.

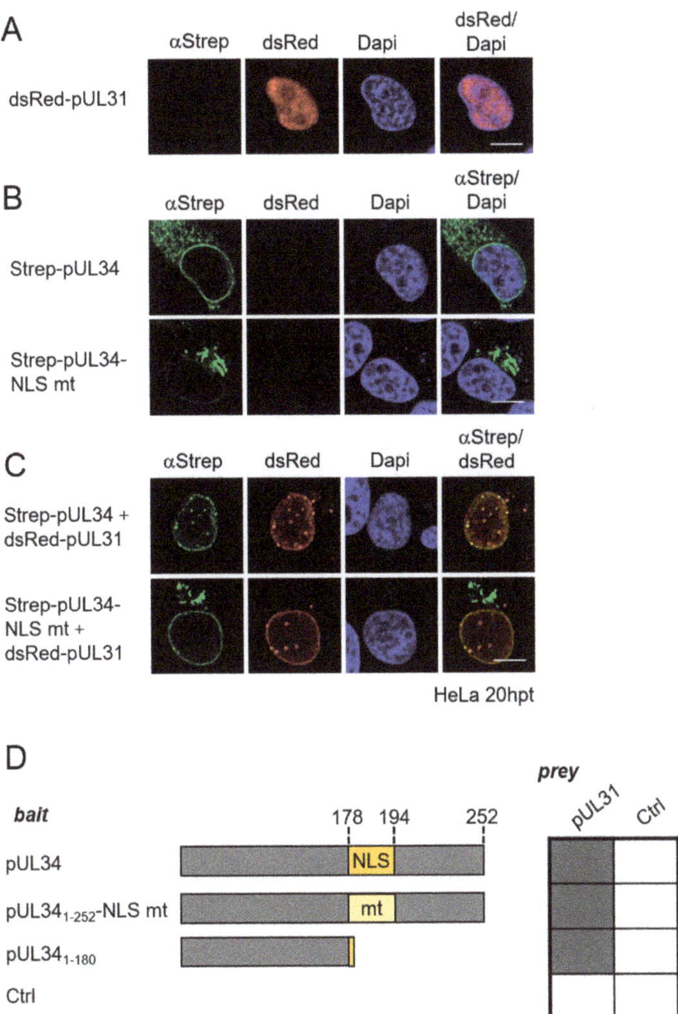

Figure 3. Targeting of pUL34 to the nuclear rim combines active transport and retention by its NEC partner pUL31. To determine the importance of the pUL34-NLS for subcellular localization of pUL34 and NEC formation, (**A**) dsRed-pUL31 and (**B**) Strep-UL34 or Strep-pUL34-NLS mt were expressed in HeLa cells for 20 h either (**A**,**B**) alone or (**C**) in combinations. Indirect immunofluorescence was performed using anti-Strep tag II antibodies followed by Alexa Fluor 488-conjugated secondary reagents. dsRed-pUL31 was visualized directly and nuclei were detected by Dapi. Scale bars correspond to 10 μm. (**D**) To determine whether mutagenesis of the pUL34-NLS plays a role in interaction of pUL34 with pUL31, the bait plasmids $pUL34_{1-252}$, $pUL34_{1-252}$-NLS mt and $pUL34_{1-180}$ (schematically depicted on the left) were co-expressed with the pUL31 prey plasmid. Negative controls (Ctrl) were represented by the respective empty plasmids. Interaction was analyzed by Y2H using the HIS3 reporter gene activation. Grey squares display positive results and white squares display negative results.

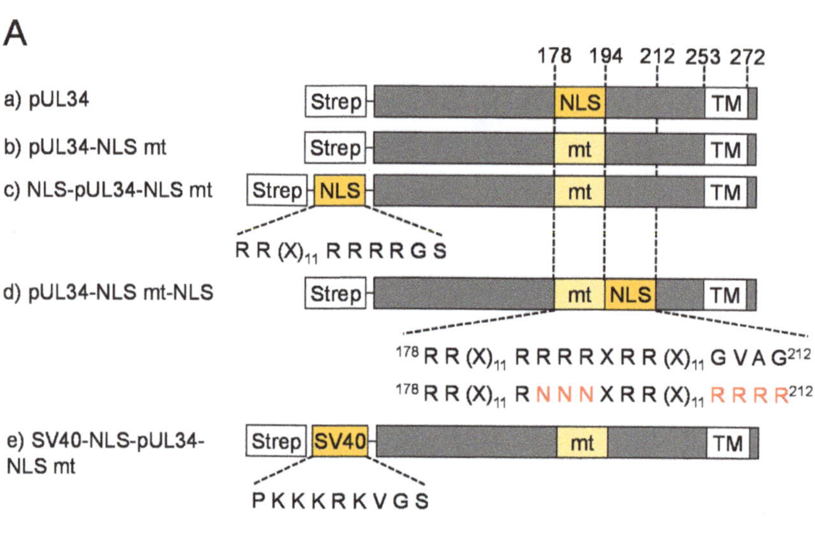

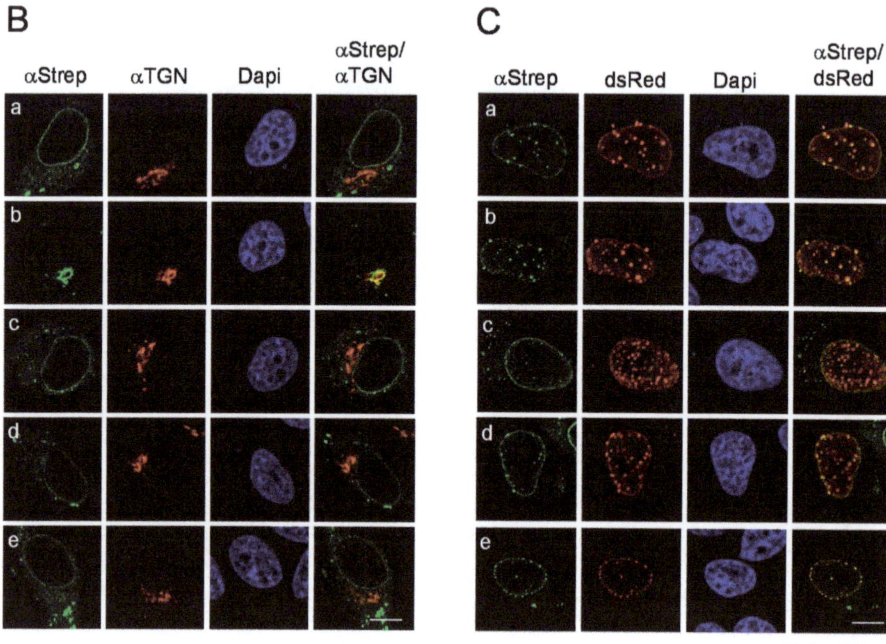

Figure 4. The pUL34-NLS functions independent of its position and can be replaced by an SV40-NLS. (**A**) Graphical depiction of Strep-pUL34, where the pUL34-NLS was replaced by the authentic NLS or a mimic in various positions. (**B**,**C**): (**a**) Strep-UL34, (**b**) Strep-pUL34-NLS mt, (**c**) Strep-NLS-pUL34-NLS mt and (**d**) Strep-pUL34-NLS mt-NLS as well as (**e**) SV40-NLS-pUL34-NLS mt were expressed in HeLa cells for 20 h either (**B**) alone or (**C**) in combination with dsRed-pUL31. Indirect immunofluorescence was performed using anti-Strep tag II and anti-TGN46 antibodies followed by Alexa Fluor 488- and 555-conjugated secondary reagents. dsRed-pUL31 was visualized directly and nuclei were detected by Dapi. Scale bars correspond to 10 μm.

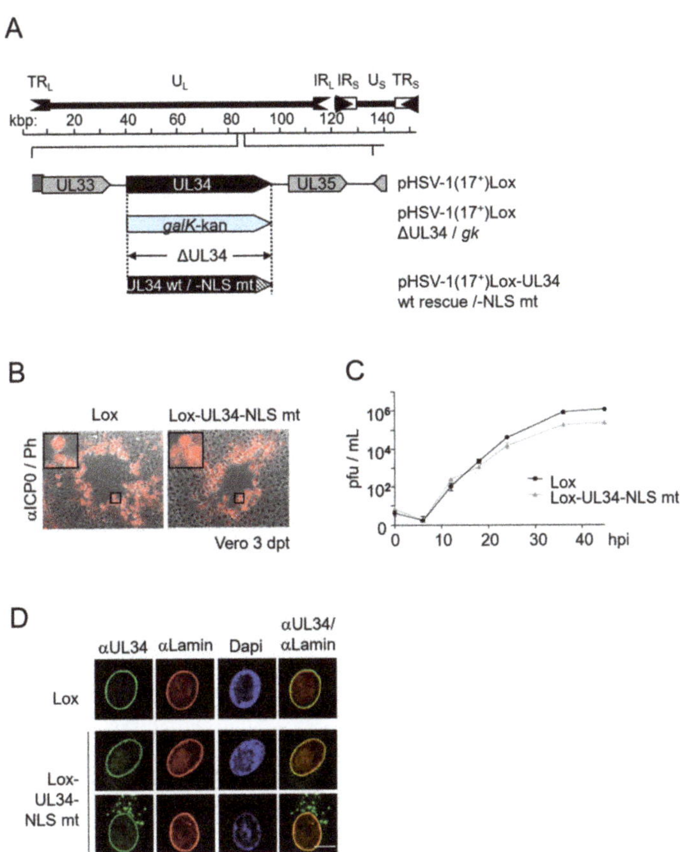

Figure 5. pHSV1(17$^+$)Lox-UL34-NLS mt is modestly compromised in viral replication. (**A**) Schematic depiction of the pHSV1(17$^+$)Lox genome, the localization of UL34 within the genome and the strategy for BAC mutagenesis. During BAC mutagenesis, UL34 is replaced by the *galK*-kan selection cassette (pHSV1(17$^+$)Lox-ΔUL34/*gK*), followed by replacement of this cassette by the wild-type gene UL34 to reach pHSV1(17$^+$)Lox-UL34 wt rescue, or by UL34 harboring the NLS mutation to generate pHSV1(17$^+$)Lox-UL34-NLS mt. (**B**) To analyze virus propagation and plaque formation, Vero cells were transfected with BAC-DNA pHSV1(17$^+$)Lox and pHSV1(17$^+$)Lox-UL34-NLS mt, fixed 3 dpt and examined by indirect immunofluorescence using monoclonal anti-ICP0 antibodies followed by Alexa Fluor 595-labelled secondary antibodies. Black squares represent magnified sections. (**C**) Growth properties of Lox-UL34-NLS mt compared to Lox were determined by infection of Vero cells with an MOI of 0.1 in triplicates, the harvest of the cell culture supernatant at a different hpi and the titration on Vero cells using plaque assay. Graphical depiction includes error bars, and the statistical analysis of the time points t12, t24, t36 and t45 showed that the growth difference between wt and mt is significant with a *p*-value < 0.05. (**D**) To determine the pUL34-NLS mt localization in the viral context, Vero cells were infected at an MOI of 1 with Lox and Lox-UL34-NLS mt and analyzed at 12 hpi by IF using a combination of monoclonal anti-Lamin A/C with anti-pUL34 antibodies, followed by Alexa Fluor 555- and Alexa Fluor 488-conjugated secondary reagents. Nuclei were visualized by DAPI staining and confocal microscopy was applied for analysis. The scale bar corresponds to 10 μm.

4. Discussion

HSV1 infection is an attractive model system to analyze the transport of integral membrane proteins to the INM for a function essential to viral propagation. There, progeny nucleocapsids assembled in the nuclear interior, associated with the INM, where a membrane-budding process called nuclear egress is required for their release to the cytoplasm and eventually to the cell surface. Two essential and conserved viral proteins, the TA membrane protein pUL34 and the nucleo-phosphoprotein pUL31, form the NEC at the INM required for release of capsids from the nucleus. Recent data showed that pUL31 is actively imported through the central nuclear pore channel [12]. With this study using several pUL34 constructs containing differently positioned tags to enlarge the cyto-/nucleoplasmic domain, we found that pUL34 and pUL31 do not interact in the cytoplasm. Furthermore, our data strongly support that membrane insertion of pUL34 already occurs at the cytoplasmic membranes prior to nuclear import. Together, these data provide strong evidence that pUL34 and pUL31 take separate routes to the nucleus, where pUL31 uses the central pore channel, while pUL34 is transported along the POM.

Bioinformatic analysis of HSV1 pUL34 revealed two clusters of positively charged residues composed of RR and RRRRR within residues 178–194, which resemble a bipartite NLS. We show here that this sequence is able and sufficient to target an unrelated soluble cytoplasmic protein to the nucleus and that it forms a functional nuclear import complex with transport factors of the importin α/β family, fulfilling the criteria of a functional NLS [33]. Physical interaction of pUL34 with various α importins required both clusters of basic residues. Consistent with previous reports [11,39], transiently expressed pUL34$_{1\text{-}252}$ lacking its transmembrane domain accumulated in the nucleus, a process that required a functional NLS, particularly if the protein was enlarged by fusion to MBP. Together, our data show that pUL34 contains a functional bipartite NLS that provides access to the classical nuclear import system.

Further analysis of this NLS revealed its impact on the subcellular distribution of a full-length TA membrane protein. Unlike the wild-type pUL34, which upon transient expression primarily locates to ER-like membranes, pUL34-NLS mt was predominantly located at the TGN. pUL34-NLS mt was redirected to the ER upon insertion of the authentic or a mimic NLS at different sites of pUL34, irrespective of the NLS sequence and position within pUL34. In the presence of its NEC partner pUL31, however, all variants of pUL34 were efficiently, although not completely, recruited to the nuclear rim. Together, these data indicate that the trafficking of pUL34 to the INM combines both an active NLS-dependent transport mechanism and retention at the INM by its NEC partner pUL31.

Different models have been proposed for transport of integral membrane proteins to the INM, including the diffusion-retention model and the transport factor-mediated model ([25] and references therein). Peripheral NPC channels are thought to be ~10 nm in width limiting the nucleoplasmic domain of an INM protein to ~60 kDa. A complex formed between the pUL34 domain of 27 kDa and transport factors of the importin α/β family would clearly exceed the peripheral pore channel. To overcome these limitations, a partial unfolding of the respective proteins is required, allowing them to stretch through the core NPC and expose their NLS to the central channel, thereby enabling the NLS to bind to transport receptors. Since the NLS of pUL34 is located at the C-terminal end of the cyto-/nucleoplasmic domain and in close vicinity to the peripheral pore membrane, unfolding is unlikely to expose the NLS to the pore channel. Thus, the underlying mechanism for active and classical transport of integral membrane proteins may be more complex, requiring further detailed studies.

Remarkably, a mutant virus encoding pUL34-NLS mt was replicating but modestly attenuated. Thus, while the NLS of pUL34 is not essential for viral replication, its presence seems to promote HSV1 growth. Interestingly, this viral phenotype correlates with partial mislocalization of pUL34-NLS mt to the cytoplasmic membranes during infection, suggesting an incomplete transport of pUL34-NLS mt to the nuclear periphery and likely to the INM, potentially explaining the attenuated phenotype. The importance of the NLS-dependent nuclear import could vary depending on the highly complex situation during infection, e.g., regarding expression and nuclear import dynamics of pUL31, or the availability of soluble transport factors. Thus, depending on the cellular and viral context, an NLS within pUL34 may be of advantage to viral replication and thus viral fitness. Interestingly, not all pUL34 orthologs seem to contain an NLS [40]. In addition to HSV1 pUL34, clusters of basic residues predicted to serve as NLS were identified in pUL34 orthologs of HSV2, HHV7, MCMV and EBV but were lacking in others [40]. This clearly indicates that the NLS is not a conserved feature of pUL34 and thus unlikely to be essential for nuclear egress and viral replication. This notion is supported by recent structural and functional data supported by the present study, where pUL34 lacking the second half of the pUL34-NLS forms stable NEC complexes [14,19]. Interestingly, while exposed in the cytoplasm, the pUL34-NLS may be partially masked by the binding of pUL31 upon NEC formation in the nucleus, making the NLS inaccessible for transport factors and finalizing pUL34 recruitment to the INM. Overall, it is thus conceivable that this NLS within pUL34 may promote the recruitment of pUL34 to the INM and thus NEC formation and capsid nuclear egress, depending on the dynamics of viral infection. This is particularly interesting in light of the capsid escort model, where pUL31 associates with nucleocapsids already in the nucleoplasm prior to escorting them to sites of primary envelopment [6,12,41]. In this scenario, part of the pUL34 molecules may actively reach the INM, there awaiting the pUL31 escorting the capsid from the nuclear interior to the INM. In parallel, pUL31 may readily interact with integral pUL34 to form NEC seeds, thereby retaining pUL34 at the INM. Thus, during HSV1 infection, both mechanisms of the active NLS-mediated import of pUL34 and retention by pUL31 may cooperate for a concerted nuclear egress, together promoting viral replication.

TA membrane proteins are a diverse group of eukaryotic membrane proteins targeted to various subcellular membranes, particularly to membranes of the secretory pathway and the nuclear envelope, among others. HSV1 encodes three TA membrane proteins, pUL34, pUL56 and pUS9. While pUL34 is located to the ER if expressed alone, pUL56 and pUS9 are targeted to the TGN [10], similar to pUL34-NLS mt, as observed in this study. Interestingly, unlike pUL34, pUL56 and pUS9 do not contain an active NLS. This could indicate that following post-translational insertion into ER membranes: TA membrane proteins are transported to the TGN by default and the mechanisms for retention or redirection to the ER are favorable for targeting to the nuclear periphery. Interestingly, the pUL34 mutant of pseudorabies (PRV) lacks an NLS [40] but contains the motif RQR 173–175 proposed to act as a Golgi retrieval signal. Mutagenesis to RQG results in the mislocalization of PRV pUL34 to the Golgi ([1] and references therein), resembling the subcellular distribution of HSV1-pUL34 lacking an NLS (Figures 3 and 4). Since either the HSV1 or PRV pUL34 mutant is attenuated in viral replication, the ER may represent a favorable starting point for transport to the INM, since it is continuous with the ONM, POM and INM. At present, the mechanisms of such a retrieval or selection process for transport to the INM is unclear. However, together these data may indicate that different mechanisms exist, one of which involves classical transport factors. For further analysis, HSV1 could serve as a functional reporter system to gain insight into a protein transport process that is vital for nuclear functions far beyond herpesviral infections.

5. Conclusions

- The nuclear import of the HSV1 NEC partners pUL34 and pUL31 ensues separately from each other.
- HSV1 pUL34 contains a functional bipartite NLS that mediates the physical interaction with transport factors of the importin α/β family and the nuclear import of an unrelated cytoplasmic protein.
- In the absence of its NLS, pUL34 is mislocalized to the TGN but retargeted to the authentic localization at the ER upon the insertion of the authentic or a mimic NLS, independent of the insertion site.
- If co-expressed with pUL31, the pUL34-NLS mt is efficiently, although not completely, targeted to the nuclear rim, where NEC formation and membrane budding seemed to occur.

- A viral mutant expressing pUL34-NLS mt is modestly attenuated but viably associated with the partial mislocalization of pUL34-NLS mt in the cytoplasm.
- HSV1 pUL34 reaches the nuclear rim and most likely the INM by an NLS-dependent active process combined with retention by its NEC partner pUL31, thereby promoting viral replication.

Author Contributions: C.F., D.M.d.S.eS., M.O. and V.R. performed all experiments. C.F., D.M.d.S.eS., M.O. and V.R. generated graphs and figures. C.F. and S.M.B. wrote the text. S.M.B. directed and guided the project. All authors have read and agreed to the published version of the manuscript.

Funding: This work was supported by grants of the Deutsche Forschungsgemeinschaft.

Institutional Review Board Statement: Not applicable.

Informed Consent Statement: Not applicable.

Data Availability Statement: Not applicable.

Acknowledgments: The HSV1 strain 17+ was kindly provided by Duncan J. McGeoch (University of Glasgow, UK). The BAC HSV1(17⁺)lox was provided by Beate Sodeik (Medizinische Hochschule Hannover, Germany). We are grateful to Jens von Einem (Institut für Virologie, Universitätsklinikum Ulm, Germany) and Richard Roller (Department of Microbiology, The University of Iowa, USA) for generously providing antibodies. We would like to thank Thomas Fries (University of the Saarland, Germany) for kindly providing purified GST and for advice during the biochemical experiments. Dirk Görlich (Max Planck-Institute Göttingen, Germany), Ulrike Kutay (Eidgenössische Technische Hochschule Zürich, Swiss) and Zsolt Ruzsics (Universitätsklinikum Freiburg, Germany) kindly provided plasmids. We would like to thank Caroline Friedel (Ludwig-Maximilians-Universität München, Germany) for the bioinformatic prediction of the pUL34-NLS and Gabriele Vacun (Fraunhofer IGB, Stuttgart, Germany) for statistical analysis. We are grateful to Markus Morrison and Stephan Eisler for support and advice in use of the confocal microscope at the University of Stuttgart (Germany).

Conflicts of Interest: The authors declare no conflict of interest.

Abbreviations

BAC	bacterial artificial chromosome
ER	endoplasmic reticulum
HSV1	Herpes simplex virus type 1
INM	inner nuclear membrane
MBP	Maltose binding protein
NEC	nuclear egress complex
NLS	nuclear localization sequence
ONM	outer nuclear membrane
POM	pore membrane
TA	tail-anchor
TGN	trans Golgi network
Y2H	yeast two-hybrid
YFP	yellow fluorescent protein

References

1. Hellberg, T.; Paßvogel, L.; Schulz, K.S.; Klupp, B.G.; Mettenleiter, T.C. Nuclear Egress of Herpesviruses: The Prototypic Vesicular Nucleocytoplasmic Transport. *Adv. Virus Res.* **2016**, *94*, 81–140. [PubMed]
2. Mettenleiter, T.C.; Klupp, B.G.; Granzow, H. Herpesvirus assembly: A tale of two membranes. *Curr. Opin. Microbiol.* **2006**, *9*, 423–429. [CrossRef]
3. Mettenleiter, T.C.; Müller, F.; Granzow, H.; Klupp, B.G. The way out: What we know and do not know about herpesvirus nuclear egress. *Cell Microbiol.* **2013**, *15*, 170–178. [CrossRef] [PubMed]
4. Roller, R.J.; Baines, J.D. Herpesvirus Nuclear Egress. *Adv. Anat. Embryol. Cell Biol.* **2017**, *223*, 143–169. [PubMed]
5. Crump, C. Virus Assembly and Egress of HSV. *Adv. Exp. Med. Biol.* **2018**, *1045*, 23–44.
6. Bailer, S.M. Venture from the Interior-Herpesvirus pUL31 Escorts Capsids from Nucleoplasmic Replication Compartments to Sites of Primary Envelopment at the Inner Nuclear Membrane. *Cells* **2017**, *6*, 46. [CrossRef]

7. Bigalke, J.M.; Heldwein, E.E. Have NEC Coat, Will Travel: Structural Basis of Membrane Budding during Nuclear Egress in Herpesviruses. *Adv. Virus Res.* **2017**, *97*, 107–141. [CrossRef]
8. Häge, S.; Sonntag, E.; Borst, E.M.; Tannig, P.; Seyler, L.; Bäuerle, T.; Bailer, S.M.; Lee, C.P.; Müller, R.; Wangen, C.; et al. Patterns of Autologous and Nonautologous Interactions Between Core Nuclear Egress Complex (NEC) Proteins of α-, β- and γ-Herpesviruses. *Viruses* **2020**, *12*, 303. [CrossRef] [PubMed]
9. Marschall, M.; Häge, S.; Conrad, M.; Alkhashrom, S.; Kicuntod, J.; Schweininger, J.; Kriegel, M.; Lösing, J.; Tillmanns, J.; Neipel, F.; et al. Nuclear Egress Complexes of HCMV and Other Herpesviruses: Solving the Puzzle of Sequence Coevolution, Conserved Structures and Subfamily-Spanning Binding Properties. *Viruses* **2020**, *12*, 683. [CrossRef]
10. Ott, M.; Marques, D.; Funk, C.; Bailer, S.M. Asna1/TRC40 that mediates membrane insertion of tail-anchored proteins is required for efficient release of Herpes simplex virus 1 virions. *Virol. J.* **2016**, *13*, 175. [CrossRef]
11. Ott, M.; Tascher, G.; Hassdenteufel, S.; Zimmermann, R.; Haas, J.; Bailer, S.M. Functional characterization of the essential tail anchor of the herpes simplex virus type 1 nuclear egress protein pUL34. *J. Gen. Virol.* **2011**, *92 Pt 12*, 2734–2745. [CrossRef]
12. Funk, C.; Ott, M.; Raschbichler, V.; Nagel, C.H.; Binz, A.; Sodeik, B.; Bauerfeind, R.; Bailer, S.M. The Herpes Simplex Virus Protein pUL31 Escorts Nucleocapsids to Sites of Nuclear Egress, a Process Coordinated by Its N-Terminal Domain. *PLoS Pathog.* **2015**, *11*, e1004957. [CrossRef] [PubMed]
13. Klupp, B.G.; Granzow, H.; Fuchs, W.; Keil, G.M.; Finke, S.; Mettenleiter, T.C. Vesicle formation from the nuclear membrane is induced by coexpression of two conserved herpesvirus proteins. *Proc. Natl. Acad. Sci. USA* **2007**, *104*, 7241–7246. [CrossRef] [PubMed]
14. Bigalke, J.M.; Heuser, T.; Nicastro, D.; Heldwein, E.E. Membrane deformation and scission by the HSV-1 nuclear egress complex. *Nat. Commun.* **2014**, *5*. [CrossRef] [PubMed]
15. Zeev-Ben-Mordehai, T.; Weberruß, M.; Lorenz, M.; Cheleski, J.; Hellberg, T.; Whittle, C.; El Omari, K.; Vasishtan, D.; Dent, K.C.; Harlos, K.; et al. Crystal Structure of the Herpesvirus Nuclear Egress Complex Provides Insights into Inner Nuclear Membrane Remodeling. *Cell Rep.* **2015**, *13*, 2645–2652. [CrossRef] [PubMed]
16. Luitweiler, E.M.; Henson, B.W.; Pryce, E.N.; Patel, V.; Coombs, G.; McCaffery, J.M.; Desai, P.J. Interactions of the Kaposi's Sarcoma-associated herpesvirus nuclear egress complex: ORF69 is a potent factor for remodeling cellular membranes. *J. Virol.* **2013**, *87*, 3915–3929. [CrossRef]
17. Hagen, C.; Dent, K.C.; Zeev-Ben-Mordehai, T.; Grange, M.; Bosse, J.B.; Whittle, C.; Klupp, B.G.; Siebert, C.A.; Vasishtan, D.; Bäuerlein, F.J.; et al. Structural Basis of Vesicle Formation at the Inner Nuclear Membrane. *Cell* **2015**, *163*, 1692–1701. [CrossRef] [PubMed]
18. Desai, P.J.; Pryce, E.N.; Henson, B.W.; Luitweiler, E.M.; Cothran, J. Reconstitution of the Kaposi's sarcoma-associated herpesvirus nuclear egress complex and formation of nuclear membrane vesicles by coexpression of ORF67 and ORF69 gene products. *J. Virol.* **2012**, *86*, 594–598. [CrossRef]
19. Bigalke, J.M.; Heldwein, E.E. Structural basis of membrane budding by the nuclear egress complex of herpesviruses. *Embo. J.* **2015**, *34*, 2921–2936. [CrossRef] [PubMed]
20. Walzer, S.A.; Egerer-Sieber, C.; Sticht, H.; Sevvana, M.; Hohl, K.; Milbradt, J.; Muller, Y.A.; Marschall, M. Crystal Structure of the Human Cytomegalovirus pUL50-pUL53 Core Nuclear Egress Complex Provides Insight into a Unique Assembly Scaffold for Virus-Host Protein Interactions. *J. Biol. Chem.* **2015**, *290*, 27452–27458. [CrossRef]
21. Lye, M.F.; Sharma, M.; El Omari, K.; Filman, D.J.; Schuermann, J.P.; Hogle, J.M.; Coen, D.M. Unexpected features and mechanism of heterodimer formation of a herpesvirus nuclear egress complex. *Embo. J.* **2015**, *34*, 2937–2952. [CrossRef]
22. Leigh, K.E.; Sharma, M.; Mansueto, M.S.; Boeszoermenyi, A.; Filman, D.J.; Hogle, J.M.; Wagner, G.; Coen, D.M.; Arthanari, H. Structure of a herpesvirus nuclear egress complex subunit reveals an interaction groove that is essential for viral replication. *Proc. Natl. Acad. Sci. USA* **2015**, *112*, 9010–9015. [CrossRef] [PubMed]
23. Marinko, J.T.; Huang, H.; Penn, W.D.; Capra, J.A.; Schlebach, J.P.; Sanders, C.R. Folding and Misfolding of Human Membrane Proteins in Health and Disease: From Single Molecules to Cellular Proteostasis. *Chem. Rev.* **2019**, *119*, 5537–5606. [CrossRef] [PubMed]
24. Katta, S.S.; Smoyer, C.J.; Jaspersen, S.L. Destination: Inner nuclear membrane. *Trends Cell Biol.* **2014**, *24*, 221–229. [CrossRef]
25. Mudumbi, K.C.; Czapiewski, R.; Ruba, A.; Junod, S.L.; Li, Y.; Luo, W.; Ngo, C.; Ospina, V.; Schirmer, E.C.; Yang, W. Nucleoplasmic signals promote directed transmembrane protein import simultaneously via multiple channels of nuclear pores. *Nat. Commun.* **2020**, *11*, 2184. [CrossRef]
26. Sandbaumhuter, M.; Dohner, K.; Schipke, J.; Binz, A.; Pohlmann, A.; Sodeik, B.; Bauerfeind, R. Cytosolic herpes simplex virus capsids not only require binding inner tegument protein pUL36 but also pUL37 for active transport prior to secondary envelopment. *Cell Microbiol.* **2013**, *15*, 248–269. [CrossRef]
27. Nygardas, M.; Paavilainen, H.; Muther, N.; Nagel, C.H.; Roytta, M.; Sodeik, B.; Hukkanen, V. A herpes simplex virus-derived replicative vector expressing LIF limits experimental demyelinating disease and modulates autoimmunity. *PLoS ONE* **2013**, *8*, e64200. [CrossRef] [PubMed]
28. Nagel, C.H.; Dohner, K.; Fathollahy, M.; Strive, T.; Borst, E.M.; Messerle, M.; Sodeik, B. Nuclear egress and envelopment of herpes simplex virus capsids analyzed with dual-color fluorescence HSV1(17+). *J. Virol.* **2008**, *82*, 3109–3124. [CrossRef]
29. Schmidt, T.; Striebinger, H.; Haas, J.; Bailer, S.M. The heterogeneous nuclear ribonucleoprotein K is important for Herpes simplex virus-1 propagation. *FEBS Lett.* **2010**, *584*, 4361–4365. [CrossRef] [PubMed]

30. Striebinger, H.; Koegl, M.; Bailer, S.M. A high-throughput yeast two-hybrid protocol to determine virus-host protein interactions. *Methods Mol. Biol. (Clifton N. J.)* **2013**, *1064*, 1–15. [CrossRef]

31. Fossum, E.; Friedel, C.C.; Rajagopala, S.V.; Titz, B.; Baiker, A.; Schmidt, T.; Kraus, T.; Stellberger, T.; Rutenberg, C.; Suthram, S.; et al. Evolutionarily conserved herpesviral protein interaction networks. *PLoS Pathog.* **2009**, *5*, e1000570. [CrossRef] [PubMed]

32. Striebinger, H.; Zhang, J.; Ott, M.; Funk, C.; Radtke, K.; Duron, J.; Ruzsics, Z.; Haas, J.; Lippé, R.; Bailer, S.M. Subcellular trafficking and functional importance of herpes simplex virus type 1 glycoprotein M domains. *J. Gen. Virol.* **2015**, *96*, 3313–3325. [CrossRef]

33. Damelin, M.; Silver, P.A.; Corbett, A.H. Nuclear protein transport. *Methods Enzymol.* **2002**, *351*, 587–607.

34. Timney, B.L.; Raveh, B.; Mironska, R.; Trivedi, J.M.; Kim, S.J.; Russel, D.; Wente, S.R.; Sali, A.; Rout, M.P. Simple rules for passive diffusion through the nuclear pore complex. *J. Cell Biol.* **2016**, *215*, 57–76. [CrossRef]

35. Reynolds, A.E.; Ryckman, B.J.; Baines, J.D.; Zhou, Y.; Liang, L.; Roller, R.J. UL31 and UL34 proteins of herpes simplex virus type 1 form a complex that accumulates at the nuclear rim and is required for envelopment of nucleocapsids. *J. Virol.* **2001**, *75*, 8803–8817. [CrossRef] [PubMed]

36. Yamauchi, Y.; Shiba, C.; Goshima, F.; Nawa, A.; Murata, T.; Nishiyama, Y. Herpes simplex virus type 2 UL34 protein requires UL31 protein for its relocation to the internal nuclear membrane in transfected cells. *J. Gen. Virol.* **2001**, *82*, 1423–1428. [CrossRef]

37. Ryckman, B.J.; Roller, R.J. Herpes simplex virus type 1 primary envelopment: UL34 protein modification and the US3-UL34 catalytic relationship. *J. Virol.* **2004**, *78*, 399–412. [CrossRef] [PubMed]

38. Bjerke, S.L.; Roller, R.J. Roles for herpes simplex virus type 1 UL34 and US3 proteins in disrupting the nuclear lamina during herpes simplex virus type 1 egress. *Virology* **2006**, *347*, 261–276. [CrossRef]

39. Bjerke, S.L.; Cowan, J.M.; Kerr, J.K.; Reynolds, A.E.; Baines, J.D.; Roller, R.J. Effects of charged cluster mutations on the function of herpes simplex virus type 1 UL34 protein. *J. Virol.* **2003**, *77*, 7601–7610. [CrossRef] [PubMed]

40. Schmeiser, C.; Borst, E.; Sticht, H.; Marschall, M.; Milbradt, J. The cytomegalovirus egress proteins pUL50 and pUL53 are translocated to the nuclear envelope through two distinct modes of nuclear import. *J. Gen. Virol.* **2013**, *94 Pt 9*, 2056–2069. [CrossRef]

41. Milbradt, J.; Sonntag, E.; Wagner, S.; Strojan, H.; Wangen, C.; Lenac Rovis, T.; Lisnic, B.; Jonjic, S.; Sticht, H.; Britt, W.J.; et al. Human Cytomegalovirus Nuclear Capsids Associate with the Core Nuclear Egress Complex and the Viral Protein Kinase pUL97. *Viruses* **2018**, *10*, 35. [CrossRef] [PubMed]

Phenotypical Characterization of the Nuclear Egress of Recombinant Cytomegaloviruses Reveals Defective Replication upon ORF-UL50 Deletion but Not pUL50 Phosphosite Mutation

Sigrun Häge [1,†], Eric Sonntag [1,†], Adriana Svrlanska [1], Eva Maria Borst [2], Anne-Charlotte Stilp [3], Deborah Horsch [1], Regina Müller [1], Barbara Kropff [1], Jens Milbradt [1], Thomas Stamminger [3], Ursula Schlötzer-Schrehardt [4] and Manfred Marschall [1,*]

[1] Institute for Clinical and Molecular Virology, Friedrich-Alexander University of Erlangen-Nürnberg (FAU), 91054 Erlangen, Germany; sigrun.haege@fau.de (S.H.); ericsonntag@web.de (E.S.); adriana.svrlanska@gmx.de (A.S.); de.horsch@gmx.de (D.H.); mueller.regina@uk-erlangen.de (R.M.); barbara.kropff@uk-erlangen.de (B.K.); jens.milbradt@lgl.bayern.de (J.M.)

[2] Institute of Virology, Hannover Medical School (MHH), 30625 Hannover, Germany; borst.eva@mh-hannover.de

[3] Institute for Virology, Ulm University Medical Center, 89081 Ulm, Germany; anne-charlotte.stilp@uniklinik-ulm.de (A.-C.S.); thomas.stamminger@uniklinik-ulm.de (T.S.)

[4] Department of Ophthalmology, University Medical Center Erlangen (FAU), 91054 Erlangen, Germany; Ursula.Schloetzer-Schrehardt@uk-erlangen.de

* Correspondence: manfred.marschall@fau.de; Tel.: +49-9131-8526089

† These authors contributed equally to this study.

Citation: Häge, S.; Sonntag, E.; Svrlanska, A.; Borst, E.M.; Stilp, A.-C.; Horsch, D.; Müller, R.; Kropff, B.; Milbradt, J.; Stamminger, T.; et al. Phenotypical Characterization of the Nuclear Egress of Recombinant Cytomegaloviruses Reveals Defective Replication upon ORF-UL50 Deletion but Not pUL50 Phosphosite Mutation. *Viruses* **2021**, *13*, 165. https://doi.org/10.3390/v13020165

Academic Editors: Shan-Lu Liu and Jennifer Corcoran

Received: 20 November 2020
Accepted: 19 January 2021
Published: 22 January 2021

Publisher's Note: MDPI stays neutral with regard to jurisdictional claims in published maps and institutional affiliations.

Abstract: Nuclear egress is a common herpesviral process regulating nucleocytoplasmic capsid release. For human cytomegalovirus (HCMV), the nuclear egress complex (NEC) is determined by the pUL50-pUL53 core that regulates multicomponent assembly with NEC-associated proteins and capsids. Recently, NEC crystal structures were resolved for α-, β- and γ-herpesviruses, revealing profound structural conservation, which was not mirrored, however, by primary sequence and binding properties. The NEC binding principle is based on hook-into-groove interaction through an N-terminal hook-like pUL53 protrusion that embraces an α-helical pUL50 binding groove. So far, pUL50 has been considered as the major kinase-interacting determinant and massive phosphorylation of pUL50-pUL53 was assigned to NEC formation and functionality. Here, we addressed the question of phenotypical changes of ORF-UL50-mutated HCMVs. Surprisingly, our analyses did not detect a predominant replication defect for most of these viral mutants, concerning parameters of replication kinetics (qPCR), viral protein production (Western blot/CoIP) and capsid egress (confocal imaging/EM). Specifically, only the ORF-UL50 deletion rescue virus showed a block of genome synthesis during late stages of infection, whereas all phosphosite mutants exhibited marginal differences compared to wild-type or revertants. These results (i) emphasize a rate-limiting function of pUL50 for nuclear egress, and (ii) demonstrate that mutations in all mapped pUL50 phosphosites may be largely compensated. A refined mechanistic concept points to a multifaceted nuclear egress regulation, for which the dependence on the expression and phosphorylation of pUL50 is discussed.

Keywords: human cytomegalovirus; core nuclear egress complex; ORF-UL50 deletion; pUL50 phosphosite mutants; phenotypical changes; differential functional relevance

1. Introduction

Human cytomegalovirus (HCMV) represents a ubiquitous pathogen that causes severe sequelae in immunocompromised patients and newborns. During lytic replication, HCMV hijacks cellular processes to allow the efficient progression of its lytic replication cycle. A crucial step is the well-orchestrated transition process of viral capsids through the nuclear envelope. Hereby, viral capsids have to overcome the nuclear lamina, a proteinaceous meshwork that limits viral budding through the inner and outer nuclear membranes. HCMV-induced disassembly of the nuclear lamina is mediated by a defined nuclear egress complex (NEC) that associates both viral and host proteins [1]. Two virus-encoded NEC proteins have homologs in all known herpesviruses and have been considered as functionally essential for a balanced regulation of herpesviral nuclear capsid egress [1–5]. This core NEC, i.e., HCMV pUL50-pUL53, provides a scaffold that is crucial for NEC multicomponent protein assembly, capsid binding and a multifaceted functionality. On this basis, at least three main functions of the multicomponent NEC have been analyzed on a molecular basis so far: (i) to mediate the multimeric recruitment of NEC-associated effectors, (ii) to promote the reorganization of the nuclear lamina as well as membranes, and (iii) to provide a docking platform for nuclear capsids. As far as the latter aspect is concerned, we identified by immuno-gold electron microscopy (EM) a high portion of nuclear capsids being associated with pUL53 [6]. It appears generally accepted that both viral and cellular protein kinases are responsible for site-specific phosphorylation of the core NEC proteins themselves as well as of nuclear lamins and lamina-associated proteins. Recent studies demonstrated that HCMV pUL50-pUL53 core NEC formation at the nuclear rim depends on the activity of the viral kinase pUL97 [7] and additionally on cellular protein kinases such as CDK1 [8]. Of note, phosphorylation-specific mass spectrometry analysis revealed five major and nine minor phosphosites in pUL50 [8], but their functional importance remained elusive so far.

The property of being phosphorylated by viral and cellular protein kinases is shared by all analyzed members of the core NEC gene family [1,3,9]. Moreover, several studies revealed strongly conserved structural features of core NECs, particularly comparing the 3D crystal structures of herpes simplex virus type 1 (HSV-1), pseudorabies virus (PrV), HCMV, murine cytomegalovirus (MCMV) and Epstein-Barr virus [10–16]. Although core NEC homologs only share low to moderate degrees of sequence identity, their globular structures are remarkably conserved [3], thus comprising (i) a hook-like N-terminal heterodimerization domain in pUL53 homologs, (ii) a C-terminal α-helical binding groove in pUL50 homologs and (iii) a pUL53-specific zinc finger. Most strikingly, a common hook-into-groove principle of core NEC heterodimerization has been identified for α-, β- and γ-herpesviruses [1,3,5,9,10,14]. However, limited data are available addressing the question whether core NEC homologs are strictly functionally conserved [1,17] or whether they may display virus-specific differences. As far as the latter point is concerned, recent investigations revealed that binding properties of core NEC proteins in terms of nonautologous heterodimeric interactions are limited within herpesviral subfamilies [2,16,17], whereas those in terms of multimeric interaction with additional NEC-associated proteins can differ substantially [1]. Using bacterial artificial chromosome-derived (BAC) virus recombinants carrying NEC gene exchanges, HCMV pUL50 and pUL53 were able to rescue viral replication of MCMV mutants defective in the expression of NEC proteins in vitro [17] or in vivo [2].

Our recent analyses of NEC protein-protein interaction and phosphorylation-specific mass spectrometry pointed to pUL50 representing the multi-interacting, phospho-regulated NEC determinant [8,18]. Since the question about functional relevance of core NEC protein phosphorylation has not been fully answered, the impact of alterations in the phosphosite patterns on viral nuclear egress should be investigated. In this study, we addressed these central questions by the use of an ORF-UL50 deletion virus, using pUL50-complementing cells for virus reconstitution, as well as by introducing pUL50 phosphosite replacement mutations. The experimental design covered an analysis of viral protein expression patterns, investigation of alternative phosphosites, coimmuno-precipitation (CoIP) analysis of altered protein-protein interactions, confocal imaging of NEC protein localization and EM as well as qPCR methods to characterize parameters of viral replication. Profound functional differences were seen between the ORF-UL50 deletion virus and the recombinants harboring phosphosite mutations and their relevance for the mechanistic understanding of herpesviral NECs is discussed.

2. Materials and Methods

2.1. Cell Culture, Virus Infection and Transient Transfection

Primary human foreskin fibroblasts (HFFs, own repository of primary cell cultures) and HFF-UL50 [19] were propagated as described previously [8,19]. HeLa and 293T cells were cultivated at 37 °C, 5% CO_2 and 80% humidity using Dulbecco's modified Eagle's medium (DMEM, 11960044, Thermo Fisher Scientific, Waltham, MA, USA) supplemented with 1× GlutaMAX™ (35050038, Thermo Fisher Scientific), 10 µg/mL gentamicin (22185.03, SERVA, Heidelberg, Germany) and 10% fetal bovine serum (FBS, F7524, Sigma Aldrich, St. Louis, MO, USA). HCMV strain AD169 variant UK, HCMV ΔUL50 (described in [19]) and phosphosite mutants generated in this study were propagated as stocks and titrated in the respective cell lines as described previously [20,21]. Infection experiments were performed at indicated multiplicity of infection (MOI) using parental or recombinant HCMVs. After incubation for 90 min at 37 °C, virus inoculi were removed and replaced by fresh growth medium. Transient transfection of expression plasmids was performed as described before [2].

2.2. Plasmids

For the generation of recombinant HCMVs, universal transfer constructs (UTCs) carrying phosphosite mutations on the template UL50-HA were generated in vector pcDNA3.1. In addition to the wild-type-like revertant pcDNA3.1-UL50(Rev1)-HA, major phosphosites (S188, T214, S216, S300 and S324) were substituted with alanine resulting in pcDNA3.1-UL50(P×5)-HA or pcDNA3.1-UL50(S216A)-HA or with glutamic acid resulting in pcDNA3.1-UL50(M×5)-HA by site-directed mutagenesis (GeneArt™ Site-Directed Mutagenesis System, Thermo Fisher Scientific). An I-*SceI*-*aphAI* cassette (kanamycin cassette) from plasmid pEP-*S/aphAI* (kindly provided by B. K. Tischer; [22]) was inserted into wild-type or mutated UL50-HA UTCs via the *AflII* restriction site (Figure S1). UTCs harboring mutations at 14 phosphorylation sites (S95, T174, T183, S188, T214, S216, S225, S227, T229, S251, S268, S300, S324 and S330) were constructed by ShineGene Molecular Biotech Inc., Shanghai, China. Expression plasmids coding for pUL50-HA, pUL53-Flag, PKCα-Flag and p32(50-282)-Flag were described previously [23–25].

2.3. Generation of Recombinant HCMV

For the generation of HCMV UL50 phosphosite mutations, the BAC HB15/AD169ΔUL50 (described in [19]) (Figure S1A) was used for the insertion of UL50 phosphosite UTCs by a two-step markerless Red recombination system as described previously (Figure S1C; [19,26]). In a first step the pUL50 phosphosite UTCs were inserted into HB15/AD169ΔUL50 as described before and in antisense direction (Figure S1C; [22,26,27]). During the second recombination step, the coding sequence for the resistance cassette was deleted resulting in the recombinant BACs (Figure S1B). Recombinant viruses were reconstituted by transfection of the generated BACs into HFFs. The ΔUL50 genome, which did not yield progeny virus after transfection into normal HFFs, could be successfully used for the reconstitution of rescue virus by the generation of pUL50-complementing cells as described earlier [19]. Virus stock was grown in doxycylin-induced (+Dox), pUL50-complementing cells and was then used for infection experiments in cultures of induced, uninduced HFF-UL50 or normal HFFs. The correctness of gene sequences and the inserted mutations was verified by repeated sequence analyses of PCR fragments derived from both, templates of BAC DNA as well as reconstituted virion DNA (summarized in Figure S7).

2.4. Antibodies

Antibodies used in this study are mAb-IE1p72, mAb-MCP, mAb-pp28 (all kindly provided by William Britt, University of Alabama, Birmingham, AL, USA), mAb-UL44 (kindly provided by Bodo Plachter, University of Mainz, Mainz, Germany), mAb-UL50.01, mAb-UL53.01, mAb-UL97.01 (all kindly provided by Stipan Jonjic and Tihana Lenac Rovis, University of Rijeka, Rijeka, Croatia), mAb-β-Actin (A5441, Sigma Aldrich, St. Louis,

MO, USA), mab-lamin A/C (ab108595, Abcam, Cambridge, UK), pAb-lamin A/C pSer22 (ABIN1532183, Antibodies online, Aachen, Germany), pAb-Pin1 (10495-1-AP, Proteintech, Rosemont, IL, USA), anti-Cytomegalovirus-Alexa Fluor 488 (IE1/MAB810X, Merck, Darmstadt, Germany), mAb-HA (Clone 7, H9658, Sigma Aldrich), mAb-Flag (F1804, Sigma Aldrich), mAb-emerin (Sc-25284, Santa Cruz, Dallas, TX, USA), anti-mouse Alexa 555 (A-21422, Thermo Fisher Scientific, Waltham, MA, USA) and anti-rabbit Alexa 488 (A-11008, Thermo Fisher Scientific).

2.5. CoIP, Phosphate Affinity (Phos-Tag) SDS-PAGE and Western Blot (Wb) Analyses

For the investigation of expression patterns, HCMV-infected HFFs were harvested and lysed at the time points indicated. To analyze protein-protein interactions in a transient expression system, 293T cells were seeded into 10-cm dishes with a density of 5.0×10^6 cells and transfected with expression plasmids. Three days post transfection (d p.t.), CoIP was performed as described previously [18]. Immunoprecipitation control samples of approximately one-tenth of total volumes were taken prior to CoIP reactions. Samples were subjected to SDS–PAGE/Wb procedure as described previously [28]. For the identification of phosphorylated protein varieties, HFFs were infected with HCMVs at an MOI of 0.3, harvested 4 days post infection (d.p.i.). and lysed in CoIP buffer without EDTA for 20 min on ice. For lambda phosphatase (PP) treatment, samples were supplemented with $1 \times$ PMP buffer, 1 mM $MnCl_2$ and 0.5 µL of lambda phosphatase (P0753S, New England Biolabs, Ipswich, MA, USA), incubated at 30 °C for 45 min and subsequently denatured at 95 °C for 10 min. Samples were analyzed by standard SDS-PAGE, supplemented with Phos-tag reagent [29] (AAL-107, Wako PureChemical Industries, Osaka, Japan) according to manufacturer's instructions, and a subsequent Wb immunostaining.

2.6. Indirect Immunofluorescence Assay and Confocal Laser-Scanning Microscopy

Transiently transfected HeLa cells or HCMV-infected HFFs were grown on coverslips, fixed at 2 d p.t. or 6 d p.i. with 10% formalin solution (10 min, room temperature) and permeabilized by incubation with 0.2% Triton X-100 solution (15 min, 4 °C). Indirect immunofluorescence staining was performed by incubation with primary antibodies as indicated for 60 min at 37 °C, followed by incubation with dye-conjugated secondary antibodies for 30 min at 37 °C. Cells were mounted with Vectashield Mounting Medium containing DAPI (H-1700, Vector Laboratories, Burlingame, CA, USA) and analyzed using a TCS SP5 confocal laser-scanning microscope (Leica Microsystems, Wetzlar, Germany). Images were processed using the LAS AF software (Leica Microsystems) and Photoshop CS5 (Adobe Inc., San José, CA, USA).

2.7. Cytomegalovirus Multistep Replication Curve Analysis

Infection experiments were performed at indicated MOIs using parental or recombinant HCMVs. Viral replication kinetics were analyzed by quantitative real-time PCR (qPCR) or digital droplet PCR (ddPCR) as described previously [21,30]. In the ddPCR procedure, DNA samples are individually fractionated in 20,000 droplets, each droplet containing either no template, one template or more templates. Thus, ddPCR is a binary end-point measurement resulting in either a positive or a negative signal [31,32]. On this basis, one single copy can be detected, so that per definition the ddPCR limit of detection was set as 10^0 copies. The standard containing 10 HCMV DNA copies reached the cycle threshold at cycle 38 and was therefore defined as the limit of detection for qPCR.

2.8. Transmission Electron Microscopy (TEM)

HFFs were plated on Nunc™ Thermanox™ coverslips (174977, Thermo Fisher Scientific) with a density of $3.0–4.0 \times 10^5$ and infected with parental or recombinant HCMVs at indicated MOI. At 4 (phosphosite mutants) or 7 d p.i. (ΔUL50), cells were fixed with 2.5% glutaraldehyde in 0.1M phosphate buffer, postfixed in 2% buffered osmium tetroxide, dehydrated in graded alcohol concentrations, and embedded in epoxy resin according to standard protocols. Ultrathin sections were stained with uranyl acetate and lead citrate and examined with a transmission electron microscope (LEO 906E, Carl Zeiss Microscopy GmbH, Oberkochen, Germany).

3. Results

3.1. Construction of Expression Plasmids and Recombinant Viruses Encoding Mutant Versions of ORF-UL50

For the analysis in various expression and infection systems, a series of constructs was generated. Transient expression was used as a basis for Wb detection of mutant pUL50, CoIP experiments and confocal immunofluorescence imaging. To this end, plasmid constructs were generated by cloning WT ORF-UL50-HA into vector pcDNA3.1 and subsequent site-directed mutagenesis to produce replacement mutations in all 14 phosphosites reported in our earlier study [8]. These phosphosites, i.e., serine or threonine residues, were categorized in five major (S188, T214, S216, S300 and S324) and nine minor sites (S95, T174, T183, S225, S227, S229, S251, S268 and S330) on the basis of their abundance of detectability by phosphorylation-specific mass spectrometry analyses. Replacement mutations were either performed through alanine (loss-of function, P×5 and P×14) or glutamic acid exchanges (gain-of-function, M×5 and M×14). The latter mutants were intended to mimic the phosphorylated state on the basis of the amino acids negative charge. Revertants to the original phosphosite pattern were also produced in the five major positions (Rev1) or all 14 positions (Rev2). Recombinant viruses were generated using two-step traceless BAC mutagenesis of the HB15 genome referring to the HCMV laboratory strain AD169 [22]. The constructs and procedures are described in Figure 1 and Figure S1. Sequencing of relevant regions was performed, so that the correctness of ORF-UL50 deletion and phosphosite mutations in the respective BACs as well as the genomes of reconstituted viruses could be verified. In order to address the question of putative compensatory mutations in other, functionally relevant viral coding sequences, also ORFs UL53 and UL97 were included in this analysis. With the exception of one identical amino acid replacement in P×5, M×5 and Rev1, no additional mutations were detected (Figure S7). Thus, we did not find an indication of a selective pressure towards compensatory mutations induced by ORF-UL50 mutagenesis in these virus recombinants.

3.2. Patterns of Expression and Protein-Protein Interaction of ORF-UL50 Deletion and Phosphosite Mutants

In order to analyze the regulatory relevance of deletion and phosphosite mutations, patterns of viral protein expression were determined under various conditions. The rescue virus of HCMV ΔUL50 was used for infection experiments performed in pUL50-complementing cells, i.e., the HFF-UL50 cells induced with doxycyclin (+Dox; [19]), in parallel to pUL50-negative HFF-UL50 (−Dox) or normal HFFs (Figure 2). The impact of +Dox induction on HFF-UL50 cells for successful pUL50 complementation and rescue of the defective ΔUL50 phenotype had been described before [19]. In the present experiment, infection of the different cells (HFF-UL50+Dox, −Dox and normal HFFs) was performed with infectious supernatant samples freshly produced under the following conditions: normal HFFs were used for the production of primary infectious supernatants of parental HCMV AD169 (WT) and recombinant HCMV ΔUL50. The supernatants were harvested at the time point when virus-induced cytopathic effect had occurred in approximately 50% of the monolayers. These supernatants were then used as inocula for the infection of normal HFF or HFF-UL50−Dox/+Dox cells to determine the differential viral expression patterns. Note the marked quantitative differences of infectivity of ΔUL50 virus compared to WT (Figure 2). Cells were harvested at consecutive time points between 24 and 96 h p.i. (h p.i.) and viral proteins of all replicative stages, i.e., immediate early (IE1p72), early (pUL44 and pUL50) and late (major capsid protein, MCP), were detected by Wb analysis. Notably, HCMV ΔUL50 showed a delay and massive quantitative reduction in the production of viral proteins, but surprisingly late-period expression was not completely abolished. Note the detectable signals of IE1p72 and pUL44 expression at time points of 48–72 h p.i. or

72–96 h p.i., respectively (Figure 2, lanes 11, 14, 20, 23, 29 and 32). It was also striking to find a very low level or even the absence of expression of the true-late protein MCP for HCMV ΔUL50. In addition, a lack of continuous presence of IE1p72 was noted at late time points of infection in the absence of Dox (96 h p.i., upper panel). Thus for the ΔUL50 virus, a very low level of infectivity and viral protein production was observed compared to WT, particularly close to the detection limit in uninduced cells (uninduced *, −Dox; induced >, +Dox). This difference was detectable for immediate early, early as well as late proteins. The finding strongly suggested that the ΔUL50 mutant, although not being entirely defective in the production of viral proteins, showed a limited and transient mode of immediate early/early protein expression. Infection periods later than 96 h were analyzed by PCR-based measurement of viral replication kinetics as described below (up to 14 d p.i.).

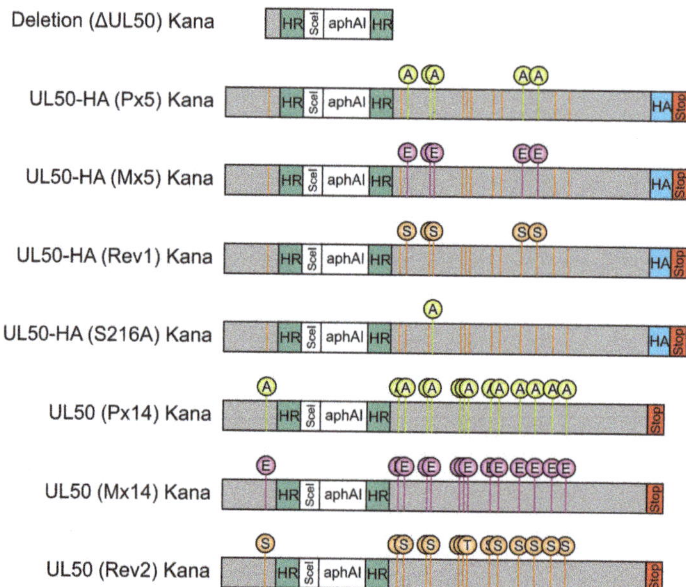

Figure 1. Schematic view of ORF-UL50 constructs used in this study. For the construct termed 'Deletion (ΔUL50) Kana', ORF-UL50 was deleted from the HB15 BACmid using Red recombination without affecting the pUL49/pUL50 overlapping region (nt 1151–1194 highlighted in light grey; [19]). The construction of universal transfer constructs (UTCs, inserted into pcDNA3.1 plasmid vector) was based on wild-type UL50-HA, containing the homologous regions (HR) required for Red recombination, an I-SceI restriction site and a kanamycin cassette (aphAI). The serine/threonine replacement mutants (five major phosphosites S188, T214, S216, S300, S324 and nine minor phosphosites S95, T174, T183, S225, S227, S229, S251, S268, S330), as generated by site-directed mutagenesis, were termed as P×5 and P×14 (5 or 14 alanine replacement mutations) or M×5 and M×14 (5 or 14 phosphorylation mimetic mutations, carrying glutamic acid replacements at identical positions). Two corresponding revertants, termed Rev1 and Rev2, were used as control constructs for the P×5/M×5 and P×14/M×14 mutants, respectively, and a single-site mutant, S216A, was additionally produced for comparison. HA-tag is highlighted in blue and stop codons in red. Details for BAC recombination using these UTCs is described by Figure S1.

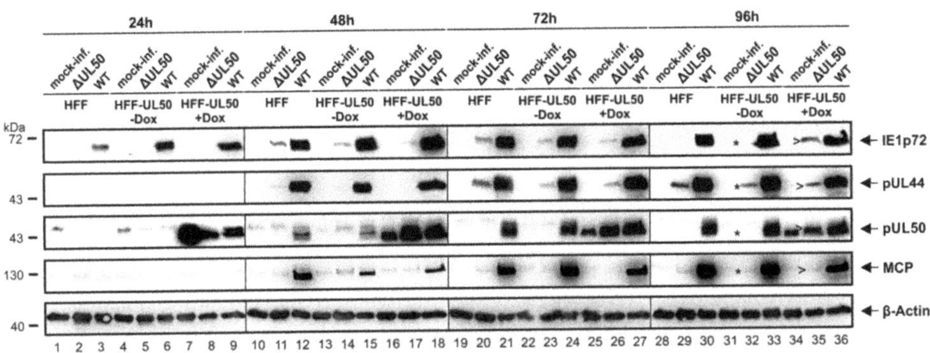

Figure 2. Analysis of viral protein expression of recombinant HCMV ΔUL50. Normal HFFs were grown in 6-well plates (4×10^5 cells/well) and used for the production of primary infectious supernatants of parental HCMV AD169 (WT) and recombinant HCMV ΔUL50. These supernatants were used as inocula (identical volumes of 1 mL/well) for the infection of normal HFF or HFF-UL50 −Dox/+Dox cells as indicated, to determine the expression patterns of HCMV WT in comparison to ΔUL50. Protein expression was analyzed at the time points indicated (h p.i.) and Western blotting (Wb) was performed using the respective antibodies. Note the marked quantitative differences, i.e., the low level of infectivity and viral protein production, even after 96 h p.i., obtained with the ΔUL50 inoculum, particularly in uninduced (*, −Dox) compared to pUL50-induced cells (>, +Dox).

In contrast to the ΔUL50 virus, HCMV pUL50 phosphosite mutants (MOI of 0.3, Figure 3), showed mostly unaltered patterns of expression of viral proteins, which was basically indistinguishable from the parental WT virus (Figure 3). The expression patterns of IE1p72, pUL44, pUL50, pUL53, pUL97 and MCP, although showing some minor variations, did not reveal a drastic change for the virus recombinants P×5, M×5, S216A at 4 d p.i. (compared to WT, the GFP-expressing reporter version AD169-GFP or the ORF-UL50 revertants Rev1 or Rev2). This finding of poorly altered expression patterns was mostly consistent for all mutants, i.e., including the major phosphosite mutants (P×5, M×5; Figure 3A), the single-site mutant S216A (Figure 3A), and the minor phosphosite mutants (P×14, M×14; Figure 3B). The finding for S216A was surprising, since a previous report demonstrated a negative effect of pUL50 mutation S216A on viral nuclear egress. In the reported case, however, the mutant virus carried a second mutation, namely pUL53 S19A [7]. This suggests that S216A alone may be compensated while double-mutations in pUL50/pUL53 may produce such phenotypical effect. Interestingly in the present experiments, the phosphosite mutants showed some variation in the quantities and kinetics of individual viral proteins, including pUL53 and/or pUL50 themselves, but this could not be clearly correlated with genotypic differences. Such altered expression levels appeared to be compensated, since the overall viral replication kinetics (see below) did not indicate detectable replicative defects. In this context, it should also be mentioned that all HCMVs expressing HA-tagged pUL50 (Figure 3A, lanes 4–7) showed somewhat reduced levels of pUL50 and pUL53 compared to WT, an observation already described before [27]. Moreover, also phosphorylated (pS22) lamin A/C and Pin1, did not show a clear-cut virus-specific alteration in most cases analyzed here, and did not markedly differ from the house-keeping protein β-actin. A general phenomenon of a virus-induced effect on the cellular level, observed already earlier [33–35], was confirmed here, namely a slight upregulation of lamin A/C pS22 and Pin1 as well as downregulation of lamin A/C by HCMV infection. Concerning the mutants' protein characteristics in SDS-PAGE separation, P×14 and M×14 showed pronounced faster or slower migration, respectively, when compared to WT. This was found in the settings of both recombinant viruses and transient expression (Figure 3B and Figure S2), and such altered migration behavior was in part also observed for the other mutants (Figure 3A). This may be explained by the massive loss of overall charge and the change in molecular masses upon replacement of serine/threonine residues to alanine or glutamic acid.

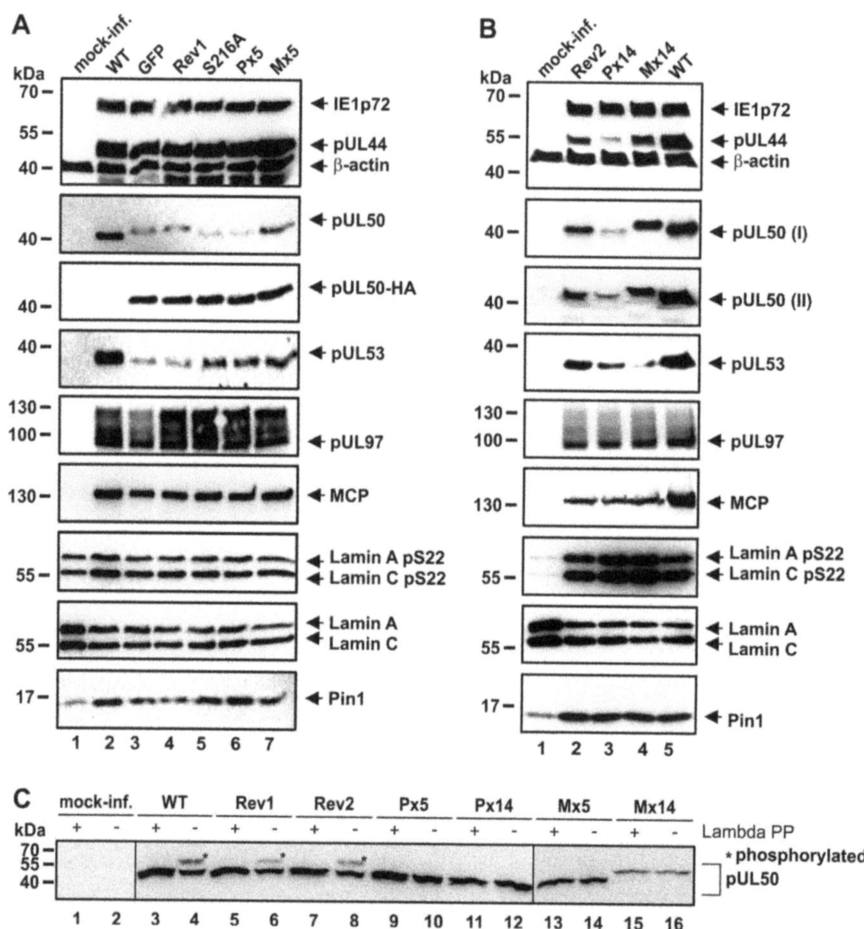

Figure 3. Analysis of viral protein expression of HCMV pUL50 phosphosite mutants. HFFs were infected with parental or recombinant HCMVs at a MOI of 0.3. Protein expression was analyzed after 4 d p.i. and Wb was performed using the indicated antibodies. (**A**) Focus was given either to major phosphosite mutants P×5 and M×5, or (**B**) the mutants including major plus minor phosphosite replacements in P×14 and M×14. Note the reduced pUL53 and pUL50 levels, as explained in the text, the latter depicted by two independent experimental replicates, labeled pUL50 (I) and (II). (**C**) Total lysates of HCMV-infected HFFs were optionally treated with lambda phosphatase (PP) for 45 min at 30 °C and subjected to Phos-tag SDS-PAGE and Wb; *, phosphorylated form of pUL50.

Given the fact that a previous publication demonstrated the functional importance of HCMV pUL50/pUL53 phosphorylation sites [7], it appeared plausible that alternative patterns of phosphorylation may replace the originally identified patterns [18]. Possibly, the phosphorylation of pUL50 may carry some intrinsic flexibility, so that the preferred positions of phosphorylation should not be as decisive as the overall negative charge of subdomains and the mimic options of alternative amino acids. Of note, HCMV pUL50 and pUL53 in total contain 59 and 42 serine/threonine residues, respectively, within their entire amino acid sequences. In order to test the possibility of alternative phosphorylation, a Phos-tag-specific Wb analysis was performed (Figure 3C). A comparison between samples of total cellular protein lysates derived from the individual virus mutants showed that the mutants P×5, M×5, P×14 and M×14 lack the phosphorylation signals (asterisks) detected for WT and Rev viruses. This may indicate that alternative pUL50 phosphorylation of the mutants was either not detectable or pUL50 phosphorylation may not be functionally important.

When analyzing the interaction capacity of the pUL50 mutants P×14 and M×14 upon transient expression, all interactions with the known interaction partners, i.e., pUL53, PKCα, p32/gC1qR and emerin, were detectable by CoIP analysis (Figure S2). Considering the quantitative levels of interaction, referring to 100% defined for pUL50-HA as the non-mutated parental protein, reduced CoIP signals were found in most cases for both of the two types of pUL50 mutants, P×14 and M×14 (Figure S2B–D; note the exception of pUL53 interaction which was found to be increased, Figure S2A).

3.3. Confocal Imaging of Mutant pUL50 Expressed by Plasmids or Recombinant HCMVs: Localization of Viral Proteins and Nuclear Lamina Components

The intracellular localization of nuclear egress-relevant proteins was investigated by confocal imaging, using both transiently transfected HeLa cells and HCMV-infected primary fibroblasts. Upon transient plasmid transfection of mutant pUL50, the question of relocalization of coexpressed pUL53 was specifically addressed. Generally, the single expression of pUL53 shows an even distribution throughout the nucleus (Figure S3A, panel i), whereas pUL50-pUL53 coexpression is typically characterized by a prominent nuclear rim colocalization (Figure S3A, panels d–f), so that this phenotype was analyzed for all phosphosite mutants. As a clear result, none of the mutants, P×5, M×5, P×14 or M×14, showed a substantial difference compared to the parental pUL50-HA, i.e., the relocalization of pUL53 to the nuclear rim was detected in all cases (Figure S3B,C, panels g–r). Some variations in the speckled pattern of rim formation were visible, but these were not based on mutant-specific phenotypical alterations. Next, HCMV-infected HFFs (MOI of 0.05, late-stage infection at 7 d p.i.) were analyzed by confocal imaging in an identical manner. In this case, the HCMV ΔUL50 showed very pronounced differences towards the parental WT strain: (i) pUL53 completely lacked the nuclear rim colocalization, but showed an even distribution throughout the nucleus, (ii) the viral DNA polymerase processivity factor pUL44 lacked its typical accumulation in viral nuclear replication compartments, but showed a mostly diffuse staining pattern, and (iii) MCP similarly lacked a replication compartment-specific accumulation (normally enhanced at their edge areas), but was also mostly diffuse (Figure 4A–C, upper four lines of panels). Thus, these three viral proteins were drastically delocalized in cells infected with the rescue HCMV ΔUL50. In this context, it should be emphasized that there was a clear difference between infection of normal HFF and HFF-UL50 +Dox cells in the localization patterns of these three proteins (Figure 4A–C, panels 9–16 each), concluding that the pUL50-complementing cells rescued the phenotype of the ΔUL50 mutant. Of note, other viral proteins did not show similar changes, like IE1 and pUL97 (both nuclear) or pp28 (cytoplasmic cVAC structures; Figure 4D,E, upper four lines of panels). This indicates that this mutant is able to transiently produce viral proteins belonging to all replicative stages from immediate early to late, but that the regulation of intracellular protein localization and transport is massively disturbed.

As far as cellular egress-relevant proteins were concerned, we analyzed lamin A/C, Ser22-phosphorylated lamin A/C and emerin (Figure S4A–C, upper four lines of panels). Interestingly none of these proteins showed virus-specific alterations when comparing HCMV ΔUL50 and the phosphosite mutants with the WT virus. It was striking to see that infection with P×14 and M×14 resulted in protein localization patterns, for all viral and cellular proteins analyzed, that we considered uneventful and indistinguishable from WT or the respective revertant Rev2 (Figure 4A–E and Figure S4A–C, lower three lines of panels). Combined these confocal imaging data, in the case of HCMV ΔUL50 mutant revealed marked characteristics of altered protein localization patterns, specifically for three viral proteins, but surprisingly not for the mutants carrying 14 phosphosite replacements.

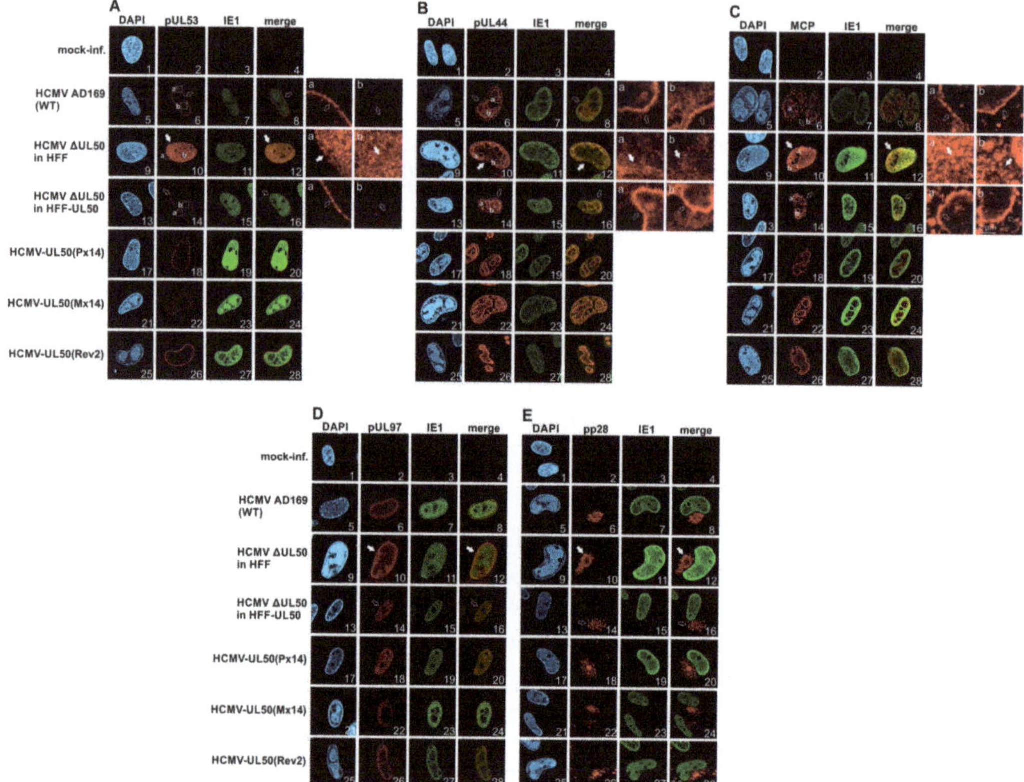

Figure 4. Confocal imaging analysis of the localization of viral proteins produced by HCMV ΔUL50 or pUL50 phosphosite mutations. Normal HFFs or pUL50-complementing cells (HFF-UL50 +Dox) were used for infection with the recombinant HCMVs at a MOI of 0.05 and harvested at 7 d p.i. Immunofluorescence staining was performed with antibodies against the indicated proteins and representative panels of confocal imaging are given (see scale bar in panel E, picture 28). (**A–C**) Viral proteins pUL53, pUL44 and MCP, showing altered localization patterns for HCMV ΔUL50 in HFF (see insets a and b and the phenotype of altered localization marked by filled white arrows, compared to normal/unaltered localization marked by framed white arrows). (**D,E**) Viral proteins pUL97, IE1 and pp28, showing no alteration in localization patterns. Additional information is provided by Figure S4.

3.4. Characteristic Differences in the Replication Kinetics between Recombinant HCMVs Carrying ORF-UL50 Deletion or pUL50 Phosphosite Mutations

The investigation of viral replication kinetics revealed some unexpected findings. Firstly, we used the ΔUL50 virus and performed a multistep replication curve analysis, using both qPCR and ddPCR measurements of the same samples, derived either from infectious media supernatants or cellular lysates after HCMV infection at different MOIs. When using HFF-UL50 complementing cells, comparing the conditions of induced (+Dox, pUL50-positive) and uninduced (−Dox, pUL50-negative), no difference was observed for the parental WT by measuring supernatant samples with ddPCR. However, a clear difference between the two conditions was seen for the ΔUL50 virus (Figure 5A). The induced sample showed a strong increase in signals compared to the uninduced setting, thus indicating a rise of viral genomic load and viral replication over time. Both ddPCR curves were significantly lower than the WT reference curve. Interestingly, however, the curve of the ΔUL50 virus did not indicate a complete lack of replication, but instead showed a slight and transient increase of viral load between 2 and 10 d p.i., which came only to a plateau at later time points (14 d p.i.; Figure 5A). Data could be confirmed for these samples by using qPCR (Figure S5A,B). A very similar finding was obtained by using normal HFFs for infection, either at MOI 0.01 or 0.001 (Figure 5B, ddPCR; see reproduction of data by qPCR in Figure S5C,D), and likewise supernatants and cellular lysates from MOI 0.01 in parallel (Figure 5C, qPCR). In all cases, signals of the ΔUL50 virus remained significantly lower than those of the parental WT, but nevertheless showed at least a slight increase of genomic loads after 4 d p.i. (Figure 5C). Thus, the ddPCR and qPCR data further illustrated a finding described already earlier [19]. The ΔUL50 rescue virions, even in the absence of pUL50-complementing cells, are not fully replication incompetent, but show some residual capacity of transient genome replication over a period of two to three viral replication cycles.

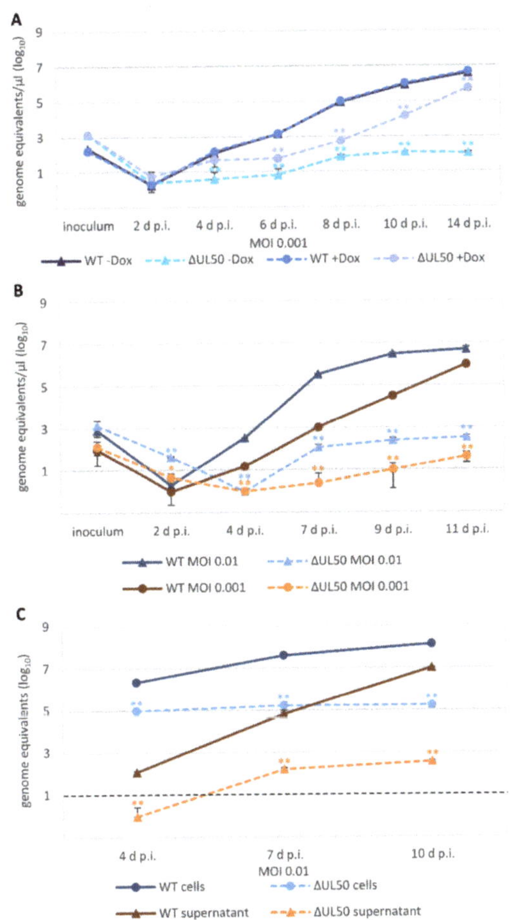

Figure 5. HCMV replication kinetics of the ORF-UL50 deletion mutant. HFF-UL50 cells (**A**) or HFFs (**B**,**C**) were infected with parental HCMV AD169 (WT) or recombinant HCMV ΔUL50 at a MOI of 0.01 or 0.001 as indicated. The expression of pUL50 in HFF-UL50 cells was either uninduced (−Dox) or induced (+Dox). Viral supernatants (**A**–**C**) or cells (**C**) were harvested at the indicated time points and viral genome equivalents were determined by IE1-specific ddPCR (**A**,**B**) or qPCR (**C**). Each infection was performed at least in triplicates; mean values and standard deviations are shown. The significance is calculated relating to WT (solid lines). The standard containing 10 HCMV DNA copies reached the cycle threshold at cycle 38 and was therefore defined as the limit of detection for qPCR, as shown by the black dashed line. ddPCR is a binary end-point measurement and can detect one single copy; *, $p \leq 0.05$; **, $p \leq 0.01$.

All the ORF-UL50 phosphosite mutants showed almost WT-like replication kinetics and efficiency, indistinguishable from the control virus qPCR curves, so that no statistically significant impairment of progeny virus production was determined (Figure 6). This was true for both the P×5 and M×5 pair as well as the P×14 and M×14 pair (Figure 6A–C), for MOIs of 0.01 and 0.001 (reproduced by ddPCR with very similar results; data not shown), and for the use of supernatants parallel to cellular lysates as a PCR template (Figure 6D). Thus, the combined findings for the viral phosphosite replacement mutants do not suggest phenotypic alteration of viral replication kinetics.

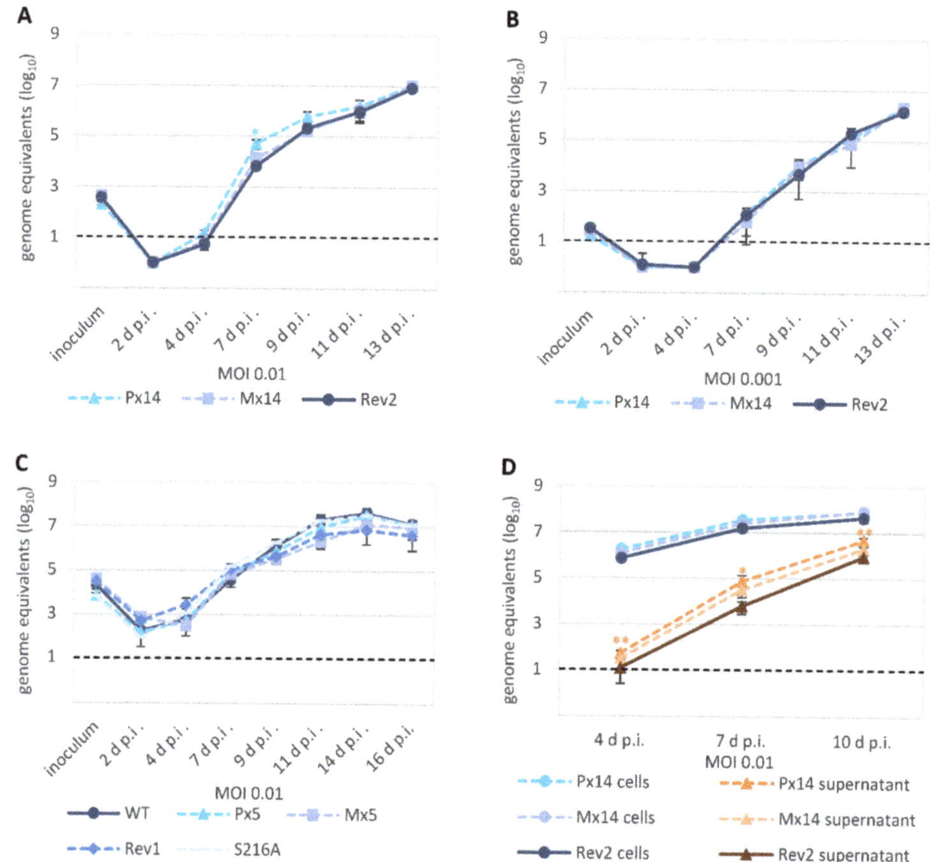

Figure 6. HCMV replication kinetics of the pUL50 phosphosite mutants. HFFs were infected with parental HCMV AD169 (WT) or recombinant HCMV phosphosite mutants P×14, M×14, Rev2, P×5, M×5, Rev1 or S216A at a MOI of 0.01 or 0.001 as indicated. Viral supernatants (**A–D**) or cells (**D**) were harvested at the indicated time points and viral genome equivalents were determined by IE1-specific qPCR. Each infection was performed at least in triplicates; mean values and standard deviations are shown. The significance is calculated relating to Rev2 or WT (solid lines). The standard containing 10 HCMV DNA copies reached the cycle threshold at cycle 38 and was therefore defined as the limit of detection for qPCR, as shown by the black dashed line. *, $p \leq 0.05$; **, $p \leq 0.01$.

3.5. Electron Microscopic Analysis of HFFs Infected with the Recombinant HCMVs

In order to assess the phenotype of viral nuclear capsid formation for WT, mutant and revertant HCMVs, transmission electron microscopy (TEM) analysis was performed (Figure 7 and Figure S6). Ultrathin sections of HCMV-infected HFFs were stained with uranyl acetate and lead citrate to visualize the nuclear types A, B and C of viral capsids (Figure 7A). Analysis was performed in two experimental settings (I, II) and a quantitation of capsid numbers was conducted by counting several nuclear sections for each virus (Figure 8A,B). In the majority, intermediate type B capsids were detected for WT, revertants as well as all phosphosite mutants (ranging between $33 \pm 10\%$ and $61 \pm 5\%$ for the individual viruses). Lower percentages were obtained for immature type A capsids ($5 \pm 0\%$ to $26 \pm 9\%$) and mature type C capsids (19 ± 4 to $45 \pm 3\%$). Some degree of variation of percentages was noted between the two experimental settings, as expressed by the values of WT (I), WT (II), Rev1 and Rev2. No clear phenotypical alteration was found for any of the phosphosite mutants $P \times 5$, $M \times 5$, S216A, $P \times 14$ or $M \times 14$, neither in quantitative nor in qualitative terms (Figure 7, Figure 8 and Figure S6). Strikingly, however, the deletion mutant ΔUL50 showed a profoundly different pattern of capsids than WT. While the conditions of pUL50 complementation (+Dox) yielded capsid numbers for ΔUL50 in the range of other viruses, conditions of lack of pUL50 ($-$Dox) showed a drastic change in two aspects: first, the ΔUL50 mutant was characterized by a strong accumulation and entrapment of capsids at the nucleoplasmic proximity of the nuclear envelope (Figure 7C, NE); second, the quantitation revealed a strong abundance of immature A type capsids (Figure 8A,B, $88 \pm 5\%$) and a drastic reduction of both B and C types ($9 \pm 4\%$ and $4 \pm 3\%$, respectively). This finding clearly confirmed the replicative defect of the ΔUL50 mutant, i.e., a lack of efficient nuclear capsid egress. The defect could be relieved by pUL50 complementation in +Dox cells. Obviously, the entrapment of intranuclear A and B capsids, aligned along the NE, was linked to an inefficiency of capsid maturation. No similar defect was observed for any of the phosphosite mutants (Figures 7 and 8). Referring to the report by Krosky et al. [36], in which an EM-based quantitation of nuclear versus cytoplasmic capsids was described, we were not able to perform a similar quantitative assessment due to the very low number of detectable capsids in the cytoplasm of all samples (Figure S6). Combined our data strongly supported the conclusion that HCMV nuclear egress is massively hampered by ORF-UL50 deletion, but is poorly affected by the pUL50 phosphosite mutations analyzed in this study.

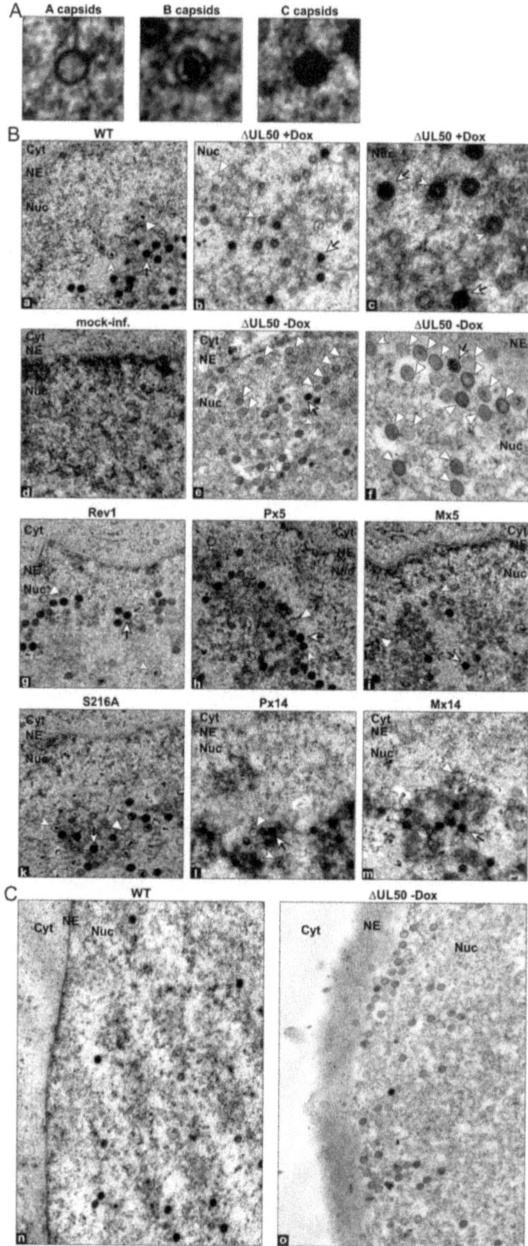

Figure 7. Electron microscopic analysis of viral nuclear capsid formation. HFFs were infected with parental WT, mutant or revertant HCMVs as indicated at a MOI adapted to maintain cell viability for harvesting and fixation at 4–7 d p.i., to be subjected to sectioning and negative staining in TEM. (**A**) Representatives of A-, B- and C-capsids. (**B**) Representative nuclei of infected cells at a 16,700-fold or 33,400-fold (c,f) magnification. The distribution of A- (white triangle), B- (white arrowhead) and C-capsids (white arrowhead with black tail) is depicted. NE, nuclear envelope; Cyt, cytoplasm; Nuc; nucleus. (**C**) Accumulation of capsids of HCMV ΔUL50 in close proximity to the nuclear envelope in contrast to a more homogeneous distribution of the WT (25,050-fold magnification).

A Quantitation of nuclear capsids in HCMV-infected fibroblasts

	Virus	Total no. of capsids (analyzed pictures)	% of A, B, C capsids		
			A	**B**	**C**
(I)	WT (AD169)	270 ± 6 (9)	5 ± 0	50 ± 3	45 ± 3
	ΔUL50 + Dox	274 ± 13 (9)	16 ± 1	61 ± 5	22 ± 5
	ΔUL50 - Dox	232 ± 23 (7)	88 ± 5	9 ± 4	4 ± 3
	Px14	431 ± 22 (4)	19 ± 3	49 ± 9	32 ± 10
	Mx14	315 ± 32 (3)	21 ± 2	50 ± 9	29 ± 8
	Rev2	274 ± 22 (3)	24 ± 3	57 ± 7	19 ± 4
(II)	WT (AD169)	231 ± 47 (4)	26 ± 9	33 ± 10	41 ± 1
	Px5	221 ± 48 (4)	24 ± 7	41 ± 16	35 ± 9
	Mx5	144 ± 14 (4)	21 ± 8	53 ± 9	26 ± 1
	Rev1	202 ± 28 (4)	15 ± 2	52 ± 7	34 ± 5
	S216A	272 ± 29 (4)	23 ± 7	42 ± 17	35 ± 10

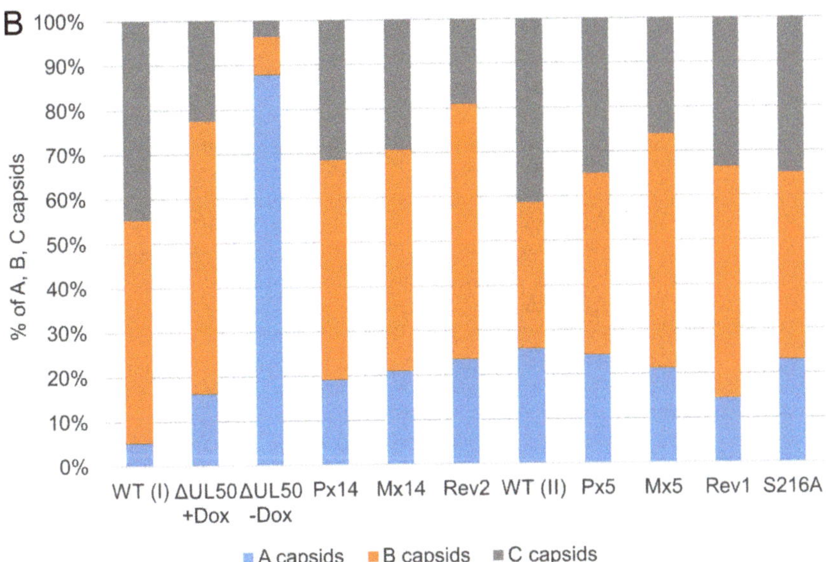

Figure 8. Quantitative evaluation of intranuclear capsid types. For the details of infection (4–7 d.p.i.) and sample preparation see Figure 7. (**A**) TEM counting was performed individually by researchers in duplicate (*n* = 4–8), using up to nine nuclear sections for each virus. Mean values ± SD of A, B and C capsids are given, as set in percentage to the total number of capsids counted. (**B**) Diagram representation of the mean values given above.

4. Discussion

In the present study we focused on the investigation of newly generated recombinant HCMVs carrying ORF-UL50 deletion or pUL50 phosphosite mutations. Their putative phenotypical changes in nuclear egress should be addressed and the rationale behind this, i.e., the question whether pUL50 phosphosite mutations would markedly alter viral replication efficiencies, originated from our previous proteomics-based identification of 14 distinct detectable phosphosites. The main findings are the following: (i) only the ORF-UL50 deletion rescue virus showed a block of genome synthesis at late infection periods (viral replication kinetics determined by qPCR/ddPCR), whereas the phosphosite mutants displayed largely unchanged kinetics, (ii) neither ORF-UL50 deletion nor phosphosite replacements led to drastic changes in viral protein expression patterns, (iii) confocal imaging of NEC colocalization and CoIP-based analysis of NEC interaction revealed three distinct protein examples (pUL53, pUL44, MCP) of differences in intracellular localization compared to WT for ORF-UL50 deletion, but not for phosphosite mutants, and (iv) TEM analysis revealed a strong abundance and accumulation of immature A-capsids for ΔUL50 in contrast to WT and pUL50 phosphosite mutants. On this basis, the current data underline the strong phenotypical changes of the ΔUL50 virus, whereas viruses carrying multiple pUL50 phosphosite mutations were found largely unaltered.

Generally, it is widely accepted knowledge that nuclear egress is determined by the functionality of herpesviral NEC proteins and represents an essential and rate-limiting step of viral replication efficiency (reviewed by [1,3–5,37–41]). A number of molecular characteristics have been identified for herpesviral core NEC proteins, among which both identical features and differences could be pointed out [1,5,9,10,42]. Especially for the example of HCMV pUL50-pUL53, at least three important functions have been experimentally described or postulated so far, namely the recruitment of NEC-associated effectors, the reorganization of the nuclear envelope and the NEC-docking and transition of nuclear capsids. Hereby, the role of regulatory phosphorylation of NEC-associated proteins moved more and more into the focus of our interest [43]. It should be stressed that most, if not all, of the HCMV core NEC and NEC-associated proteins are either subject to massive phosphorylation or represent NEC-/lamina-phosphorylating protein kinases themselves. A previous report described a replacement of S216A in pUL50 together with S19A in pUL53 resulting in the loss of efficient NEC rim location and viral replication [7]. Although this study showed some methodological differences compared to our data (e.g., different cell types used for transient transfection or HCMV infection, respectively, in addition pUL50/pUL53 double-mutant instead of pUL50 single-mutant), divergent results could not be explained at this stage. In order to address the apparent discrepancy, we reconstituted five BAC-derived HCMVs harboring various phosphosite patterns of pUL50. Confirming our initial results, we could not identify a functional impact of the pUL50 phosphosites under conditions of HCMV multiround replication in cultured primary fibroblasts. Although three parameters were intensively studied for these mutants, i.e., viral replication kinetics and protein production, NEC protein association and nuclear lamin phosphorylation and viral nuclear capsid egress (TEM analysis), no functional impairment during any stage of the viral replication cycle could be detected. Thus, we conclude that although pUL50 is heavily phosphorylated by viral and cellular kinases, our findings so far cannot substantiate the importance of phosphorylation at the analyzed 14 phosphosites in pUL50 for NEC functionality [7,8,23,44].

Combined, the findings suggest that alternative pUL50 phosphorylation was either not detectable or phosphorylation may generally not be functionally essential. This situation may prompt to speculate about the importance of additional minor phosphosites in pUL50 or, even more likely, on the basis of current data, in pUL53 [7,8]. The observations published so far strongly suggest that NEC is regulated by phosphorylation [1], even though phosphorylation of pUL50 itself had no effect on viral replication as described in this report. As a commonly accepted feature, the main regulatory protein kinase of the HCMV-specific NEC is the viral kinase pUL97. Its main NEC-associated viral and cellular phosphorylated substrates have been identified as the pUL50-pUL53 complex, pUL97 autophosphorylation, pp65 with unknown NEC impact, the multiligand-binding bridging factor p32/gC1qR, emerin and nuclear lamins A/C [45]. Additional cellular kinases, such as CDK1, can contribute to the phosphorylation of the core NEC and NEC-associated proteins [8,23,25,44]. To date, however, with the exception of lamin A/C phosphorylation, the regulatory impact of NEC-specific phosphorylation activities has not been clarified.

Considering this aspect of functional relevance of phosphosites beyond NEC regulation, multiple events of protein phosphorylation during human herpesviral and non-herpesviral infections were described for a variety of viruses, some of these with or without functional importance [46–50]. Radestock and colleagues demonstrated that the predominant HIV-1 phosphoprotein 6 (p6) comprised several phosphosites, but their substitution had no effect on p6 functions including virus release [48]. Regarding the functional relevance of multiple-site phosphorylation of nuclear proteins more related to our study, examples have been described for two human and animal α-herpesviruses. The bovine herpesvirus 1 (BoHV-1) nuclear protein VP8 was found to be phosphorylated by both cellular and viral kinases at various sites, thereby promoting viral replication [50]. With regard to core NEC proteins, the phosphorylation of six N-terminal serine residues of HSV-1 pUL31 by viral kinase pUS3 was described. These sites were characterized as indispensable for capsid release from the nucleus and thus their functional relevance during HSV-1 infection was postulated [51]. More recent data implied that some of these residues within the extreme N-terminus of pUL31 might be crucial for nuclear egress. Substitutions of four phosphorylated residues resulted in capsid accumulations adjacent to the inner nuclear membrane, whereas single substitutions had no detectable effect [52].

Intriguingly, the major phosphosites are not positionally conserved throughout the pUL50 protein family in α-, β- and γ-herpesviruses. Alignments revealed that HCMV major phosphosites seem to be exclusively present in pUL50 but not in other herpesviral homologs [14]. However, our sequence analysis indicated some partial homologies; i.e., S216 of HCMV pUL50 was found in human herpesvirus 6 and 7 (HHV-6 and HHV-7; S216) homologs but not in MCMV. Some additional alignment patterns such as the proline-directed kinase motif, comprising a serine residue with a subsequent proline (Ser-Pro), have been identified in VZV (S209) and PrV (S194) homologs. These short sequence sections point to a possible substitutional relevance of alternative serine residues within the respective pUL50 homologs. Moreover, the relevant phosphosites identified so far are predominantly located in non-globular regions of pUL50; thus, positions of putative phosphosites within pUL50 homologs might differ and are not easily detectable by alignment approaches.

Combined, this report strongly suggests the conserved functional importance of the cytomegalovirus core NEC proteins. In particular, our data underline the central, albeit not fully essential, role of pUL50 as a multi-interacting determinant of the HCMV core NEC, at least considering the efficiency of multiple rounds of viral lytic replication. The study suggests a refined model of pUL50 functionality, in which the regulation of viral replication, specifically the nuclear egress, seems to allow some compensatory measures to balance alterations in the expression and phosphorylation levels of pUL50. It has not escaped our notice that pUL50 is not only present at low copy numbers in infectious virions [53], but also comprises additional, NEC-independent activities [54,55]. Thus, more detailed studies on the virion-associated state of pUL50 may help to understand this multifaceted regulatory aspect in the near future.

Supplementary Materials: The following are available online at https://www.mdpi.com/1999-4915/13/2/165/s1. Figure S1, Schematic representation of the generation of recombinant HCMVs; Figure S2, CoIP-based interaction analysis of pUL50 phophosite mutants; Figure S3, Confocal imaging analysis of the localization of pUL50 and pUL53 of transfected phosphosite mutants; Figure S4, Confocal imaging analysis of the localization of cellular proteins produced by HCMV ΔUL50 or pUL50 phosphosite mutations; Figure S5, HCMV replication kinetics of the ORF-UL50 deletion mutant; Figure S6, Electron microscopic investigation of recombinant viruses; Figure S7: Sequence alignments of ORFs UL50, UL53 and UL97 of reconstituted viruses.

Author Contributions: Conceptualization, M.M., E.S., J.M., S.H.; methodology, S.H., E.S., A.S., E.M.B., A.-C.S., D.H., R.M., B.K., U.S.-S., M.M.; validation, S.H., E.S., D.H., R.M., U.S.-S., M.M.; formal analysis, S.H., E.S., D.H., R.M., U.S.-S., M.M.; investigation, S.H., E.S., A.S., E.M.B., A.-C.S., D.H., R.M., B.K., J.M., T.S., U.S.-S., M.M.; resources, M.M., E.M.B., T.S., U.S.-S.; data curation, S.H., E.S., D.H., R.M., U.S.-S., M.M.; writing—original draft preparation, M.M., S.H., E.S.; writing—review and editing, S.H., E.S., T.S., M.M.; supervision, M.M., E.M.B., J.M., T.S., U.S.-S.; project administration,

M.M.; funding acquisition, M.M. All authors have read and agreed to the published version of the manuscript.

Funding: This research was funded by Deutsche Forschungsgemeinschaft (DFG grants SFB796/C3, MA1289/11-1, MI2143/2-1) and Deutscher Akademischer Austauschdienst (DAAD-Go8 Marschall/Rawlinson 2017–18, 2020–21).

Data Availability Statement: Not Applicable.

Acknowledgments: The authors express their gratitude to the research groups of Jutta Eichler, Heinrich Sticht and Yves Muller (all FAU, Erlangen-Nürnberg) for very helpful scientific exchange and support in the joint nuclear egress project, to Jintawee Kicuntod, Martin Schütz, Josephine Lösing and Christina Wangen (all research group M.M.) for help in the evaluation of EM and confocal images, to William Rawlinson (Serology and Virology Division, Prince of Wales Hospital, Sydney, Australia) and his research group, particularly Stuart Hamilton, Diana Wong and Ece Egilmezer, for long-term cooperation, Patrick König (Virology, Univ. Ulm), Bodo Plachter (Virology, Univ. Mainz), Michael Mach (Virology, FAU Erlangen-Nürnberg) and Peter Lischka (AiCuris Anti-infective Cures GmbH, Wuppertal, Germany) for methodological contributions, long-term support and stimulating scientific discussion, to Anne Wang and colleagues, Shanghai ShineGene Molecular Biotech, Inc. (Shanghai, China) for generating UTCs including site-directed mutagenesis, to Stipan Jonjic, Tihana Lenac Roviš (Dept. Histology & Embryology, Univ. Rijeka, Croatia) and William Britt (Virology, Univ. Birmingham, AL, USA) for the generous supply with HCMV-specific monoclonal antibodies, and to Victoria Jackiw (Language Center, FAU Erlangen-Nürnberg) for reading the manuscript.

Conflicts of Interest: The authors declare no conflict of interest. The funders had no role in the design of the study; in the collection, analyses, or interpretation of data; in the writing of the manuscript, or in the decision to publish the results.

References

1. Marschall, M.; Muller, Y.A.; Diewald, B.; Sticht, H.; Milbradt, J. The human cytomegalovirus nuclear egress complex unites multiple functions: Recruitment of effectors, nuclear envelope rearrangement, and docking to nuclear capsids. *Rev. Med. Virol.* **2017**, *27*. [CrossRef] [PubMed]
2. Hage, S.; Sonntag, E.; Borst, E.M.; Tannig, P.; Seyler, L.; Bauerle, T.; Bailer, S.M.; Lee, C.P.; Muller, R.; Wangen, C.; et al. Patterns of Autologous and Nonautologous Interactions Between Core Nuclear Egress Complex (NEC) Proteins of alpha-, beta- and gamma-Herpesviruses. *Viruses* **2020**, *12*, 303. [CrossRef] [PubMed]
3. Marschall, M.; Häge, S.; Conrad, M.; Alkhashrom, S.; Kicuntod, J.; Schweininger, J.; Kriegel, M.; Lösing, J.; Tillmanns, J.; Neipel, F.; et al. Nuclear Egress Complexes of HCMV and Other Herpesviruses: Solving the Puzzle of Sequence Coevolution, Conserved Structures and Subfamily-Spanning Binding Properties. *Viruses* **2020**, *12*, 683. [CrossRef] [PubMed]
4. Draganova, E.B.; Thorsen, M.K.; Heldwein, E.E. Nuclear Egress. *Curr. Issues Mol. Biol.* **2020**, *41*, 125–170. [CrossRef]
5. Lye, M.F.; Wilkie, A.R.; Filman, D.J.; Hogle, J.M.; Coen, D.M. Getting to and through the inner nuclear membrane during herpesvirus nuclear egress. *Curr. Opin. Cell Biol.* **2017**, *46*, 9–16. [CrossRef]
6. Milbradt, J.; Sonntag, E.; Wagner, S.; Strojan, H.; Wangen, C.; Lenac Rovis, T.; Lisnic, B.; Jonjic, S.; Sticht, H.; Britt, W.J.; et al. Human Cytomegalovirus Nuclear Capsids Associate with the Core Nuclear Egress Complex and the Viral Protein Kinase pUL97. *Viruses* **2018**, *10*, 35. [CrossRef]
7. Sharma, M.; Bender, B.J.; Kamil, J.P.; Lye, M.F.; Pesola, J.M.; Reim, N.I.; Hogle, J.M.; Coen, D.M. Human cytomegalovirus UL97 phosphorylates the viral nuclear egress complex. *J. Virol.* **2015**, *89*, 523–534. [CrossRef]
8. Sonntag, E.; Milbradt, J.; Svrlanska, A.; Strojan, H.; Hage, S.; Kraut, A.; Hesse, A.M.; Amin, B.; Sonnewald, U.; Coute, Y.; et al. Protein kinases responsible for the phosphorylation of the nuclear egress core complex of human cytomegalovirus. *J. Gen. Virol.* **2017**, *98*, 2569–2581. [CrossRef]
9. Mettenleiter, T.C.; Muller, F.; Granzow, H.; Klupp, B.G. The way out: What we know and do not know about herpesvirus nuclear egress. *Cell Microbiol.* **2013**, *15*, 170–178. [CrossRef]
10. Bigalke, J.M.; Heldwein, E.E. Structural basis of membrane budding by the nuclear egress complex of herpesviruses. *EMBO J.* **2015**, *34*, 2921–2936. [CrossRef]
11. Hagen, C.; Dent, K.C.; Zeev-Ben-Mordehai, T.; Grange, M.; Bosse, J.B.; Whittle, C.; Klupp, B.G.; Siebert, C.A.; Vasishtan, D.; Bauerlein, F.J.; et al. Structural Basis of Vesicle Formation at the Inner Nuclear Membrane. *Cell* **2015**, *163*, 1692–1701. [CrossRef] [PubMed]
12. Leigh, K.E.; Sharma, M.; Mansueto, M.S.; Boeszoermenyi, A.; Filman, D.J.; Hogle, J.M.; Wagner, G.; Coen, D.M.; Arthanari, H. Structure of a herpesvirus nuclear egress complex subunit reveals an interaction groove that is essential for viral replication. *Proc. Natl. Acad. Sci. USA* **2015**, *112*, 9010–9015. [CrossRef] [PubMed]
13. Lye, M.F.; Sharma, M.; El Omari, K.; Filman, D.J.; Schuermann, J.P.; Hogle, J.M.; Coen, D.M. Unexpected features and mechanism of heterodimer formation of a herpesvirus nuclear egress complex. *EMBO J.* **2015**, *34*, 2937–2952. [CrossRef] [PubMed]

14. Walzer, S.A.; Egerer-Sieber, C.; Sticht, H.; Sevvana, M.; Hohl, K.; Milbradt, J.; Muller, Y.A.; Marschall, M. Crystal Structure of the Human Cytomegalovirus pUL50-pUL53 Core Nuclear Egress Complex Provides Insight into a Unique Assembly Scaffold for Virus-Host Protein Interactions. *J. Biol. Chem.* **2015**, *290*, 27452–27458. [CrossRef]

15. Zeev-Ben-Mordehai, T.; Weberruss, M.; Lorenz, M.; Cheleski, J.; Hellberg, T.; Whittle, C.; El Omari, K.; Vasishtan, D.; Dent, K.C.; Harlos, K.; et al. Crystal Structure of the Herpesvirus Nuclear Egress Complex Provides Insights into Inner Nuclear Membrane Remodeling. *Cell Rep.* **2015**, *13*, 2645–2652. [CrossRef]

16. Muller, Y.A.; Hage, S.; Alkhashrom, S.; Hollriegl, T.; Weigert, S.; Dolles, S.; Hof, K.; Walzer, S.A.; Egerer-Sieber, C.; Conrad, M.; et al. High-resolution crystal structures of two prototypical beta- and gamma-herpesviral nuclear egress complexes unravel the determinants of subfamily specificity. *J. Biol. Chem.* **2020**. [CrossRef]

17. Schnee, M.; Ruzsics, Z.; Bubeck, A.; Koszinowski, U.H. Common and specific properties of herpesvirus UL34/UL31 protein family members revealed by protein complementation assay. *J. Virol.* **2006**, *80*, 11658–11666. [CrossRef]

18. Sonntag, E.; Hamilton, S.T.; Bahsi, H.; Wagner, S.; Jonjic, S.; Rawlinson, W.D.; Marschall, M.; Milbradt, J. Cytomegalovirus pUL50 is the multi-interacting determinant of the core nuclear egress complex (NEC) that recruits cellular accessory NEC components. *J. Gen. Virol.* **2016**, *97*, 1676–1685. [CrossRef]

19. Häge, S.; Horsch, D.; Stilp, A.C.; Kicuntod, J.; Müller, R.; Hamilton, S.T.; Egilmezer, E.; Rawlinson, W.D.; Stamminger, T.; Sonntag, E.; et al. A quantitative nuclear egress assay to investigate the nucleocytoplasmic capsid release of human cytomegalovirus. *J. Virol. Methods* **2020**, *283*, 113909. [CrossRef]

20. Klenovsek, K.; Weisel, F.; Schneider, A.; Appelt, U.; Jonjic, S.; Messerle, M.; Bradel-Tretheway, B.; Winkler, T.H.; Mach, M. Protection from CMV infection in immunodeficient hosts by adoptive transfer of memory B cells. *Blood* **2007**, *110*, 3472–3479. [CrossRef]

21. Lorz, K.; Hofmann, H.; Berndt, A.; Tavalai, N.; Mueller, R.; Schlotzer-Schrehardt, U.; Stamminger, T. Deletion of open reading frame UL26 from the human cytomegalovirus genome results in reduced viral growth, which involves impaired stability of viral particles. *J. Virol.* **2006**, *80*, 5423–5434. [CrossRef] [PubMed]

22. Tischer, B.K.; von Einem, J.; Kaufer, B.; Osterrieder, N. Two-step red-mediated recombination for versatile high-efficiency markerless DNA manipulation in Escherichia coli. *Biotechniques* **2006**, *40*, 191–197. [PubMed]

23. Milbradt, J.; Auerochs, S.; Marschall, M. Cytomegaloviral proteins pUL50 and pUL53 are associated with the nuclear lamina and interact with cellular protein kinase C. *J. Gen. Virol.* **2007**, *88*, 2642–2650. [CrossRef] [PubMed]

24. Milbradt, J.; Auerochs, S.; Sticht, H.; Marschall, M. Cytomegaloviral proteins that associate with the nuclear lamina: Components of a postulated nuclear egress complex. *J. Gen. Virol.* **2009**, *90*, 579–590. [CrossRef] [PubMed]

25. Marschall, M.; Marzi, A.; aus dem Siepen, P.; Jochmann, R.; Kalmer, M.; Auerochs, S.; Lischka, P.; Leis, M.; Stamminger, T. Cellular p32 recruits cytomegalovirus kinase pUL97 to redistribute the nuclear lamina. *J. Biol. Chem.* **2005**, *280*, 33357–33367. [CrossRef]

26. Tischer, B.K.; Smith, G.A.; Osterrieder, N. En passant mutagenesis: A two step markerless red recombination system. *Methods Mol. Biol.* **2010**, *634*, 421–430. [CrossRef]

27. Schmeiser, C.; Borst, E.; Sticht, H.; Marschall, M.; Milbradt, J. The cytomegalovirus egress proteins pUL50 and pUL53 are translocated to the nuclear envelope through two distinct modes of nuclear import. *J. Gen. Virol.* **2013**, *94*, 2056–2069. [CrossRef]

28. Webel, R.; Milbradt, J.; Auerochs, S.; Schregel, V.; Held, C.; Nöbauer, K.; Razzazi-Fazeli, E.; Jardin, C.; Wittenberg, T.; Sticht, H.; et al. Two isoforms of the protein kinase pUL97 of human cytomegalovirus are differentially regulated in their nuclear translocation. *J. Gen. Virol.* **2011**, *92*, 638–649. [CrossRef]

29. Kinoshita, E.; Kinoshita-Kikuta, E.; Takiyama, K.; Koike, T. Phosphate-binding tag, a new tool to visualize phosphorylated proteins. *Mol. Cell Proteom.* **2006**, *5*, 749–757. [CrossRef]

30. Sonntag, E.; Hahn, F.; Bertzbach, L.D.; Seyler, L.; Wangen, C.; Muller, R.; Tannig, P.; Grau, B.; Baumann, M.; Zent, E.; et al. In Vivo proof-of-concept for two experimental antiviral drugs, both directed to cellular targets, using a murine cytomegalovirus model. *Antivir. Res* **2019**, *161*, 63–69. [CrossRef]

31. Hindson, B.J.; Ness, K.D.; Masquelier, D.A.; Belgrader, P.; Heredia, N.J.; Makarewicz, A.J.; Bright, I.J.; Lucero, M.Y.; Hiddessen, A.L.; Legler, T.C.; et al. High-throughput droplet digital PCR system for absolute quantitation of DNA copy number. *Anal. Chem.* **2011**, *83*, 8604–8610. [CrossRef] [PubMed]

32. Pinheiro, L.B.; Coleman, V.A.; Hindson, C.M.; Herrmann, J.; Hindson, B.J.; Bhat, S.; Emslie, K.R. Evaluation of a droplet digital polymerase chain reaction format for DNA copy number quantification. *Anal. Chem.* **2012**, *84*, 1003–1011. [CrossRef] [PubMed]

33. Milbradt, J.; Hutterer, C.; Bahsi, H.; Wagner, S.; Sonntag, E.; Horn, A.H.; Kaufer, B.B.; Mori, Y.; Sticht, H.; Fossen, T.; et al. The Prolyl Isomerase Pin1 Promotes the Herpesvirus-Induced Phosphorylation-Dependent Disassembly of the Nuclear Lamina Required for Nucleocytoplasmic Egress. *PLoS Pathog.* **2016**, *12*, e1005825. [CrossRef] [PubMed]

34. Milbradt, J.; Webel, R.; Auerochs, S.; Sticht, H.; Marschall, M. Novel mode of phosphorylation-triggered reorganization of the nuclear lamina during nuclear egress of human cytomegalovirus. *J. Biol. Chem.* **2010**, *285*, 13979–13989. [CrossRef] [PubMed]

35. Schütz, M.; Thomas, M.; Wangen, C.; Wagner, S.; Rauschert, L.; Errerd, T.; Kießling, M.; Sticht, H.; Milbradt, J.; Marschall, M. The peptidyl-prolyl cis/trans isomerase Pin1 interacts with three early regulatory proteins of human cytomegalovirus. *Virus Res.* **2020**, *285*, 198023. [CrossRef]

36. Krosky, P.M.; Baek, M.C.; Coen, D.M. The human cytomegalovirus UL97 protein kinase, an antiviral drug target, is required at the stage of nuclear egress. *J. Virol.* **2003**, *77*, 905–914. [CrossRef] [PubMed]

37. Bigalke, J.M.; Heldwein, E.E. Have NEC Coat, Will Travel: Structural Basis of Membrane Budding During Nuclear Egress in Herpesviruses. *Adv. Virus Res.* **2017**, *97*, 107–141. [CrossRef]
38. Bailer, S.M. Venture from the Interior-Herpesvirus pUL31 Escorts Capsids from Nucleoplasmic Replication Compartments to Sites of Primary Envelopment at the Inner Nuclear Membrane. *Cells* **2017**, *6*, 46. [CrossRef]
39. Hellberg, T.; Passvogel, L.; Schulz, K.S.; Klupp, B.G.; Mettenleiter, T.C. Nuclear Egress of Herpesviruses: The Prototypic Vesicular Nucleocytoplasmic Transport. *Adv. Virus Res.* **2016**, *94*, 81–140. [CrossRef]
40. Mettenleiter, T.C. Breaching the Barrier-The Nuclear Envelope in Virus Infection. *J. Mol. Biol.* **2016**, *428*, 1949–1961. [CrossRef]
41. Lee, C.P.; Chen, M.R. Escape of herpesviruses from the nucleus. *Rev. Med. Virol.* **2010**, *20*, 214–230. [CrossRef] [PubMed]
42. Mettenleiter, T.C.; Klupp, B.G.; Granzow, H. Herpesvirus assembly: An update. *Virus Res.* **2009**, *143*, 222–234. [CrossRef]
43. Marschall, M.; Feichtinger, S.; Milbradt, J. Regulatory roles of protein kinases in cytomegalovirus replication. *Adv. Virus Res.* **2011**, *80*, 69–101. [CrossRef] [PubMed]
44. Oberstein, A.; Perlman, D.H.; Shenk, T.; Terry, L.J. Human cytomegalovirus pUL97 kinase induces global changes in the infected cell phosphoproteome. *Proteomics* **2015**, *15*, 2006–2022. [CrossRef] [PubMed]
45. Steingruber, M.; Marschall, M. The Cytomegalovirus Protein Kinase pUL97: Host Interactions, Regulatory Mechanisms and Antiviral Drug Targeting. *Microorganisms* **2020**, *8*, 515. [CrossRef] [PubMed]
46. Keating, J.A.; Striker, R. Phosphorylation events during viral infections provide potential therapeutic targets. *Rev. Med. Virol.* **2012**, *22*, 166–181. [CrossRef]
47. Keck, F.; Ataey, P.; Amaya, M.; Bailey, C.; Narayanan, A. Phosphorylation of Single Stranded RNA Virus Proteins and Potential for Novel Therapeutic Strategies. *Viruses* **2015**, *7*, 5257–5273. [CrossRef]
48. Radestock, B.; Morales, I.; Rahman, S.A.; Radau, S.; Glass, B.; Zahedi, R.P.; Muller, B.; Krausslich, H.G. Comprehensive mutational analysis reveals p6Gag phosphorylation to be dispensable for HIV-1 morphogenesis and replication. *J. Virol.* **2013**, *87*, 724–734. [CrossRef]
49. Yu, M.; Summers, J. Multiple functions of capsid protein phosphorylation in duck hepatitis B virus replication. *J. Virol.* **1994**, *68*, 4341–4348. [CrossRef]
50. Zhang, K.; Afroz, S.; Brownlie, R.; Snider, M.; van Drunen Littel-van den Hurk, S. Regulation and function of phosphorylation on VP8, the major tegument protein of bovine herpesvirus 1. *J. Virol.* **2015**, *89*, 4598–4611. [CrossRef]
51. Mou, F.; Wills, E.; Baines, J.D. Phosphorylation of the U(L)31 protein of herpes simplex virus 1 by the U(S)3-encoded kinase regulates localization of the nuclear envelopment complex and egress of nucleocapsids. *J. Virol.* **2009**, *83*, 5181–5191. [CrossRef] [PubMed]
52. Klupp, B.G.; Hellberg, T.; Rönfeldt, S.; Franzke, K.; Fuchs, W.; Mettenleiter, T.C. Function of the Nonconserved N-Terminal Domain of Pseudorabies Virus pUL31 in Nuclear Egress. *J. Virol.* **2018**, *92*. [CrossRef] [PubMed]
53. Couté, Y.; Kraut, A.; Zimmermann, C.; Büscher, N.; Hesse, A.M.; Bruley, C.; De Andrea, M.; Wangen, C.; Hahn, F.; Marschall, M.; et al. Mass Spectrometry-Based Characterization of the Virion Proteome, Phosphoproteome, and Associated Kinase Activity of Human Cytomegalovirus. *Microorganisms* **2020**, *8*, 820. [CrossRef] [PubMed]
54. Lee, M.K.; Kim, Y.J.; Kim, Y.E.; Han, T.H.; Milbradt, J.; Marschall, M.; Ahn, J.H. Transmembrane Protein pUL50 of Human Cytomegalovirus Inhibits ISGylation by Downregulating UBE1L. *J. Virol.* **2018**, *92*. [CrossRef]
55. Lee, M.K.; Hyeon, S.; Ahn, J.H. The Human Cytomegalovirus Transmembrane Protein pUL50 Induces Loss of VCP/p97 and Is Regulated by a Small Isoform of pUL50. *J. Virol.* **2020**, *94*. [CrossRef]

Article

Human Cytomegalovirus Nuclear Egress Complex Subunit, UL53, Associates with Capsids and Myosin Va, but Is Not Important for Capsid Localization towards the Nuclear Periphery

Adrian R. Wilkie [1,2], Mayuri Sharma [1], Margaret Coughlin [3,4], Jean M. Pesola [1], Maria Ericsson [4], Jessica L. Lawler [1,2], Rosio Fernandez [1] and Donald M. Coen [1,2,*]

[1] Department of Biological Chemistry and Molecular Pharmacology, Blavatnik Institute, Harvard Medical School, Boston, MA 02115, USA; adrian.r.wilkie@gmail.com (A.R.W.); mmayuri@gmail.com (M.S.); jean_pesola@hms.harvard.edu (J.M.P.); jlawler46@gmail.com (J.L.L.); rof36@pitt.edu (R.F.)

[2] Committee on Virology, Harvard University, Cambridge, MA 02138, USA

[3] Department of Systems Biology, Blavatnik Institute, Harvard Medical School, Boston, MA 02115, USA; margaret_coughlin@hms.harvard.edu

[4] Electron Microscopy Core Facility, Department of Cell Biology, Blavatnik Institute, Harvard Medical School, Boston, MA 02115, USA; maria_ericsson@hms.harvard.edu

* Correspondence: don_coen@hms.harvard.edu

Citation: Wilkie, A.R.; Sharma, M.; Coughlin, M.; Pesola, J.M.; Ericsson, M.; Lawler, J.L.; Fernandez, R.; Coen, D.M. Human Cytomegalovirus Nuclear Egress Complex Subunit, UL53, Associates with Capsids and Myosin Va, but Is Not Important for Capsid Localization towards the Nuclear Periphery. *Viruses* **2022**, *14*, 479. https://doi.org/10.3390/v14030479

Academic Editor: Graciela Andrei

Received: 27 January 2022
Accepted: 23 February 2022
Published: 26 February 2022

Abstract: After herpesviruses encapsidate their genomes in replication compartments (RCs) within the nuclear interior, capsids migrate to the inner nuclear membrane (INM) for nuclear egress. For human cytomegalovirus (HCMV), capsid migration depends at least in part on nuclear myosin Va. It has been reported for certain herpesviruses that the nucleoplasmic subunit of the viral nuclear egress complex (NEC) is important for this migration. To address whether this is true for HCMV, we used mass spectrometry and multiple other methods to investigate associations among the HCMV NEC nucleoplasmic subunit, UL53, myosin Va, major capsid protein, and/or capsids. We also generated complementing cells to derive and test HCMV mutants null for UL53 or the INM NEC subunit, UL50, for their importance for these associations and, using electron microscopy, for intranuclear distribution of capsids. We found modest associations among the proteins tested, which were enhanced in the absence of UL50. However, we found no role for UL53 in the interactions of myosin Va with capsids or the percentage of capsids outside RC-like inclusions in the nucleus. Thus, UL53 associates somewhat with myosin Va and capsids, but, contrary to reports regarding its homologs in other herpesviruses, is not important for migration of capsids towards the INM.

Keywords: capsid migration; human cytomegalovirus; UL53; UL50; myosin Va; major capsid protein; mass spectrometry; virus genetics; complementing cells; null mutants

1. Introduction

Newly assembled herpesvirus capsids translocate from the nucleus to the cytoplasm in a complicated process known as nuclear egress. Nuclear egress includes four distinct steps: (1) migration of capsids from the nuclear interior, where viral genome replication and encapsidation occur within discrete replication compartments (RCs), to the nuclear rim; (2) disruption of the nuclear lamina providing access to the inner nuclear membrane (INM): (3) budding of capsids through the INM (primary envelopment); and (4) de-envelopment at the outer nuclear membrane (reviewed in [1–4]).

Step 1 of this process—migration from the nuclear interior to the nuclear rim—is poorly understood. In one model, as initially proposed for an alphaherpesvirus, herpes simplex virus 1 (HSV-1) [5] capsids move from the nuclear interior towards the nuclear rim by actomyosin-directed transport. For a betaherpesvirus, human cytomegalovirus (HCMV), this model is supported for at least some intranuclear capsid migration by data showing that the virus induces the formation of actin filaments in the nucleus that associate with capsids and myosin Va, that treatment with an actin-depolymerizing drug disrupts these filaments and impairs migration to the nuclear rim and nuclear egress, and that siRNA and a nuclear localized dominant negative mutant that antagonize myosin Va also impair these processes [6,7]. However, how the capsid would interact with the actomyosin machinery to be transported from the nuclear interior to the nuclear rim remains unclear.

Another poorly understood aspect of nuclear egress is the relationship between this first step of nuclear egress and the viral nuclear egress complex (NEC). The NEC consists of two virus-encoded subunits, one that is anchored in the INM, and the other that binds to its nucleoplasmic face. For HCMV, the INM-anchored subunit is UL50 and the nucleoplasmic subunit is UL53. Both subunits are essential for viral replication and nuclear egress [8–10]. The HCMV NEC recruits the viral protein kinase, UL97, to the nuclear rim for phosphorylation and thus disruption of the nuclear lamina [10,11]. Based on work with other herpesvirus NECs, the HCMV NEC also likely orchestrates capsid budding during primary envelopment (reviewed in [1–4]). However, there is evidence that nucleoplasmic subunits of certain herpesvirus NECs participate in processes upstream of events at the nuclear rim, such as DNA packaging, and capsid migration to the nuclear periphery [12–16]. Moreover, homologs of UL53, such as the HSV-1 and pseudorabies virus nucleoplasmic NEC subunit, UL31, have been found to associate with intranuclear capsid proteins [16–18], and recently, some evidence that HCMV UL53 associates with intranuclear capsids was provided [19]. We were intrigued by these reports, as they raised the possibility that HCMV UL53 might interact with both capsids and the actomyosin machinery to mediate migration of HCMV capsids to the nuclear rim.

This study began with a proteomics investigation to identify candidate proteins that interact with UL53. This identified possible associations of UL53 with myosin Va and with capsid proteins, and we found several lines of evidence for modest associations of these components that were consistent with the hypothesis that UL53 might serve as a bridge between the actomyosin machinery and capsids during their migration to the nuclear rim. We then generated complementing cell lines to allow us to derive stocks of viral mutants with sufficiently high titers to allow us to test whether UL53 was required for associations of myosin Va with capsids, and for capsid migration towards the nuclear rim.

2. Materials and Methods

2.1. Cells and Viruses

Human foreskin fibroblasts (HFF) cells (ATCC, CRL-1684) and human embryonic kidney (293T) cells (ATCC (Manassas, VA, USA), CRL-11268) were propagated in DMEM containing 10% fetal bovine serum (FBS). The HCMV laboratory strain AD169 was used in all experiments. AD169-RV encoding a FLAG-tagged version of UL53 (53-F), and bacterial artificial chromosomes (BACs) encoding this virus and UL53-null (53N), UL50-null (50N), and UL53-null rescue-derivative (53NR) viruses have been described previously [10]. The production of HFFs expressing UL50 or UL53 and infectious 50N, 53N, and 53NR viruses is described below. Viruses were propagated and titrated as described previously [10,20].

2.2. Mass-Spectrometry

Immunoprecipitation (IP) of UL53-FLAG from nuclear lysates of infected cells for mass spectrometry was carried out as described previously [21]. Briefly, HFFs were infected with 53-F or WT HCMV (MOI 3) and cells were harvested at 72 h post-infection (hpi). Nuclear fractions were then isolated and subjected to α-FLAG IP. For mass spectrometry, eluates in Laemmli buffer were run on a 4–20% SDS-polyacrylamide gel. Extracted bands were submitted to the Taplin Mass-Spectrometry Facility (Harvard Medical School, Boston, MA, USA) for liquid chromatography-tandem mass-spectrometry (LC-MS/MS) analysis.

2.3. Immunoprecipitation

For immunoprecipitation (IP) of transfected cells, 5×10^6 293T cells/plate were seeded in 100 mm plates. Cells were then transfected with a pcDNA vector encoding a FLAG-tagged version of UL53 (UL53-FLAG) [22] and either a pcDNA-based plasmid encoding an HA-tagged version of UL50 (UL50-HA) [22] a lentiviral vector encoding green fluorescent protein with a nuclear localization sequence (GFP-NLS, [7]), or a lentiviral vector encoding the long tail of myosin Va fused to GFP-NLS (LT-GFP-NLS, [7]) (total of 10 μg DNA/plate; 1 μg/mL doxycycline was added to induce expression from the lentiviral vectors). At 48 h post-transfection, the cell monolayers were washed with Dulbecco's phosphate-buffered saline (DPBS) and whole cell lysates were harvested in EBC buffer (50 mM Tris (pH 8.0), 150 mM NaCl, 0.5% NP-40) containing one Complete EDTA-free protease inhibitor tablet (Roche, Indianapolis, IN, USA) per 50 mL. For IP, 25 μL of EZ-View α-FLAG M2 affinity gel (Sigma, St. Louis, MO, USA) was added to lysates and rotated overnight at 4 °C. Resin was centrifuged and washed 4 times in 750 μL ice-cold EBC buffer by rotating for 20 min at 4 °C between washes. After the final spin, the resin pellet was mixed with 25 μL of EBC buffer, and protein was eluted from resin by incubation with 50 μL of 2x Laemmli buffer at 95 °C for 5 min and analyzed by Western blot as described below.

For IP of UL53-FLAG from infected cells, 2.5×10^6 HFFs were infected with either 53-F or WT HCMV (MOI 1). At 72 hpi, cells were harvested, and nuclei were isolated and lysed using a Nuclear Complex Co-IP Kit (Active Motif, Carlsbad, CA, USA). The nuclear lysate was mixed with 1 mL EBC buffer containing one Complete EDTA-free protease inhibitor tablet (Roche) per 50 mL and precleared with 100 μL of mouse IgG-agarose (Sigma) by rotating at 4 °C for 5 h. For IP, 40 μL of EZ-View α-FLAG M2 affinity gel (Sigma) was added to precleared lysate and rotated overnight at 4 °C. Resin was centrifuged and washed 4 times in 750 μL ice-cold EBC buffer by rotating for 20 min at 4 °C between washes. After the final spin, the resin pellet was mixed with 40 μL of EBC buffer, and protein was eluted from resin by incubation with 80 μL of 2x Laemmli buffer at 95 °C for 5 min and analyzed by Western blot as described below.

2.4. Western Blotting

For Western blotting of IPs, lysates and eluates in Laemmli buffer were separated on a 4–20% SDS-polyacrylamide gel (Bio-Rad, Hercules, CA, USA). Proteins were then transferred onto a PVDF membrane, blocked with 5% milk in DPBS-T (DPBS with 0.5% Tween-20), and probed with primary antibodies (see below for sources and dilutions) overnight at 4 °C with rocking. Membranes were washed 3x with DPBS-T for 10 min at room temperature (RT) with rocking. Membranes were then incubated with TrueBlot secondary antibodies conjugated to horseradish peroxidase (HRP) (Rockland, Limerick, PA, USA) at 1:1000 for 1 h at RT with rocking, followed by washing. Finally, Pierce chemiluminescence solution (ThermoFisher, Waltham, MA, USA) was added to membranes and signal was detected with film.

For all other Western blotting, cells were harvested by washing with DPBS followed by the addition of 2x Laemmli buffer with protease inhibitors (ThermoFisher) directly to the monolayer. Lysates were scraped off the plate, boiled at 95 °C for 5 min, and processed as described above.

Primary antibody dilutions were as follows: rabbit α-myosin Va (Cell Signaling Technology, Danvers, MA, USA, #3402), 1:1000; mouse α-β actin (Sigma A5441), 1:5000; mouse α-FLAG M2 (Sigma F1804), 1:100; rabbit α-UL53 [10], 1:500; rabbit α-UL50 [10], 1:500; rabbit α-PCNA (Abcam, Cambridge, UK, ab18197), 1:700; mouse α-MCP (a kind gift from William Britt, University of Alabama, Birmingham, AL, USA), 1:1000; rabbit α-GFP (ThermoFisher A11122), 1:1000.

2.5. Immunofluorescence Analysis

For immunofluorescence, 1×10^5 HFFs/well were seeded on glass coverslips in a 24-well plate followed by either mock infection or infection with WT, 53-F, 50N/53F

HCMV (MOI 1) as indicated in the text. At the time-points indicated, cells were fixed at RT in 3.7% formaldehyde/DPBS. Cells were then permeabilized at RT in 0.1% Triton X-100/DPBS, washed 3x with DPBS, and blocked overnight in a mixture of 1% bovine serum albumin (Sigma) and 5% human serum (Sigma) in DPBS. The following antibodies and dilutions were used for primary staining: rabbit α-myosin Va (Abcam, ab11094), 1:50; mouse α-FLAG M2 (Sigma F1804), 1:500; mouse α-MCP (a kind gift from William Britt, University of Alabama, Birmingham, AL, USA), 1:250. Antibodies were diluted in a mixture of 1% BSA/5% human serum in DPBS and added to coverslips for 1 h at RT with rocking. Primary antibodies were removed and coverslips washed 3x with DPBS for 5 min with rocking at RT. The staining procedure was repeated with the appropriate fluorescently labeled Alexa-fluor secondary antibodies (ThermoFisher), and DAPI was applied in the last 10 min of the secondary antibody incubation. After the final washes, coverslips were mounted on glass slides using ProLong Anti-fade (ThermoFisher). Imaging was carried out at the Nikon Imaging Center (NIC) at Harvard Medical School using a Nikon Ti spinning-disk confocal laser microscope. Postacquisition image analysis was conducted using Metamorph and ImageJ software packages.

2.6. Immunoelectron Microscopy

Here, 2×10^5 HFFs/well were seeded in 12-well plates and infected with either WT or 53-F HCMV (MOI 1). At 72 hpi, cells were washed with DPBS, trypsinized, and harvested. The cell suspension was layered on top of a cushion of 4% paraformaldehyde and 0.1% glutaraldehyde in DPBS and pelleted for 3 min at 3000 rpm. The supernatant was carefully removed and fresh 4% paraformaldehyde and 0.1% glutaraldehyde were added. After fixation at RT for 2 h, the fixative was replaced with DPBS. Prior to freezing in liquid nitrogen the cell pellets were infiltrated with 2.3 M sucrose in DPBS (containing 0.2 M glycine to quench free aldehyde groups) for 15 min. Frozen samples were sectioned at -120 °C, the sections were transferred to formvar-carbon coated copper grids. Grids were floated on DPBS or stored on 2% gelatin dishes at 4 °C until immunogold labeling. The gold labeling was carried out at RT on a piece of parafilm. Antibodies and protein A gold were diluted in 1% BSA in PBS. The diluted primary antibody solution (α-myosin Va 1:10) was centrifuged 1 min at 14,000 rpm prior to labeling to avoid possible aggregates. Grids were floated on drops of 1% BSA for 10 min to block for unspecific labeling, transferred to 5 μL drops of primary antibody and incubated for 30 min. The grids were then washed in 4 drops of DPBS for a total of 15 min, transferred to 5 μL drops of 10 nm Protein A gold for 20 min, washed in 4 drops of DPBS for 15 min and 6 drops of double-distilled water. Contrasting/embedding of the labeled grids was carried out on ice in 0.3% uranyl acetate in 2% methyl cellulose for 10 min. Grids were picked up with metal loops (diameter slightly larger than the grid) and the excess liquid was removed by streaking on filter paper, leaving a thin coat of methylcellulose.

The grids were examined on a JEOL 1200EX electron microscope and images were recorded with an AMT 2k CCD camera. Labeled and unlabeled capsids were counted, and the percentage of capsids associated with at least one gold particle was calculated for each condition, and analyzed by Fisher's exact test using GraphPad Prism Version 7 software, GraphPad Software, San Diego, CA, USA.

2.7. Generation of Infectious 53N and 50N Viruses

To generate HFFs stably expressing either UL53 or UL50, each viral gene was amplified from WT HCMV BAC DNA with the following primers: UL53, forward: 5′-TAAGCAGCG GCCGCATGTCTAGCGTGAGCGGC GTGCGCA-3′; UL53, reverse: 5′-TGCTTAGGATCCT CAAGGCGCACGAATGCTGTTGAGAAACAGCGG-3′; UL50, forward: 5′-TAAGCAGCG GCCGCATGGAGATGAACAAGG TTCTCCATC-3′; UL50, reverse: 5′-TGCTTAGGATCCTC AGTCGCGGTGTGCGGAGCGTGTCGGA-3′. Each PCR product was digested with Not1 and BamH1 restriction enzymes and cloned into the pLVX-eF1αlentiviral vector (generous gift from the late Gregory Pari, University of Nevada, Reno, NV, USA). Lentiviruses were produced following transfection of 293T with these pLVX-eIFα-based plasmids and used to transduce HFFs as described previously [7]. The UL53 expressing cell lines were then electroporated with WT, 53N, or 53NR BACS, and the UL50 expressing cell line was electroporated with WT or 50N HCMV BAC DNA. After several weeks, viral supernatant was harvested and used for experiments as described in the text.

2.8. Transmission Electron Microscopy

Transmission electron microscopy (TEM) was utilized to assess subnuclear capsid distribution by counting capsids in the nuclei either inside or outside of RC-like inclusions under the conditions described in the text using a TecnaiG^2 Spirit BioTWIN electron microscope equipped with an AMT 2k CCD camera. Processing for image acquisition was performed as described previously [23]. Intranuclear capsid distributions were assessed by counting capsids within or outside of electron dense inclusions in the interior of the nucleoplasm that have been considered to be RCs (e.g., [15]) in representative sections of whole nuclei [6]. We term these RC-like inclusions. Data analyses for capsid distributions were performed as described previously [23], using ordinary one-way ANOVA corrected for multiple comparisons using the Holm–Sidak test, while analyses of capsid counts were performed using the Kruskal–Wallis test and Dunn's tests for multiple comparisons. All statistical analyses used GraphPad Prism version 7 (capsid distributions) and version 9.3.1 for Mac (capsid counts).

2.9. Correlative Light Electron Microscopy

Correlative light electron microscopy (CLEM) was conducted as described previously [10]. Briefly, HFFs were electroporated with either 53N or 53NR BAC DNA. The following day, cells were seeded onto gridded glass bottom dishes, and on day 7 or 8 postelectroporation they were fixed, imaged with fluorescence and phase microscopy to visualize the electroporated cells and the grid, and processed for EM. GFP-positive cells were identified by their grid coordinates, excised, and remounted for serial sectioning. Imaging was carried out using a TecnaiG^2 Spirit BioTWIN microscope. Representative whole-cell sections from three GFP-positive cells containing capsids were analyzed for each condition. Fisher's exact test was applied to data using GraphPad Prism software version 7.

3. Results

3.1. Mass Spectrometry Identifies Potential UL53 Binding Partners

To identify possible viral and cellular binding partners of UL53 in the nuclei of infected cells, human foreskin fibroblasts (HFFs) were infected with HCMV expressing a FLAG-tagged version of UL53 (53-F; multiplicity of infection (MOI) 3), or as a control, wild type (WT) HCMV, and, at 72 hpi, nuclear lysates were prepared and immunoprecipitated using α-FLAG antibody conjugated resin [21]. Immunoprecipitates were then subjected to liquid chromatography-tandem mass-spectrometry (LC-MS/MS) analysis, which uncovered peptides derived from numerous viral and cellular proteins in nuclear lysates from 53F-infected but not WT-infected cells. Of particular interest for this study were major capsid protein (MCP; 20 peptides; 13% coverage; peptides shown in Figure 1A) the primary protein constituent of capsids, and the capsid portal protein (UL104; 15 peptides; 23% coverage; peptides shown in Figure 1B). Also of interest for this study was the cellular protein myosin Va (MyoVa; 8 peptides; 6% coverage; peptides shown in Figure 1C), an F-actin-based host motor protein that we have previously shown is important for nuclear egress at the stage of localization towards the nuclear rim [7]. A list of viral proteins identified in this study is presented in Table S1 in Supplementary Materials. A similar list of cellular proteins will be reported separately.

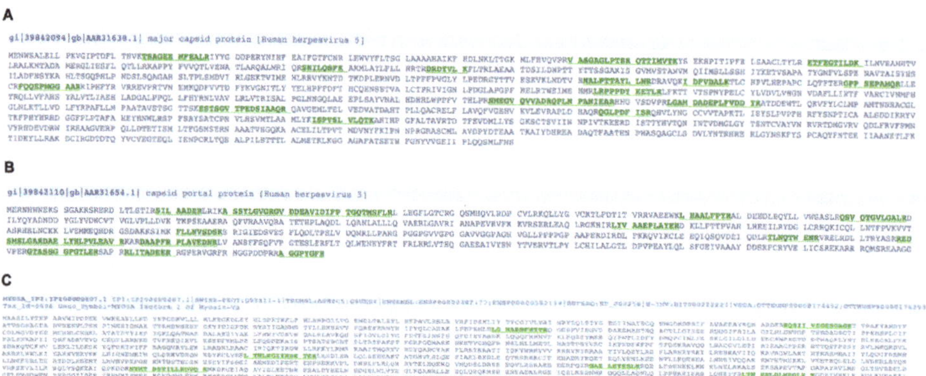

Figure 1. HFFs were infected with 53-F (MOI 3), nuclear lysates were harvested at 72 hpi, and subjected to α-FLAG IP. Proteins were resolved on an SDS-polyacrylamide gel and extracted bands were sent for LC-MS/MS analysis. Sequences of major capsid protein (**A**), capsid portal protein (**B**), and myosin Va (**C**) are shown with positions of unique peptides detected in the LC-MS/MS analysis highlighted in green.

3.2. Co-Immunoprecipitation of UL53, Myosin Va, and Major Capsid Protein

Given the reports that UL53 homologs of herpes simplex virus-1 (HSV-1), Epstein–Barr virus (EBV), and mouse CMV participate in the process of capsid migration [13,15,16], and our previous findings that myosin Va interacts with capsids and is important for such migration [7], we were intrigued by this finding. We therefore investigated whether associations of UL53 with myosin Va and MCP detected by MS (which could be direct or indirect) could also be detected using other assays. We began by asking whether UL53 can associate with the long tail (LT) region of myosin Va that contains the cargo-binding globular tail domain (GTD). To that end, we co-transfected 293T cells with a plasmid encoding UL53-FLAG together with a plasmid expressing LT fused to green fluorescent protein (GFP) with a nuclear localization signal (LT-GFP-NLS), GFP-NLS as a negative control, or UL50-HA as a positive control. UL53-FLAG was then immunoprecipitated from whole cell lysates with α-FLAG resin and immunoprecipitates were analyzed by Western blot using anti-FLAG antibodies to detect UL53, anti-HA antibodies to detect UL50, and anti-GFP antibodies to detect LT-GFP-NLS and GFP-NLS. We readily detected UL50-HA and LT-GFP-NLS in UL53-FLAG immunoprecipitates, but did not detect GFP-NLS, indicating that UL53 can associate directly or indirectly with myosin Va (Figure 2A).

We then investigated whether we could detect associations between UL53 and MCP and/or endogenous myosin Va in HCMV-infected cells. HFFs were infected with either 53-F HCMV, or as a negative control to detect nonspecific associations, WT HCMV (MOI 1). Nuclear lysates were prepared at 72 hpi and subjected to α-FLAG immunoprecipitation (IP), and lysates and eluates were analyzed by Western blot using antibodies against MCP, myosin Va, UL50 (as a positive control) and PCNA (as a negative control (Figure 2B). We found that UL53 immunoprecipitated from lysates of 53-F, but not WT infected cells, as expected. Also as expected, UL50 co-immunoprecipitated with UL53, while the host protein, PCNA, which is not known to bind UL53, did not. We detected MCP and myosin Va in Western blots of immunoprecipitates from 53-F infected nuclear lysates. We could also detect these proteins in immunoprecipitates from WT infected nuclear lysates, indicating background, nonspecific associations, but at lower levels (for myosin Va, ~3 to 4-fold lower based on a dilution series; Figure 2C), indicating that the majority of the MCP and myosin Va found in FLAG IP from nuclear lysates of 53-F infected cells arose from specific associations. These data confirm the MS results, suggesting modest associations, which could be direct or indirect, of UL53 with myosin Va and MCP in nuclear lysates.

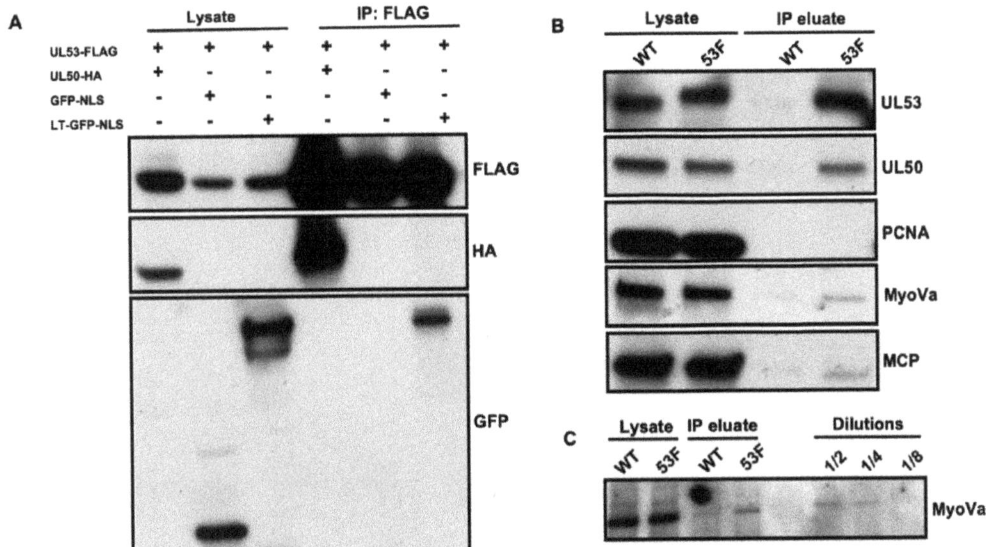

Figure 2. (**A**) 293T cells were co-transfected with a plasmid expressing UL53-FLAG together with a plasmid encoding the myosin Va long tail (LT) fused to GFP with an NLS (LT-GFP-NLS), GFP-NLS (negative control), or UL50-HA (positive control). UL53-FLAG was immunoprecipitated from whole cell lysates with α-FLAG IP and immunoprecipitates were analyzed by Western blot using the antibodies indicated at the right of each panel. (**B**) HFFs were infected with 53-F or WT HCMV (MOI 1). At 72 hpi, nuclear lysates were prepared, subjected to α-FLAG IP, and lysates and eluates were analyzed by Western blot using antibodies against the proteins indicated to the right of each panel. (**C**) The 53-F immunoprecipitates from above were diluted as indicated and analyzed by Western blot for myosin Va (MyoVa).

3.3. A Population of UL53 Colocalizes with Myosin Va and Capsids in the Nucleoplasm

To study possible associations of UL53, MCP, and myosin Va by a different method, we examined the localization of UL53, MCP, and myosin Va in infected cells, initially using immunofluorescence assays (IFA). HFFs were infected with 53-F HCMV (MOI 1) and at 72 hpi cells were fixed, stained with α-myosin Va, α-MCP, and α-FLAG antibodies, and single optical sections were imaged with confocal microscopy (Figure 3A). As expected, most UL53 was found at the nuclear rim; however, a population was also present in the nucleoplasm. We have previously shown that myosin Va concentrates in replication compartments (RCs) [7], and the staining pattern observed in this experiment was consistent with that localization. MCP also concentrated at the periphery of these structures where it colocalized with myosin Va (yellow in merged image). We observed some colocalization of UL53 with myosin Va and MCP both at the periphery of RCs and between the RCs and nuclear rim, but not at the rim (white in the merged image). Colocalization of the three proteins at the periphery of RCs was verified by measuring the fluorescence intensity of each channel (Figure 3B).

We then utilized immunoelectron microscopy (immunoEM) to assess UL53 association with capsids at the ultrastructural level. We infected HFFs with 53-F HCMV (MOI 1), fixed cells at 72 hpi, and processed cells by staining with an α-FLAG antibody followed by 10 nm protein A-gold secondary staining. Consistent with our IFA findings, most UL53 could be found at the nuclear rim where it concentrated in areas where the INM appeared slightly infolded (Figure 3C). Furthermore, these infoldings occasionally contained capsids in the perinuclear space likely in the process of capsid budding (Figure 3D), consistent with a previous report [24]. We also observed a population of UL53 in the nucleoplasm that associated with capsids in RC-like inclusions, consistent with our IFA results (Figure 3C,E). Thus, these IFA and immunoEM data provide further evidence that UL53 modestly associates with capsids in the nucleoplasm, consistent with the modest associations seen in co-immunoprecipitation experiments.

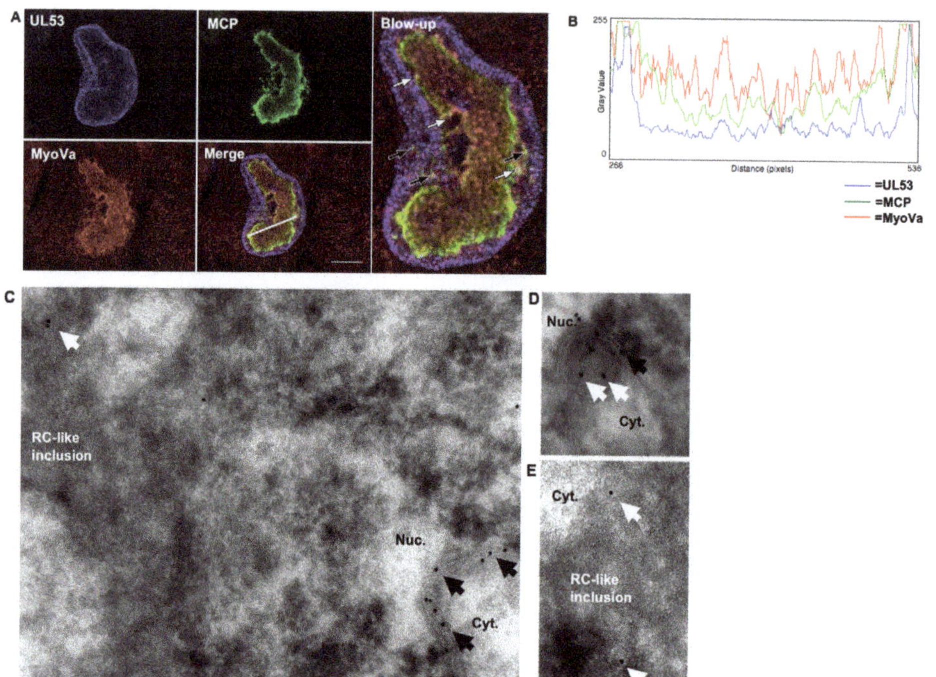

Figure 3. (**A**) HFFs were infected with 53-F HCMV (MOI 1) and fixed at 72 hpi. Cells were then stained with α-FLAG (blue), α-MCP (green), and α-myoVa (red) antibodies and imaged with spinning-disk confocal microscopy. Yellow color indicates colocalization between myosin Va (MyoVa) and MCP; purple color indicates colocalization between myosin Va and UL53; cyan color indicates colocalization between UL53 and MCP; white color and arrows indicate colocalization between UL53, MCP, and myosin Va (black arrows indicate examples of colocalization between the RC periphery and the nuclear rim; white arrows indicate examples of colocalization at the nuclear rim). Images are single optical sections. Scale bar is 10μm. (**B**) To measure colocalization between UL53, MCP, and myosin Va at the RC periphery, the fluorescence intensity of each channel was plotted across the indicated white line. (**C–E**) HFFs were infected with 53-F HCMV (MOI 1) and fixed for immunoEM at 72 hpi. The cells were further processed by primary staining with an α-FLAG antibody, followed by secondary staining with a 10nm protein A-gold secondary. Imaging was conducted using a transmission electron microscope. Black arrows indicate UL53 associated with the INM. White arrows indicate UL53 associated with capsids.

3.4. Generation of UL53 and UL50 Null Mutants Using Complementing Cell Lines

We wondered how UL53 would localize relative to myosin Va and MCP in the absence of UL50, and we especially wished to test whether UL53 is necessary for capsid localization to the nuclear rim, and for myosin Va association with capsids. We attempted to use correlative light-electron microscopy (CLEM) of HFFs electroporated with bacterial artificial chromosomes (BACs) expressing GFP and containing *UL53*-null (53N) HCMV or a rescued derivative (53NR), fixing at 7 or 8 days postelectroporation (dpe), and imaging with phase and fluorescence microscopy. (We had previously used these BACs to study infection in the absence or presence of UL53 [10]). Based on the grid coordinates of GFP-positive cells, we then traced back their location using EM and imaged capsids in representative whole cell sections (Figure S1 in Supplementary Materials). We observed a lower percentage of capsids located within 2 μm of the nuclear rim in 53N versus 53NR electroporated cells, and taken at face value, this difference was statistically significant (Figure S1 in Supplementary Materials). However, we did not trust that this difference was meaningful as we only were able to observe a total of three cells per condition, and some electroporated cells contained many fewer capsids than others and did not display obvious RC-like inclusions (Figure S1), suggesting that infection was not at the same stage for all cells.

Given these results and as CLEM is labor intensive, time consuming, and able to examine only a few cells per experiment, we instead generated infectious 50N HCMV that expresses either untagged (50N) or a FLAG-tagged UL53 (50N/53-F) and also *UL53*-null (53N) virus using complementing UL50-expressing (50HFFs) and UL53-expressing (53HFFs) cells, respectively. Previous attempts in our lab to generate complementing HFFs using a retroviral transduction system were unsuccessful; thus we opted to use the pLVX-eF1α lentiviral vector that had previously been used to complement HCMV *UL84* null virus growth in HFFs [25]. Following transduction and puromycin selection, we found that 50HFFs and 53HFFs expressed UL50 or UL53, respectively, using Western blot (Figure 4A). We then electroporated HFFs with either 53N or 50N BAC DNA and observed virus spread in complementing but not noncomplementing HFFs, as expected (Figure 4B). Infection of noncomplementing HFFs with infectious 53N or 50N HCMV confirmed that each virus did not detectably express UL53 or UL50, respectively (Figure 4A). After concentration by ultracentrifugation, we were able to generate relatively high titer stocks (~10^6 PFU/mL) of the null mutant viruses using the complementing cells. A recent paper has described the generation of cells that express UL50 and complement a *UL50* null mutant [26].

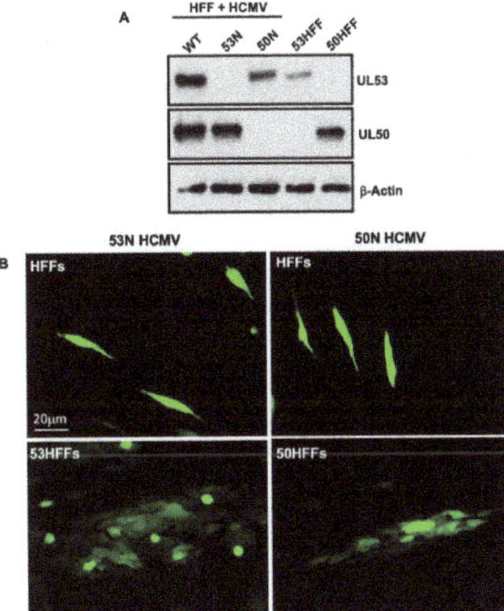

Figure 4. (**A**) HFFs were transduced with lentiviruses expressing UL53 or UL50 creating 53HFFs or 50HFFs, respectively. Expression of UL50 and UL53 in whole cell lysates of WT-infected cells (leftmost lane) was compared with that in uninfected 53HFF and 50HFF (rightmost two lanes), and with HFFs infected with 53N or 50N (second and third lanes from left, respectively) by Western blot, with β-actin as a loading control. Infected cells were harvested at 72 hpi. Cells and viruses used are indicated at the top of the image. Antibodies used are indicated to the right of the image. (**B**) Noncomplementing HFFs (top rows) and 53HFFs or 50HFFs (bottom rows) were electroporated with 53N or 50N GFP BAC DNA and imaged with widefield fluorescence microscopy to assess virus spread at 7 dpe.

3.5. Increased Association of UL53 with Major Capsid Protein and Myosin Va in the Absence of UL50

We next infected HFFs with 53-F or 50N/53-F HCMV (produced in 50HFFs) and fixed cells for confocal microscopy at 72 hpi. As both viruses also express GFP, there were not enough laser channels to image UL53, MCP, and myosin Va together in these samples. Therefore, we stained cells with DAPI, α-FLAG, and either α-myosin Va or α-MCP antibodies. As was seen in Figure 3, most UL53 was found at the nuclear rim during 53-F infection; however, a subpopulation was present in the nucleoplasm where it colocalized with myosin Va (Figure 5A) or MCP (Figure 5B), consistent with localization at the RC periphery and between the RC periphery and the nuclear rim. Notably, we found that in the absence of UL50 following infection with 50N/53F, essentially no UL53 was present at the nuclear rim and there was much more obvious colocalization of nucleoplasmic UL53 with myosin Va (Figure 5A) or MCP (Figure 5B), again consistent with all three proteins localizing at the RC periphery. Collectively, these data provide further evidence for associations of UL53 with myosin Va and MCP at the periphery of the RC.

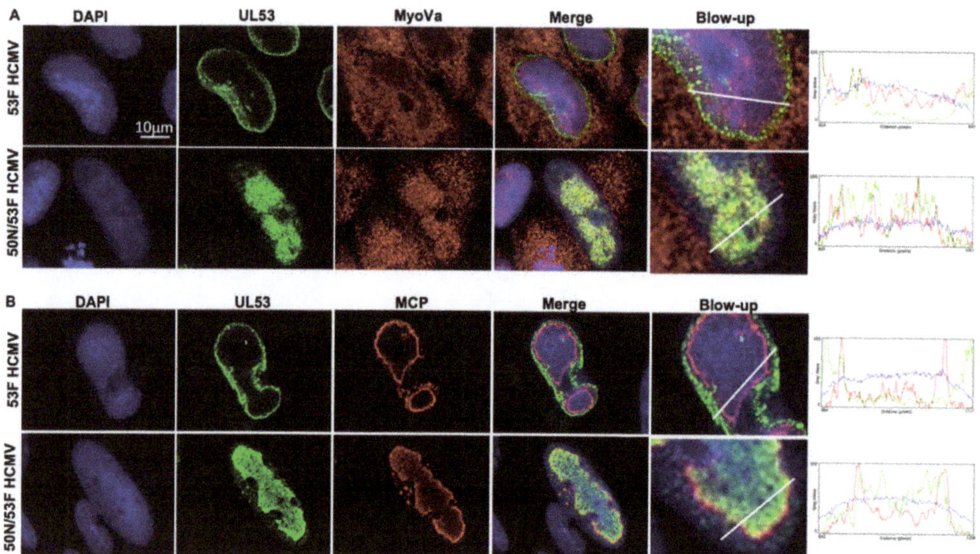

Figure 5. HFFs were infected with 53-F or 50N/53-F HCMV (MOI 1) and fixed at 72 hpi. Cells were then stained with DAPI (blue), α-FLAG (green), and in (**A**) α-myoVa (red) antibodies, or in (**B**) α-MCP (red) antibodies, and imaged with spinning-disk confocal microscopy. Yellow color indicates colocalization of UL53 with myoVa or MCP. Colocalization was measured by plotting fluorescence intensity of each channel across the indicated white lines. Images are single optical sections.

3.6. UL53 Is Not Required for the Association between Capsids and Myosin Va

The association of UL53 with capsids (Figure 3C–E) and with myosin Va (Figures 2 and 3A,B) evoked the hypothesis that this NEC subunit might serve as a bridge between capsids and the actomyosin system for intranuclear transport. We had previously shown that myosin Va interacts with capsids in HCMV-infected cells using immunoEM [7]. We therefore conducted immunoEM to determine whether UL53 is important for this association. HFFs were infected with WT or 53N HCMV (MOI 1), fixed at 72 hpi, and processed for immunoEM by staining with α-myosin Va antibodies followed by secondary staining using 10nm protein A-gold. We observed that myosin Va associated with a similar percentage (~30%) of nuclear capsids in both WT- and 53N-infected cells (Figure 6). We thus conclude that UL53 is not required for the association between capsids and myosin Va.

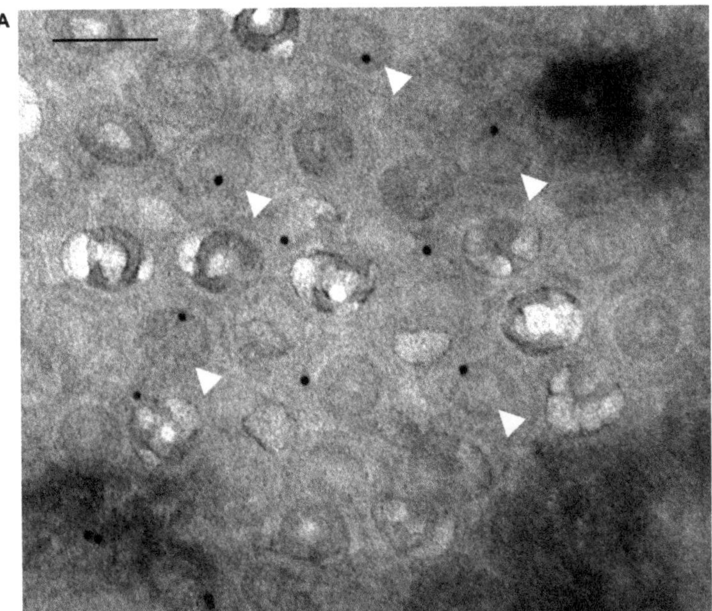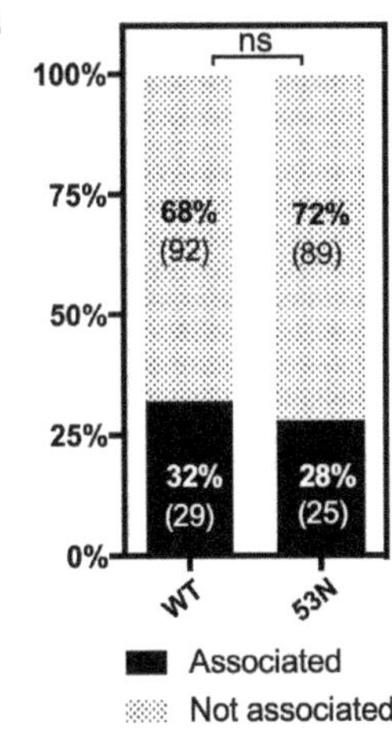

Figure 6. HFFs were infected with either WT or 53N HCMV (MOI 1) and fixed for immunoEM at 72 hpi. The cells were further processed by primary staining with α-myoVa followed by secondary staining with 10-nm protein A-gold. Imaging was conducted using a transmission electron microscope. (**A**) shows a representative image of the 53N condition. White arrowheads indicate capsids that are associated with myoVa. Scale bar is 100 nm. (**B**) The percentage of capsids associated with at least one gold particle was calculated for each condition (WT, n = 121; 53N, n = 144). The p value was calculated using Fisher's exact test (ns = not significant, p = 0.76).

3.7. UL53 Is Not Important for Capsid Localization Away from RC-like Inclusions

Our finding that UL53 was not important for myosin Va association with capsids led us to investigate whether UL53 facilitates HCMV capsid migration to the nuclear rim, as has been reported for UL53 homologs of some other herpesviruses [13,15,16] and suggested for HCMV UL53 [19]. To address this question, we infected HFFs with stocks of WT, 53N, or 50N HCMV (MOI 1). Cells were fixed and processed for EM at 72 hpi. RC-like inclusions were apparent in the nuclei of all cells analyzed (examples in Figure 7A), which allowed us to evaluate capsid distribution relative to the inclusions as we had done previously [7]. We observed no decrease and, if anything, a slight increase in the percentage of capsids located outside RC-like inclusions in cells infected with 53N or 50N virus compared to WT, although this difference was not significant (Figure 7B). The mutant-infected cells contained roughly similar numbers of nuclear capsids as WT-infected cells (Figure 7C), as we have observed with other mutants defective for nuclear egress (e.g., [23]). Thus, UL53 is not important for capsid localization away from RC-like inclusions.

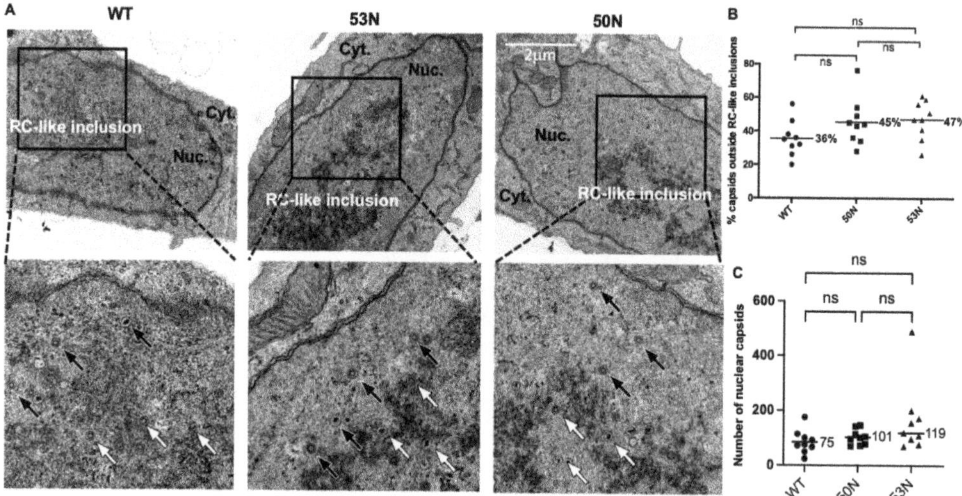

Figure 7. (**A**) HFFs were infected with WT, 50N, or 53N HCMV (MOI 1) and at 72 hpi cells were fixed and processed for EM. Black arrows point to capsids outside RC-like inclusions. White arrows point to capsids associated with RC-like inclusions (**B**) The percentage of capsids located outside RC-like inclusions was calculated in sections of 9 nuclei per condition and plotted. The horizontal bars indicate the mean percentage of capsids for each condition. *p*-values were calculated using ordinary one-way ANOVA corrected for multiple comparisons using the Holm–Sidak test (ns = not significant. WT vs. 50N, *p* = 0.18; WT vs. 53N, *p* = 0.16; 50N vs. 53N, *p* = 0.79). (**C**) The number of capsids found in the same sections analyzed in (**B**). The horizontal bars indicate the median numbers of capsids per section; the medians were calculated to obviate misleading skewing of the results by an outlying data point. Data were analyzed using a Kruskal–Wallis test with Dunn's tests to provide *p*-values corrected for multiple comparisons (ns = not significant). WT vs. 50N, *p* = 0.8346; WT vs. 53N, *p* = 0.0806; 50N vs. 53N, *p* = 0.7768).

4. Discussion

How nascent herpesvirus capsids migrate from RCs in the nuclear interior to the periphery during nuclear egress is poorly understood. It is particularly unclear whether the NEC or its subunits, which are crucial for later steps of nuclear egress, play a role in this migration. We performed a mass spectrometry study to look for proteins that associate directly or indirectly with the nucleoplasmic subunit of the HCMV NEC, UL53, and identified myosin Va and capsid proteins as candidate UL53-interacting proteins. Follow-up studies provided evidence to validate these associations, although they were rather modest. Our results together with results reported previously in various herpesvirus systems [5–7,12–16,19], led to the hypothesis that UL53 might serve as a bridge between capsids and nuclear actomyosin machinery. To test this role for UL53 and, more generally, a role in migration from the nuclear interior to the nuclear rim, we generated stocks of *UL50-* and *UL53*-null viruses using newly derived complementing cells. Experiments using these mutant viruses led to the conclusion that UL53 is neither important for associations between capsids and myosin Va nor for capsid localization away from RC-like inclusions in the nuclear interior. We discuss each of these findings below.

Our mass spectrometry, co-IP, IFA, and immunoEM results suggested associations among UL53, myosin Va, and major capsid proteins or assembled capsids in infected nuclei. By IFA and immunoEM, during WT infection, myosin Va and MCP or capsid associations were found largely within RCs or RC-like inclusions in the nuclear interior, while most UL53 could be found at the nuclear rim (Ref. [7] and this study). However, even during WT infection, a subpopulation of UL53 was also evident in the nuclear interior where it colocalized with myosin Va and MCP (as shown by IFA) or capsids (as shown by immunoEM—the IFA studies do not distinguish between capsids and unassembled MCP or partially assembled particles). This is consistent with reports that the HSV-1 homolog of UL53 (HSV-1 UL31), co-localizes with MCP hexons and associates with the portal vertex of capsids in the nucleoplasm [16,17]. The results are also consistent with a report that HCMV UL53 could be found in co-IP using antibodies against smallest capsid protein, and on intranuclear capsids by immunoEM [19]. However, the capsids shown in that report were close to the nuclear envelope rather than in RC-like inclusions. Regardless, the relatively small amount of nucleoplasmic UL53 involved in associations with myosin Va and capsids likely reflects the small proportion that is not bound to UL50 at the INM under steady state conditions. Consistent with that surmise, UL53 primarily localizes to RCs during 50N infection, where we detected more obvious colocalization of UL53 with myosin Va or MCP.

Thus, we speculate that, ordinarily, some UL53 and myosin Va associate initially with capsids in RCs, and that subsequent movement to the nuclear periphery and primary envelopment occur relatively quickly and are thus not easily detected in fixed cells. For interactions between UL53 and capsids, there is considerable evidence for association of UL53 homologs of other viruses with various capsid proteins, particularly HSV-1 UL31 with HSV-1 UL25, or inner tegument proteins [17,18,27–30]. A recent study made the exciting finding that in HSV-1, UL25, interacts with the NEC and promotes formation of pentagonal arrays, which are posited to anchor capsids to the NEC and promote NEC curvature [30]. The HCMV homolog of HSV-1 UL25, UL77, was not one of the proteins detected in our MS study, although we did detect MCP, the portal protein UL104, and several other components of intranuclear capsids including the terminase subunits, UL56 and UL89, and the inner tegument protein, UL32 (pp150), as well as TRS1, which has been reported to abet capsid assembly (Table S1 in Supplementary Materials). Our failure to detect UL77 may reflect limitations of the mass spectrometric approach, our use of soluble nuclear lysates, or differences in the HCMV and HSV-1 systems.

In contrast with the literature on associations of UL53 homologs with capsids and/or capsid proteins, there is little evidence regarding how myosin Va associates with these and with UL53. Such associations could be direct or indirect. Given these associations and previous studies indicating a role for UL53 homologs and myosin Va in intranuclear distribution of capsids [7,15,16] the hypothesis that UL53 might serve as an adaptor protein, bridging capsids and myosin Va, was attractive. However, deletion of UL53 did not discernibly affect the association between capsids and myosin Va. Thus, the details of UL53 interactions with myosin Va and nucleoplasmic capsids remain unclear.

Our finding that a population of UL53 localizes to the nucleoplasm was consistent with reports that UL53 homologs participate in events upstream of primary envelopment. For example, roles for UL53 homologs in viral DNA packaging (or stabilization of filled capsids) have been described for α-, β-, and γ-herpesviruses [12–15]. Furthermore, UL53 homologs have been suggested to facilitate capsid migration to the nuclear periphery. In one paper, it was stated in the Discussion that in cells replicating Epstein–Barr virus genomes lacking its *UL53* homolog, capsids were homogeneously distributed throughout the nucleus rather than being mostly aligned along the nuclear membrane [13]. In a second paper, in one of the figures, capsids derived from MCMV expressing a dominant-negative version of its UL53 homolog, M53, clustered in the nuclear interior away from the nuclear rim [15]. These authors went on to suggest that impaired viral DNA packaging observed with this mutant (and with another dominant-negative *M53* mutant [14]) causes capsids to stall at packaging sites and that M53 mediates MCMV capsid localization to the nuclear periphery [15]. Subsequently, it was shown in an interesting paper [16] that an HSV-1 mutant, in which two basic patches of the N-terminal segment of its UL53 homolog, UL31, were made less basic, is defective for nuclear egress. IFA of UL31 and capsid proteins, particularly in cells transfected with a BAC containing the mutant genome, led to the conclusion that the N-terminal segment is required to direct migration of nucleocapsids to sites of primary envelopment at the nuclear periphery [16].

While our initial CLEM results, using cells into which BACs containing the null mutant genome were introduced, raised the possibility that UL53 is important for capsid localization towards the nuclear periphery, subsequent analysis of a larger number of cells that were directly and synchronously infected by null mutant viruses indicated that deletion of *UL53* (or *UL50*) did not result in altered intranuclear localization of capsids. (We also observed that copious DNA-filled C capsids formed in the absence of UL53, as we had previously [10], indicating that UL53 is not essential for viral DNA packaging or for stabilization of C capsids.) It is possible that UL53 differs from its EBV, MCMV, and HSV-1 homologs in its ability to promote capsid migration to the nuclear rim. It is also possible that certain differences in results among the systems might reflect differences in how experiments were performed such as differences in preparation of infected cells (e.g., transfection vs. infection), mutants used (e.g., null vs. ones retaining certain activities), and whether and how intranuclear distributions of capsids were quantified.

What role does UL53 interactions with nucleoplasmic capsids play? One possibility is that these interactions are transient and/or infrequent and play no role. A second speculative possibility is that any UL53 that associates with capsids at the RC and migrates with them to the nuclear rim is then able to dock those capsids to any unoccupied UL50 at the INM for subsequent primary envelopment. Such a role would not depend on whether capsids migrate by the nuclear actomyosin machinery [5] or by random diffusion [31,32]. Additional studies are required to address these possibilities.

Supplementary Materials: The following are available online at https://www.mdpi.com/article/10.3390/v14030479/s1, Table S1: Viral proteins detected in FLAG-immunoprecipitates from nuclear lysates of cells infected with 53-F, but not wild-type HCMV; Figure S1: CLEM analysis of intranuclear distribution of capsids.

Author Contributions: Conceptualization, A.R.W.; formal analysis, J.M.P.; investigation, A.R.W., M.S., M.C., M.E., J.L.L. and R.F.; resources, M.E.; writing—original draft preparation, A.R.W. and D.M.C.; writing—review and editing, A.R.W., M.S., M.C., M.E., J.M.P., J.L.L., R.F. and D.M.C.; visualization, M.C. and A.R.W.; supervision, A.R.W. and D.M.C.; project administration, A.R.W. and D.M.C.; funding acquisition, A.R.W. and D.M.C. All authors have read and agreed to the published version of the manuscript.

Funding: This research was funded by National Institutes of Health grants R01 AI026077 and F31 20651.

Institutional Review Board Statement: The Institutional Review Board (IRB) of the Harvard Faculty of Medicine determined that these studies are not human subjects research as defined by DHHS or FDA regulations, because they involve human fibroblast cultures that are derived from discarded pathological materials that cannot be identifiably linked to any individual and meet the conditions regarding coded biological specimens or data.

Informed Consent Statement: Not applicable.

Data Availability Statement: Data supporting the reported results can be found in the figures and the Supplementary Materials.

Acknowledgments: Dedicated to the memory of Greg Pari, whose generosity with ideas and reagents was crucial for these studies. We are grateful to Greg and Cyprian Rossetto who kindly provided pLVX-eIF1α, William Britt for generously providing anti-MCP antibody, and staff at the Taplin Mass Spectrometry facility, the Nikon Imaging Center, and the Electron Microscopy Facility at Harvard Medical School for acquisition and analysis of mass spectrometry, spinning disk confocal microscopy, and electron microscopic data, respectively. We thank David Knipe, Samara Reck-Peterson, James Hogle, Thomas Schwarz, Frederick Wang, and Lee Gehrke, as well as Blair Strang, Jeremy Kamil, and other members of the Coen lab for helpful discussions.

Conflicts of Interest: The authors declare no conflict of interest.

References

1. Bigalke, J.M.; Heldwein, E.E. Nuclear Exodus: Herpesviruses Lead the Way. *Annu. Rev. Virol.* **2016**, *3*, 387–409. [CrossRef] [PubMed]
2. Lye, M.F.; Wilkie, A.R.; Filman, D.J.; Hogle, J.M.; Coen, D.M. Getting to and through the inner nuclear membrane during herpesvirus nuclear egress. *Curr. Opin. Cell Biol.* **2017**, *46*, 9–16. [CrossRef] [PubMed]
3. Sanchez, V.; Britt, W. Human Cytomegalovirus Egress: Overcoming Barriers and Co-Opting Cellular Functions. *Viruses* **2021**, *14*, 15. [CrossRef] [PubMed]
4. Roller, R.J.; Johnson, D.C. Herpesvirus Nuclear Egress across the Outer Nuclear Membrane. *Viruses* **2021**, *13*, 2356. [CrossRef]
5. Forest, T.; Barnard, S.; Baines, J.D. Active intranuclear movement of herpesvirus capsids. *Nat. Cell Biol.* **2005**, *7*, 429–431. [CrossRef] [PubMed]
6. Wilkie, A.R.; Lawler, J.; Coen, D.M. A Role for Nuclear F-Actin Induction in Human Cytomegalovirus Nuclear Egress. *mBio* **2016**, *7*, e01254-16. [CrossRef]
7. Wilkie, A.R.; Sharma, M.; Pesola, J.M.; Ericsson, M.; Fernandez, R.; Coen, D.M. A role for myosin Va in human cytomegalo-virus nuclear egress. *J. Virol.* **2018**, *92*, e01849-17. [CrossRef] [PubMed]
8. Dunn, W.; Chou, C.; Li, H.; Hai, R.; Patterson, D.; Stolc, V.; Zhu, H.; Liu, F. Functional profiling of a human cytomegalovirus genome. *Proc. Natl. Acad. Sci. USA* **2003**, *100*, 14223–14228. [CrossRef]
9. Yu, D.; Silva, M.C.; Shenk, T. Functional map of human cytomegalovirus AD169 defined by global mutational analysis. *Proc. Natl. Acad. Sci. USA* **2003**, *100*, 12396–12401. [CrossRef] [PubMed]
10. Sharma, M.; Kamil, J.P.; Coughlin, M.; Reim, N.I.; Coen, D.M. Human Cytomegalovirus UL50 and UL53 Recruit Viral Protein Kinase UL97, Not Protein Kinase C, for Disruption of Nuclear Lamina and Nuclear Egress in Infected Cells. *J. Virol.* **2014**, *88*, 249–262. [CrossRef]
11. Sharma, M.; Bender, B.J.; Kamil, J.P.; Lye, M.F.; Pesola, J.M.; Reim, N.I.; Hogle, J.M.; Coen, D.M. Human cytomegalovirus UL97 phosphorylates the viral nuclear egress complex. *J. Virol.* **2015**, *89*, 523–534. [CrossRef] [PubMed]
12. Chang, Y.; Van Sant, C.; Krug, P.W.; Sears, A.; Roizman, B. The null mutant of the U(L)31 gene of herpes simplex virus 1: Construction and phenotype in infected cells. *J. Virol.* **1997**, *71*, 8307–8315. [CrossRef] [PubMed]
13. Granato, M.; Feederle, R.; Farina, A.; Gonnella, R.; Santarelli, R.; Hub, B.; Faggioni, A.; Delecluse, H.-J. Deletion of Ep-stein-Barr virus BFLF2 leads to impaired viral DNA packaging and primary egress as well as to the production of defective viral particles. *J. Virol.* **2008**, *82*, 4042–4051. [CrossRef]
14. Popa, M.; Ruzsics, Z.; Lötzerich, M.; Dölken, L.; Buser, C.; Walther, P.; Koszinowski, U.H. Dominant Negative Mutants of the Murine Cytomegalovirus M53 Gene Block Nuclear Egress and Inhibit Capsid Maturation. *J. Virol.* **2010**, *84*, 9035–9046. [CrossRef] [PubMed]
15. Pogoda, M.; Bosse, J.B.; Wagner, F.M.; Schauflinger, M.; Walther, P.; Koszinowski, U.H.; Ruzsics, Z. Characterization of con-served region 2-deficient mutants of the cytomegalovirus egress protein pM53. *J. Virol.* **2012**, *86*, 12512–12524. [CrossRef]
16. Funk, C.; Ott, M.; Raschbichler, V.; Nagel, C.-H.; Binz, A.; Sodeik, B.; Bauerfeind, R.; Bailer, S.M. The Herpes Simplex Virus Protein pUL31 Escorts Nucleocapsids to Sites of Nuclear Egress, a Process Coordinated by Its N-Terminal Domain. *PLOS Pathog.* **2015**, *11*, e1004957. [CrossRef]
17. Yang, K.; Wills, E.; Lim, H.Y.; Zhou, Z.H.; Baines, J.D. Association of Herpes Simplex Virus pUL31 with Capsid Vertices and Components of the Capsid Vertex-Specific Complex. *J. Virol.* **2014**, *88*, 3815–3825. [CrossRef]
18. Leelawong, M.; Guo, D.; Smith, G.A. A Physical Link between the Pseudorabies Virus Capsid and the Nuclear Egress Complex. *J. Virol.* **2011**, *85*, 11675–11684. [CrossRef] [PubMed]
19. Milbradt, J.; Sonntag, E.; Wagner, S.; Strojan, H.; Wangen, C.; Rovis, T.L.; Lisnic, B.; Jonjic, S.; Sticht, H.; Britt, W.J.; et al. Human Cytomegalovirus Nuclear Capsids Associate with the Core Nuclear Egress Complex and the Viral Protein Kinase pUL97. *Viruses* **2018**, *10*, 35. [CrossRef]
20. Leigh, K.; Sharma, M.; Mansueto, M.S.; Boeszoermenyi, A.; Filman, D.; Hogle, J.; Wagner, G.; Coen, D.M.; Arthanari, H. Structure of a herpesvirus nuclear egress complex subunit reveals an interaction groove that is essential for viral replication. *Proc. Natl. Acad. Sci. USA* **2015**, *112*, 9010–9015. [CrossRef]
21. Sharma, M.; Kamil, J.P.; Coen, D.M. Preparation of the Human Cytomegalovirus Nuclear Egress Complex and Associated Proteins. *Methods Enzymol.* **2016**, *569*, 517–526. [CrossRef]
22. Sam, M.D.; Evans, B.T.; Coen, D.M.; Hogle, J.M. Biochemical, Biophysical, and Mutational Analyses of Subunit Interactions of the Human Cytomegalovirus Nuclear Egress Complex. *J. Virol.* **2009**, *83*, 2996–3006. [CrossRef]
23. Reim, N.I.; Kamil, J.P.; Wang, D.; Lin, A.; Sharma, M.; Ericsson, M.; Pesola, J.M.; Golan, D.E.; Coen, D.M. Inactivation of ret-inoblastoma protein does not overcome the requirement for human cytomegalovirus UL97 in lamina disruption and nu-clear egress. *J. Virol.* **2013**, *87*, 5019–5027. [CrossRef] [PubMed]
24. Monte, P.D.; Pignatelli, S.; Zini, N.; Maraldi, N.M.; Perret, E.; Prevost, M.C.; Landini, M.P. Analysis of intracellular and intraviral localization of the human cytomegalovirus UL53 protein. *J. Gen. Virol.* **2002**, *83*, 1005–1012. [CrossRef]
25. Pari, G.; Rossetto, C.; University of Nevada, Reno, NV, USA. Personal communication, 2015.
26. Häge, S.; Sonntag, E.; Svrlanska, A.; Borst, E.M.; Stilp, A.-C.; Horsch, D.; Müller, R.; Kropff, B.; Milbradt, J.; Stamminger, T.; et al. Phenotypical Characterization of the Nuclear Egress of Recombinant Cytomegaloviruses Reveals Defective Replication upon ORF-UL50 Deletion but Not pUL50 Phosphosite Mutation. *Viruses* **2021**, *13*, 165. [CrossRef] [PubMed]

27. Yang, K.; Baines, J.D. Selection of HSV capsids for envelopment involves interaction between capsid surface components pUL31, pUL17, and pUL25. *Proc. Natl. Acad. Sci. USA* **2011**, *108*, 14276–14281. [CrossRef] [PubMed]

28. Leelawong, M.; Lee, J.I.; Smith, G.A. Nuclear Egress of Pseudorabies Virus Capsids Is Enhanced by a Subspecies of the Large Tegument Protein That Is Lost upon Cytoplasmic Maturation. *J. Virol.* **2012**, *86*, 6303–6314. [CrossRef] [PubMed]

29. Takeshima, K.; Arii, J.; Maruzuru, Y.; Koyanagi, N.; Kato, A.; Kawaguchi, Y. Identification of the Capsid Binding Site in the Herpes Simplex Virus 1 Nuclear Egress Complex and Its Role in Viral Primary Envelopment and Replication. *J. Virol.* **2019**, *93*, e01290-19. [CrossRef] [PubMed]

30. Draganova, E.B.; Zhang, J.; Zhou, Z.H.; Heldwein, E. Structural basis for capsid recruitment and coat formation during HSV-1 nuclear egress. *eLife* **2020**, *9*, 56627. [CrossRef]

31. Bosse, J.B.; Virding, S.; Thiberge, S.Y.; Scherer, J.; Wodrich, H.; Ruzsics, Z.; Koszinowski, U.H.; Enquist, L.W. Nuclear herpes-virus capsid motility is not dependent on F-actin. *mBio* **2014**, *5*, e01909-14. [CrossRef] [PubMed]

32. Bosse, J.B.; Hogue, I.B.; Feric, M.; Thiberge, S.Y.; Sodeik, B.; Brangwynne, C.P.; Enquist, L.W. Remodeling nuclear architec-ture allows efficient transport of herpesvirus capsids by diffusion. *Proc. Natl. Acad. Sci. USA* **2015**, *112*, E5725–E5733. [CrossRef] [PubMed]

Article

Tegument Protein pp150 Sequence-Specific Peptide Blocks Cytomegalovirus Infection

Dipanwita Mitra [1], Mohammad H. Hasan [1], John T. Bates [1,2], Gene L. Bidwell III [3,4,5] and Ritesh Tandon [1,2,6,*]

1 Department of Microbiology and Immunology, University of Mississippi Medical Center, 2500 North State Street, Jackson, MS 39216, USA; dmitra@umc.edu (D.M.); mohammad_hasan@brown.edu (M.H.H.); jtbates@umc.edu (J.T.B.)
2 Department of Medicine, University of Mississippi Medical Center, 2500 North State Street, Jackson, MS 39216, USA
3 Department of Neurology, School of Medicine, University of Mississippi Medical Center, Jackson, MS 39216, USA; gbidwell@umc.edu
4 Department of Cell and Molecular Biology, University of Mississippi Medical Center, Jackson, MS 39216, USA
5 Department of Pharmacology and Toxicology, University of Mississippi Medical Center, Jackson, MS 39216, USA
6 Biomolecular Sciences, School of Pharmacy, University of Mississippi, Oxford, MS 38677, USA
* Correspondence: rtandon@umc.edu; Tel.: +1-601-984-1705

Abstract: Human cytomegalovirus (HCMV) tegument protein pp150 is essential for the completion of the final steps in virion maturation. Earlier studies indicated that three pp150nt (N-terminal one-third of pp150) conformers cluster on each triplex (Tri1, Tri2A and Tri2B), and extend towards small capsid proteins atop nearby major capsid proteins, forming a net-like layer of tegument densities that enmesh and stabilize HCMV capsids. Based on this atomic detail, we designed several peptides targeting pp150nt. Our data show significant reduction in virus growth upon treatment with one of these peptides (pep-CR2) with an IC_{50} of 1.33 μM and no significant impact on cell viability. Based on 3D modeling, pep-CR2 specifically interferes with the pp150–capsid binding interface. Cells pre-treated with pep-CR2 and infected with HCMV sequester pp150 in the nucleus, indicating a mechanistic disruption of pp150 loading onto capsids and subsequent nuclear egress. Furthermore, pep-CR2 effectively inhibits mouse cytomegalovirus (MCMV) infection in cell culture, paving the way for future animal testing. Combined, these results indicate that CR2 of pp150 is amenable to targeting by a peptide inhibitor, and can be developed into an effective antiviral.

Keywords: CMV; herpesviruses; peptide therapy; tegument; nuclear egress

Citation: Mitra, D.; Hasan, M.H.; Bates, J.T.; Bidwell, G.L., III; Tandon, R. Tegument Protein pp150 Sequence-Specific Peptide Blocks Cytomegalovirus Infection. *Viruses* **2021**, *13*, 2277. https://doi.org/10.3390/v13112277

Academic Editor: Donald M. Coen

Received: 31 August 2021
Accepted: 12 November 2021
Published: 15 November 2021

Publisher's Note: MDPI stays neutral with regard to jurisdictional claims in published maps and institutional affiliations.

1. Introduction

Human Cytomegalovirus (HCMV), a betaherpesvirus, infects the majority of the world's population but acute disease is manifested only in a small proportion of infected individuals [1]. Primary infection or reactivation of latent virus can cause life-threatening complications in immunocompromised individuals such as AIDS patients and transplant recipients. HCMV is also the leading cause of congenital infections and is associated with a multitude of cardiovascular diseases [2,3]. Currently, there are no commercially available vaccines for HCMV infection. Antivirals are available; however, adverse side effects and drug resistance are a growing concern.

HCMV has a prototypical herpesvirus virion (infectious virus particle) that consists of an external membranous envelope, a proteinaceous layer called the tegument and an icosahedral capsid that contains a compressed, large (>240 kb) double-stranded DNA genome. Like all herpesviruses, HCMV maturation occurs in two distinct phases, primary and secondary maturation [4]. Primary maturation begins in the host-cell nucleus where viral genome replication, capsid assembly and encapsidation take place [5]. After assembly, the nucleocapsids (NC) migrate from the nucleus to the cytoplasm. During this nuclear egress, the nucleocapsid first undergoes a primary envelopment at the inner nuclear membrane, then traverses through the nuclear envelope, followed by de-envelopment at the outer nuclear membrane, and finally reaches the cytoplasm where it accumulates within a ring-shaped perinuclear structure known as the cytoplasmic virus assembly compartment (vAC) where the secondary or final steps of virion maturation occur [4–9].

The virion maturation process is characterized by numerous virus–virus, host–host and host–virus interactions, and the tegument proteins play critical roles during this process [1,8,9]. The tegument proteins are also active during viral entry as they are released into the cells along with viral capsids and aid in viral gene expression, virus replication and evasion of host immune responses [1,10]. During late stages of infection, tegument proteins accumulate to high levels in the vAC and contribute to virion maturation and egress [8–10]. Thus, tegument proteins offer important targets for the development of antiviral therapy against HCMV infection.

The HCMV UL32 gene encodes a prominent betaherpesvirus-conserved virion tegument protein, pp150 (basic phosphoprotein/ppUL32), which is initially expressed in the nucleus but accumulates within the vAC during late stages and supports the final steps in virion maturation [11,12]. To date, several studies have been conducted to understand the structure and function of pp150 [11–21]. It is well-established that pp150 controls cytoplasmic events during virion maturation [12,14,15]. The principal role of pp150 is to stabilize and retain the nucleocapsid organization throughout the final or secondary envelopment inside the vAC [8,11,12]. The cryoEM reconstructions of HCMV virion show that pp150 is arranged in upper and lower helix bundles, which are joined by a central helix [16,18,22–24]. The HCMV capsid is an ensemble of 60 asymmetric units, with each of these units containing (i) 16 copies of the major capsid protein (MCP), which exists in penton and hexon capsomers, (ii) 16 copies of small capsid proteins (SCP) that sit atop each MCP, (iii) Five and one-third heterotrimeric triplexes (Ta, Tb, Tc, Td, Te and Tf, respectively) composed of the triplex monomer protein Tri1 (minor capsid-binding protein), which is coupled with triplex proteins Tri2A and Tri2B (also known as minor capsid proteins) and (iv) 16 copies of the pp150 molecules [1,18,22]. Three pp150 proteins cluster above each triplex and extend towards the three SCPs atop nearby MCPs, forming a net-like layer of tegument densities that enmesh HCMV capsids [22,24]. The atomic model construct for the N-terminal third of pp150 (pp150nt; residues 1 to 285) suggests that N-terminal residues 1 to 275 alone are sufficient for pp150-capsid binding [22]. Several conserved regions in the N-terminal 275 residues have been identified, including a 27 amino acid cysteine tetrad region, which is conserved across all primate cytomegaloviruses, and two betaherpesvirus-conserved regions (CR1 and CR2) [11–13]. The cysteine tetrads and CR1 lie in pp150nt's upper helix bundle, whereas CR2 lies in the lower helix bundle (Figure 1). Pp150nt–capsid interactions occur in both upper and lower helix bundles and stabilize the nucleocapsid for the production of infectious HCMV virion [18,22].

Peptide therapeutics are a promising new strategy for targeted therapy. Various studies have shown the approach of using peptides as potential antivirals [25,26]. Notable examples of peptide therapeutics against herpesvirus infections include anti-heparan sulfate (HS) peptides [27–30]. HS and its highly modified form 3-O-sulfated HS (3-OS HS) is critical for herpes simplex virus-1 (HSV-1) and CMV entry into cells [27,31], and HS removal is required when herpesviruses including CMV is released from the cells after virion maturation [32]. Twelve-mer peptides (G1 and G2) derived from a random M-13 phage display library specifically bind to HS and 3-OS HS, and block HSV-1 entry into human corneal fibroblasts and CHO-K1 cells in vitro. In vivo administration of these peptides completely blocks HSV-1 spread in the mouse cornea. G2 peptides isolated against 3-OS HS inhibit CMV entry into retinal pigment epithelial cells and thereby block CMV infection [27]. A novel synthetic polybasic peptide (p5 + 14) that binds to hypersulfated HS, effectively inhibits HS-mediated entry of MCMV, HCMV and HSV in vitro [29]. Additionally, the D-form of a p5 + 14 related peptide, known as p5R$_D$ effectively reduces CMV infection in vitro and in vivo [30]. All of the above findings indicate that peptides could be used as novel therapeutics against CMV infection.

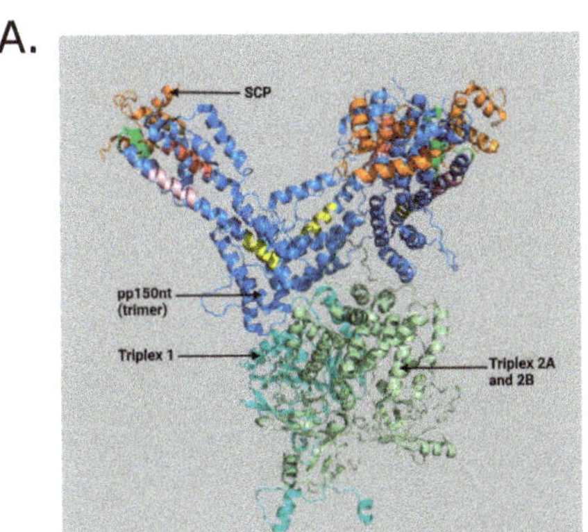

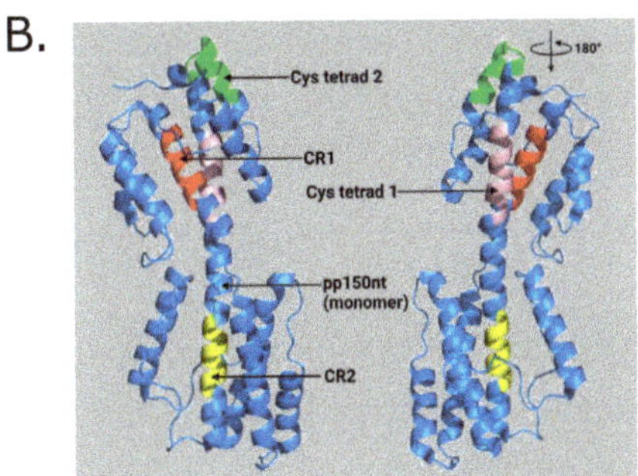

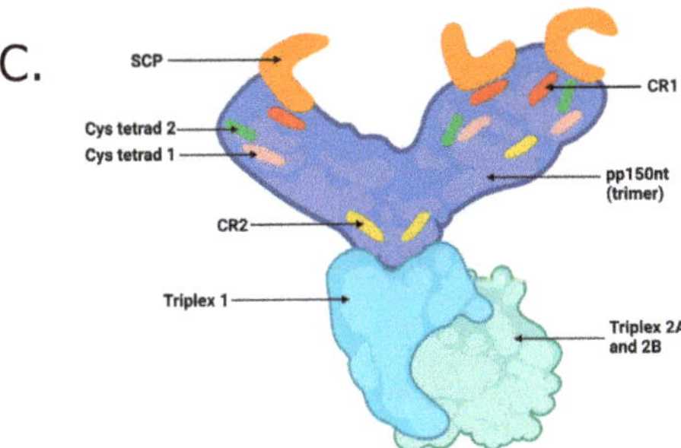

Figure 1. (**A**) Illustration of pp150nt structure and its binding interface with capsid proteins in an HCMV virion. SCP (small capsid protein) binds to MCP (major capsid protein). Triplexes (also known as minor capsid-binding protein) are heterotrimers composed of Tri1 and a Tri2A–Tri2B heterodimer. Three pp150nt residues (pp150nt trimer) cluster above triplexes and extend towards SCP atop nearby MCP, forming a net-like layer of tegument densities. (**B**) Close-up view showing conserved regions in pp150 monomer generated by PyMOL software (Schrödinger, LLC) using RSCB PDB ID: 5VKU and UNIPROT ID: Q6SW99. (**C**) Cartoon illustration of pp150nt trimer and its binding interface with capsid proteins in an HCMV virion generated with BioRender.com (access date: 10 November 2021) Orange—SCP; Blue—pp150nt; Cyan—Triplex 1; Pale green—Triplex 2A and 2B; Red—pp150 conserved region 1 (CR1); Yellow—pp150 conserved region 2 (CR2); Pink—Cysteine tetrad 1; Green—Cysteine tetrad 2 [22].

We targeted the conserved pp150nt regions with sequence-specific peptides, with the goal of developing these peptides into highly effective antivirals. The atomic details of pp150nt structure and its binding interface with capsid proteins (Figure 1A–C) guided the design of peptides targeting CR1, CR2 and cysteine tetrad regions (Table 1; Figure 2). The data show that at least one of these peptides (pep-CR2) is effective in inhibiting HCMV and MCMV growth in cell culture. Microscopic images also suggests that pep-CR2 treated HCMV-infected cells sequester pp150 in the nucleus, thereby compromising the organization of vAC and virion maturation process. Overall, the results in this study indicate that CR2 of pp150 is amenable to targeting by a peptide inhibitor, which can be developed into an effective antiviral.

Table 1. Amino acid sequence of test and control peptides [1].

Peptide	Sequence	HCMV Pp150 UNIPROT Entry (Q6SW99) Residue Numbers
Pep-control	DYKDDDDK (Flag Sequence) Control	NA
Pep-CR1	LFNELMLWL (CR1)	52–60
Pep-CR2	NKLVYTGRL (CR2)	201–209
Pep-CysTetrad1	KCLARIQERCK (Cysteine Tetrad 1)	223–233
Pep-CysTetrad2	MCLSFDSNYCR (Cysteine Tetrad 2)	241–251

[1] All peptides were synthesized with a N-terminal myristoylation by GenScript, USA.

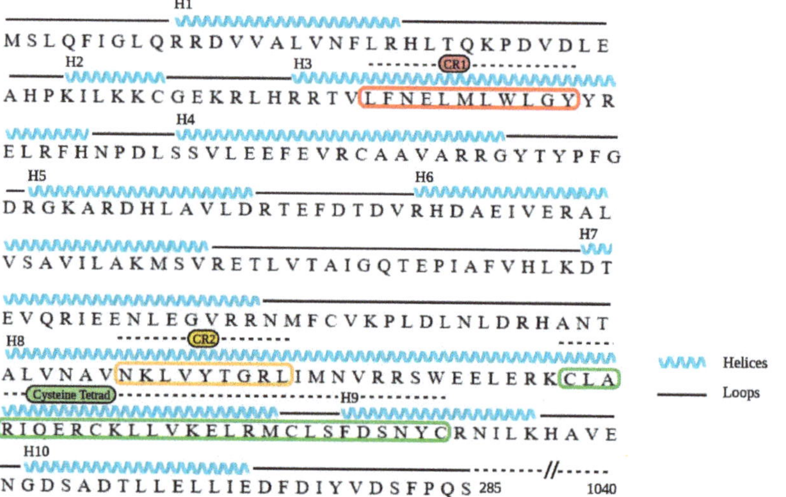

Figure 2. Amino acid sequence and structure-derived secondary structure of pp150nt. Black lines represent loops, blue lines represent helices, CR1 is boxed in red, CR2 boxed in yellow and cysteine tetrad boxed in green. Illustrated in BioRender.com (access date: 10 November 2021) using RSCB PDB ID: 5VKU and UNIPROT ID: Q6SW99 [22].

2. Materials and Methods

Preparation of peptides: Peptides were designed with a sequence identical to the amino acid sequences in the conserved regions in pp150nt (Table 1). All peptides were modified to include a N-terminal myristoyl group for better cell penetration, and were synthesized by GenScript, USA to >85% purity. Seven mg of lyophilized peptides (MW = 1558.02 daltons) was resuspended in 449.3 µL of 100% dimethyl sulfoxide (DMSO) to achieve a stock concentration of 10 mM before use; however, the working concentration of the peptides included less than 0.1% DMSO.

Cells: Human foreskin fibroblasts (HFF) and mouse endothelial fibroblasts (MEF) were cultured in Dulbecco's modified Eagle's medium (DMEM; Corning, Manassas, VA; catalog# 10-013-CM) containing 10% fetal bovine serum (Gibco, Life Technologies, Grand Island, NY, catalog# 10437-028), 4.5 g/mL glucose, 2 mM L-glutamine, 1 mM sodium pyruvate, and 100 U/mL penicillin–streptomycin (Corning, Manassas, VA; catalog# 30-002-CI) at 37 °C with 5% CO_2.

Virus: HCMV (TowneBAC and BAD32 strains) was grown on HFF cells and MCMV (K181 strain) was grown on MEF cells. Virus stock was prepared in 3X autoclaved milk, sonicated 3 times for 10 s with a 30 s gap, and stored at −80 °C. The 3X autoclaved milk is prepared from Carnation (Nestle) instant nonfat dry milk powder. Further, 10% milk was prepared in nanopure water, pH was adjusted to 7.0 and it was autoclaved three times.

Virus infection: During infection, media was removed from the wells of cell culture plates and appropriately diluted virus stock was absorbed onto the cells in DMEM without serum. Cells were incubated for 1 h with gentle shaking every 10 min followed by washing 3X with serum-free DMEM. Fresh complete medium was added, and cells were incubated until the endpoint.

Virus titers: HCMV- and MCMV-infected or mock-infected samples (in triplicates) were harvested within the medium at the designated time points post infection, and stored at −80 °C before titration. On the day of titration, harvested samples (in triplicates) were sonicated three times for 10 s each with 30 s gap. Monolayers of cells (HFFs and MEFs for HCMV and MCMV infection, respectively) were grown in 12-well tissue-culture plates, and serial dilutions of sonicated samples were absorbed onto them for 1 h (in duplicates), followed by 3X washing with serum-free DMEM. For HCMV titers, fresh DMEM containing 10% FBS was added to HFFs and cells were incubated for 9 to 10 days post infection (dpi). For MCMV titers, a carboxymethyl cellulose (CMC; EMD Millipore; catalog# 217274) overlay with complete DMEM (one-part autoclaved CMC and three parts media) was added, and the cells were incubated for 5 days. At the endpoint, DMEM/CMC+DMEM overlay was removed, and cells were washed 2X with PBS. Infected monolayers were fixed in 100% methanol for 5 min. HFFs were immediately stained with Giemsa stain, modified (Sigma-Aldrich, MilliporeSigma, US catalog# GS1L), and MEFs with 1:20 dilution of 1% crystal violet (Fisher Chemicals, Fair Lawn, NJ; catalog# C581-25) for 15 minutes. Plates were finally washed with tap water, air-dried, and plaques with clear zone (MCMV) or dark Giemsa stain (HCMV) were quantified.

IC$_{50}$ assay and percent inhibition calculation: To determine the half-maximal inhibitory concentration (IC_{50}) of pep-CR2, HFFs were pretreated for 1 h with a range of pep-CR2 concentrations (1 µM, 2.5 µM, 5 µM, 7.5 µM, 10 µM, 12.5 µM, 15 µM, 17.5 µM and 20 µM), pep-control, ganciclovir (GCV) or mock (in triplicates), and then infected with HCMV at a multiplicity of infection (MOI) of 0.1. Cells were fixed at 10 dpi, and the number of foci for each concentration in triplicate wells were enumerated. The data (# of foci) were normalized in GraphPad Prism v9.0, taking the highest number of average foci in the assay (164.66) as 0% inhibition and the lowest number of average foci (2) as 100% inhibition. The percentage (%) inhibition of pep-CR2 was plotted against a concentration range of 1 µM-20 µM on a bar graph. The calculation used in GraphPad Prism was ((value—minOfValues)/(maxOfValues—minOfValues))*100.

Microscopy: Samples were prepared using established protocols for fluorescence microscopy. Briefly, HFF cells were grown on coverslip inserts in 24-well tissue-culture plates. Cells were pretreated with pep-control and pep-CR2 at 10 µM concentration for 1 h before infection with HCMV (BAD32 strain where pp150 is GFP tagged) at an MOI of 3.0. At the endpoint (4 dpi), cells were fixed in 3.7% formaldehyde for 10 min and were incubated in 50 mM NH_4Cl in PBS for 10 min to reduce autofluorescence. Followed by 2X PBS wash, cells were incubated in 0.5% Triton X-100 for 20 min for permeabilization and washed with PBS. Finally, cells were incubated in Hoechst solution (ThermoFisher Scientific, Waltham, MA, USA, Catalog# 33342) in PBS (1:3000) for 10 min to stain the nucleus, followed by PBS wash. Coverslips were retrieved from the wells and were mounted on glass slides with a drop of mounting medium (2.5% DABCO in Fluoromont G), and air-dried for two hours before imaging. Images were acquired on an EVOS-FL epifluorescent microscope (ThermoFisher Scientific, Waltham, MA, USA).

Cell viability: HFF and MEF cells were plated in 24-well tissue-culture plates and grown to confluency (in triplicates). Cells were pretreated for 1 h with appropriate concentration of controls and test peptides and then infected with HCMV or MCMV at an MOI of 3.0 or mock-infected. At 3 dpi, HFFs received fresh, complete DMEM containing fresh peptides or controls. The medium was removed at 5 dpi, and HFFs were harvested by trypsinization. For MEFs, CMC+DMEM overlay was removed at 3 dpi before harvesting the cells by trypsinization. Cell viability was determined using trypan blue exclusion using a TC20 automated cell counter (BioRad Laboratories, Hercules, CA, USA) following the manufacturer's protocol.

Statistics: Student's t-tests were conducted in Graphpad Prism, comparing the means of different groups (GraphPad Prism version 9.0.0, GraphPad Software, San Diego, CA, USA, www.graphpad.com; access date 10 November 2021). Standard error of mean was plotted as error bars. A p value of <0.05 was considered significant. An asterisk (*) indicates significant inhibition compared to the wild type. For multiple comparisons, data were analyzed with one-way ANOVA in GraphPad Prism 9.0 and differences between groups were considered significant at a p value of <0.05.

3. Results

3.1. Inhibition of Virus Growth upon Pep-CR2 Treatment

First, we sought to establish the inhibitory efficiency of the peptides. Primary human foreskin fibroblasts (HFF) were pretreated with pep-CR1, pep-CR2, pep-CysTetrad1, pep-CysTetrad2, pep-control and DMSO at 10 µM concentration for 1 h, before infecting them with HCMV at an MOI of 3.0. Peptide concentrations were maintained in the infected cell culture. Cells were harvested at 5 days post infection (dpi), and viral plaque-forming units were enumerated by plating of serial dilutions on fibroblasts. The titers indicated >10-fold reduction in virus growth in pep-CR2-treated cells but not when treated with pep-CR1, pep-CysTetrad1, pep-CysTetrad2, pep-control or DMSO (Figure 3A). To assess the impact of pep-CR2 on virus spread, plaque sizes were calculated and compared between mock vs. pep-CR2-treated groups. Pep-CR2-treated infected cells showed significantly smaller plaque sizes compared to the mock-treated group, indicating reduction in virus spread to the adjacent cells in a foci, hence inhibiting virus growth (Figure 3B,C). Next, to determine the half-maximal inhibitory concentration (IC_{50}) of pep-CR2, HFFs were pretreated with pep-CR2 (with a concentration range of 1–20 µM), pep-control and ganciclovir (GCV; as an inhibitory control). Cells were then infected with HCMV at an MOI of 0.1, and virus yield was measured at 10 dpi by enumerating the plaque-forming units. The percentage inhibition of HCMV by pep-CR2 was plotted against the treatment concentration (Figure 4A), and the IC_{50} was calculated to be 1.33 µM with a 95% confidence interval of 1.182 to 1.474 µM. Compared to the pep-control, pep-CR2 showed 90% inhibition of virus titers at 10 µM concentration (Figure 4B), confirming previous results at higher MOI (Figure 3A). To test the cytotoxicity of pep-CR2, a cell viability assay was performed on pep-CR2 (10 µM)-treated infected and uninfected HFFs along with appropriate controls. Cell viability was similar across treatment groups in this assay (Figure 5A). Furthermore, pep-CR2-treated infected cells had significantly higher cell viability compared to the control groups, indicating that pep-CR2 efficiently protected the HCMV-infected cells from virus-induced lytic death (Figure 5B). Overall, the data from these experiments indicate a significant reduction in virus growth upon pep-CR2 treatments, without impacting cell viability.

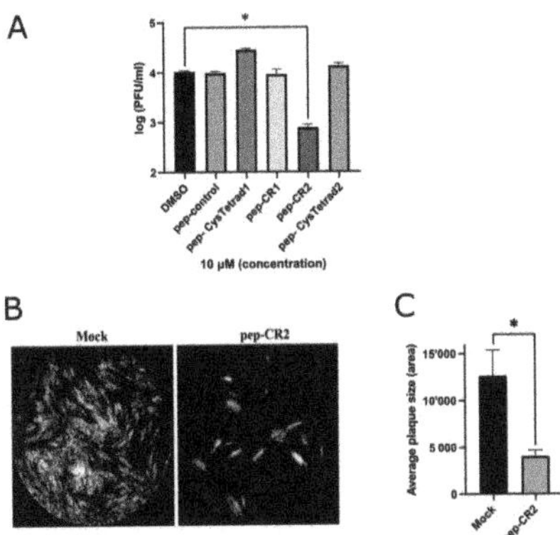

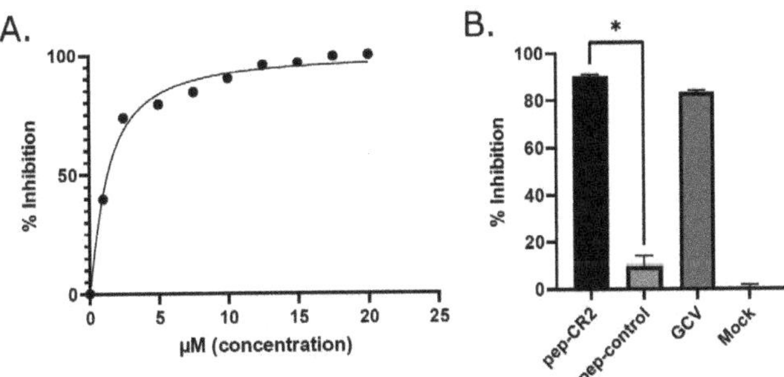

Figure 3. (**A**) Inhibition of virus growth upon pep-CR2 treatment. Cells (human foreskin fibroblasts (HFF)) were pretreated with 10 µM concentration of control (DMSO), pep-control, pep-CysTetrad1, pep-CysTetrad2, pep-CR1 and pep-CR2 (in triplicates) and infected with HCMV at an MOI of 3.0. Cells were harvested at 5 dpi (days post infection) and virus titers were assessed on HFFs. Data were analyzed by Student's *t*-tests, comparing the means of DMSO control and the peptide-treated group. (**B**) GFP+ images showing plaque size comparison in pep-CR2 and mock-treatment control group. Cells were pretreated with pep-CR2 (10 µM) and mock for 1 h (in triplicates) before infecting them with HCMV at a low MOI of 0.1. Cells were fixed at 10 dpi and virus yield was measured by counting the number of plaques. Images were acquired on a Cytation5 (BioTek Instruments, USA) cell-imaging reader with 4X magnification. (**C**) Plaque size comparison between pep-CR2 and mock-treated cells. Individual plaque sizes were measured by counting the area of the plaques in ImageJ software (National Institutes of Health, Bethesda, MD, USA). Standard error of mean was plotted as error bars. Data were analyzed by Student's t-tests, comparing the means of the test and the control group. A *p* value of <0.05 was considered significant. An asterisk (*) indicates significant inhibition compared to wild type.

Figure 4. (**A**) Determining the IC$_{50}$ (half-maximal inhibitory concentration) of pep-CR2. HFFs were pretreated for 1 h with different concentrations of pep-CR2 (1 µM, 2.5 µM, 5 µM, 7.5 µM, 10 µM, 12.5 µM, 15 µM, 17.5 µM and 20 µM), pep-control, ganciclovir (GCV) or mock (in triplicates) and then infected with HCMV at a low MOI of 0.1. Cells were fixed at 10 dpi, and virus yield was measured by counting the number of plaques. The percentage inhibition of pep-CR2 is plotted against concentration range of 1 µM-20 µM. The IC$_{50}$ of pep-CR2 was calculated to be 1.33 µM with a 95% confidence interval of 1.182 to 1.474 µM. (**B**) Comparing inhibition of virus growth by pep-CR2, pep-control, GCV and no-treatment control (Mock) at 10 µM concentration. Pep-CR2 treated cells showed ~90% inhibition in virus growth compared to the controls. Data were analyzed by Student's *t*-tests, comparing the means of control and treatment groups. Standard error of mean was plotted as error bars. A *p* value of <0.05 was considered significant. An asterisk (*) indicates significant inhibition compared to wild type.

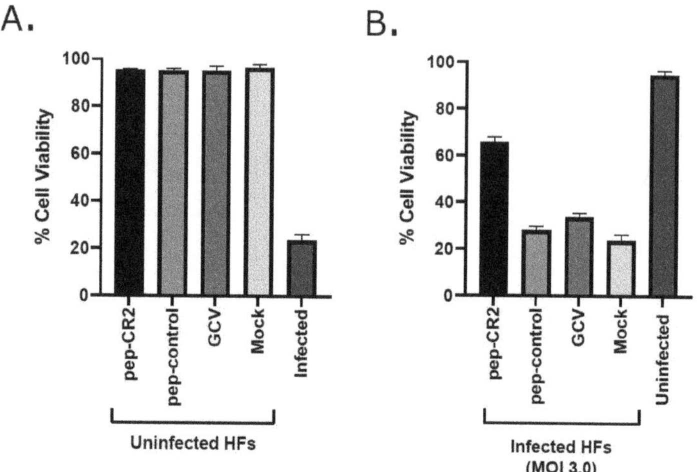

Figure 5. Cell viability (%) in (**A**) pep-CR2-treated uninfected cells vs. (**B**) pep-CR2-treated infected cells. Cells were pretreated with pep-CR2 as well as with appropriate controls (pep-control and GCV) or mock (in triplicates) for 1 h and were then either infected with HCMV at a high MOI of 3.0 or mock-infected. Cell viability was performed by trypan blue exclusion assay at 5 dpi for both groups. Results indicate that pep-CR2 protects cells from virus-induced lytic cell death. Samples were analyzed with one-way ANOVA in GraphPad Prism 9.0 and showed significant differences between groups ($p < 0.05$).

3.2. Pep-CR2 Treatment Sequesters pp150 in the Nucleus of Infected Cells

To study the localization of pp150 in pep-CR2- and pep-control-treated infected cells, HFFs were pretreated with pep-CR2 or pep-control, as described earlier, and then infected with a strain of HCMV where pp150 is fused with GFP (BAD32, MOI 3.0) [20]. Cells were fixed and imaged by epifluorescent microscopy at 4 dpi. These localization studies showed that pep-CR2-treated cells contain a diffused GFP signal in the cytoplasm of infected cells in contrast to a solid juxtanuclear sphere of GFP, which corresponds to vAC in the pep-control-treated cells (Figure 6). The intensity of the GFP signal was much higher in the nucleus of pep-CR2-treated cells compared to the nucleus of the pep-control-treated cells. These results indicate that pep-CR2 may be compromising virion maturation by sequestering pp150 in the nucleus of infected cells and interfering with the organization of the vAC.

3.3. Pep-CR2 Shows Similar Inhibitory Potential against Murine Cytomegalovirus (MCMV)

Conserved virion tegument protein pp150 is encoded by HCMV UL32 gene. The amino terminal one-third of pp150 contains CR1 and CR2, which are known to be the two most highly conserved regions among all betaherpesvirus pp150s [13]. To further strengthen our hypothesis that targeting of conserved regions of pp150 by pep-CR2 blocks CMV infection, we investigated the inhibitory potential of pep-CR2 against MCMV. Since CMV is species specific, we performed the MCMV experiments in mouse endothelial fibroblasts (MEFs). MEFs were pretreated with pep-CR2 and pep-control, as described earlier, before infecting them with MCMV (K181) at an MOI of 3.0. Peptide concentrations were maintained in the infected cell culture. Cells were harvested at 3 dpi, and virus titers were enumerated. The titers indicate a >10-fold reduction in virus growth in pep-CR2-treated cells compared with the pep-control treatment group (Figure 7A). Next, to rule out any cytotoxic effects of pep-CR2, a cell viability assay was performed on pep-CR2-treated infected and uninfected MEFs along with pep-control and GCV as appropriate controls. Cell viability was not affected at the treated concentration of pep-CR2 in uninfected cells (Figure 7B). Additionally, pep-CR2-treated infected cells had higher cell viability compared with control groups, indicating that pep-CR2 efficiently protected the MCMV-infected cells from virus-induced lytic death (Figure 7C). Overall, these results indicate that pep-CR2 shows similar inhibitory potential against MCMV compared to HCMV.

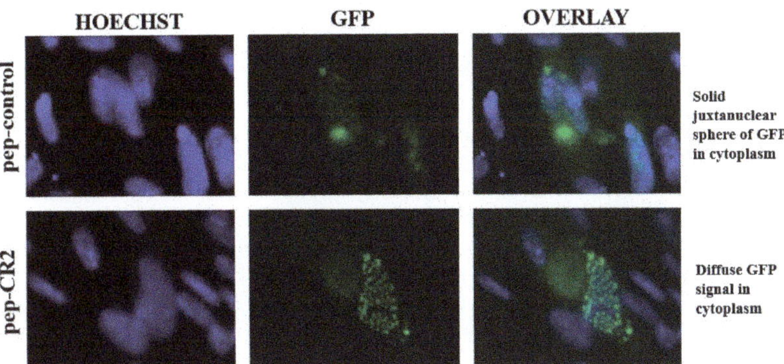

Figure 6. Epifluorescent imaging indicating pp150 localization in the nucleus and the cytoplasm of pep-CR2 and pep-control-treated cells. Cells were pretreated with pep-control and pep-CR2 at 10 µM concentration for 1 h before infecting them with HCMV (BAD32 strain where pp150 is GFP tagged) at an MOI of 3.0. Cells were fixed and imaged under an epifluorescent microscope at 4 dpi. Hoechst stains the nucleus. Results indicate that pep-CR2 may be compromising virion maturation by sequestering pp150 in the nucleus of infected cells and interfering with the organization of the virus assembly compartment (vAC).

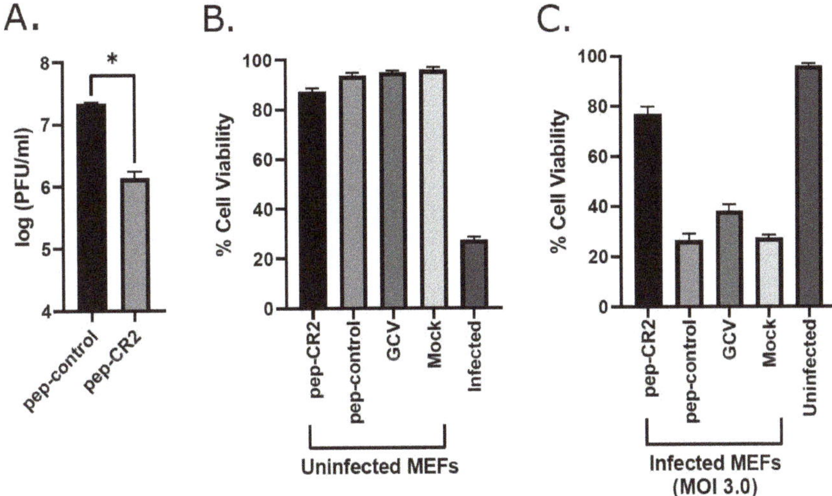

Figure 7. (**A**) Inhibition of MCMV growth upon pep-CR2 treatment. MEF cells were pretreated with 10 µM concentration of pep-CR2 and pep-control (in triplicates) and infected with MCMV at an MOI of 3.0. Cells were harvested at 3 dpi and virus titers were assessed on MEFs. Data were analyzed by Student's t-tests, comparing the means of pep-control and pep-CR2. Standard error of mean was plotted as error bars. A *p* value of <0.05 was considered significant. An asterisk (*) indicates significant inhibition compared to wild type. Cell viability (%) in (**B**) pep-CR2-treated uninfected MEFs vs. (**C**) pep-CR2-treated infected MEFs. Cells were pretreated with pep-CR2 as well as with appropriate controls (pep-control and GCV) or mock (in triplicates) for 1 h and were then either infected with MCMV at a high MOI of 3.0 or mock-infected. Cell viability was performed by trypan blue exclusion assay at 3 dpi for both groups. Results indicate that pep-CR2 protects cells from virus-induced lytic cell death. Samples were analyzed with one-way ANOVA in GraphPad Prism 9.0 and showed significant differences between groups (*p* < 0.05).

4. Discussion

In this study, we utilized multiple approaches to demonstrate that CR2 of CMV tegument protein pp150 is amenable to targeting by sequence-specific peptide pep-CR2. We first screened the inhibitory efficiency of peptides that targeted the conserved regions in pp150. The results indicated that upon treatment with peptides that target CR2 of pp150 (pep-CR2), there is a significant reduction in virus growth in cell culture (Figure 3). To further investigate the dose response of pep-CR2 on infected human fibroblasts, we performed an IC_{50} assay using a concentration range of 1 to 20 μM. The half-maximal inhibitory concentration was calculated to be 1.33 μM (Figure 4A), and pep-CR2 showed ~90% inhibition of virus growth at 10 μM concentration (Figure 4B). Pep-CR2 treatment of HFFs did not affect cell viability for the duration of treatment (Figure 5A) confirming that the observed reduction in virus titer was not due to cell death. Moreover, when HFs were pretreated and maintained in pep-CR2 throughout the course of infection, the cells resisted infection-induced cell death at late time post infection (Figure 5B). To understand the possible mechanism of inhibition of HCMV replication by pep-CR2, we looked at the cytoplasmic and nuclear localization of pp150 at late time post infection. The results indicated that pp150 is sequestered in the nucleus of infected cells that are treated with pep-CR2 (Figure 6). This would conform to a mechanistic disruption of pp150 loading onto capsids and subsequent cytoplasmic egress, which might compromise vAC organization and virion maturation processes.

Tegument proteins play a critical role in the CMV life cycle process in the host cell, and can be considered as important targets for the development of antiviral therapy. Pp150 is a late viral tegument protein which supports the final steps of the maturation process [9,11–13]. CMV infection can persist in latent, chronic and productive stages depending on the host physiology and the types of cells being infected [1]. CMV most likely establishes life-long persistence by replicating at a subclinical level and occasionally reactivating from the latent stage triggered by various biological, chemical or physical stimulations including immunological disorders, radiation and cell differentiation among others [1,33,34]. Therefore, an effective antiviral against CMV would target cell-to-cell spread of latently reactivated virus. This can be effectively achieved by targeting late stages of virus maturation.

During the primary CMV virion maturation, capsid assembly and DNA encapsidation take place inside the nucleus [1]. After the nucleocapsid is formed, it acquires some tegument proteins, including pp150, and migrates from the nucleus to the cytoplasm. Once the nucleocapsid is in the cytoplasm, it accumulates in a ring-shaped perinuclear viral assembly compartment (vAC), where the final or secondary steps of maturation take place. Pp150 is a large, conserved, CMV virion tegument protein, which hitches a ride on the capsids during the nuclear egress and accumulates in the vAC [8,9,11,12]. Pp150 is well-characterized to control cytoplasmic events during virus maturation as it stabilizes and retains the nucleocapsid organization throughout the final envelopment inside the vAC [12].

Pep-CR2 likely interferes with the pp150–capsid interaction in the nucleus of the cells, thereby blocking pp150 loading onto capsids and its translocation out of the nucleus and into the cytoplasm. Additionally, viral DNA can exit the nucleus only when is it packaged inside the capsid. In the absence of pp150, or if the pp150–capsid interaction is abolished, the nucleocapsids can still exit the nucleus but most likely will be degraded in the cytoplasm [12]. Pep-CR2 would block pp150 from capsid binding by preventing interactions with the triplexes. This interface may not be completely evident in the 2D models depicted in Figure 1; however, conformational changes may occur that could allow CR2 and the triplexes to come closer together. Alternatively, pep-CR2 could be binding to a capsid protein prior to their nuclear localization or capsid assembly that then would prevent pp150 association. Furthermore, the peptide could be preventing other host or viral protein interactions with pp150 that would also result in a similar phenotype to what is observed in this study.

Pp150-interacting host proteins determine virus maturation, and a disruption of these interactions can compromise the virus maturation process. Pp150 interacts with Bicaudal D1 (BicD1), a protein that regulates trafficking within the secretory pathway, and this interaction is required for the correct localization of pp150 in the vAC of the infected cells [19]. In the presence of pep-CR2, the BicD1–pp150 interaction will be blocked or reduced, since the exit of pp150 from the nucleus to the cytoplasm is affected. This will interfere with the formation of vAC and proper virus maturation. Other CMV proteins such as pUL96 interact with pp150 and stabilize pp150-associated nucleocapsids [14]. Interfering with the pp150–pUL96 interaction will also lead to a defect in the formation of vAC, since pUL96 and pp150 are both implicated in the formation and maintenance of vAC.

Additionally, dynamin–clathrin-mediated endocytic pathways are important for cytoplasmic stages of CMV maturation [35]. Pp150 is known to interact directly with clathrin, which affects the virion maturation process [20]. In the presence of pep-CR2, the pp150–clathrin interaction would also be compromised due to reduced availability of pp150 in the cytoplasm, which in turn would affect the vAC organization and the maturation process. BicD1 also interacts with the dynamin–clathrin-mediated membrane trafficking pathway [36]. Since pp150 is known to interact with both BicD1 and clathrin, it is possible that the presence of pep-CR2 would disrupt pp150–clathrin and/or pp150–BicD and/or BicD1–clathrin interactions, resulting in impaired trafficking and virus maturation.

Cryo-electron microscopic studies on isolated B capsids purified from HCMV- and simian cytomegalovirus (SCMV)-infected cells show that pp150 interacts with capsid proteins during virus maturation [37,38]. The amino terminal one-third of HCMV pp150 (pp150nt) as well as the SCMV homolog of UL32 are responsible for capsid binding [13]. This pp150nt–capsid interaction is important for the virion maturation process because it retains and stabilizes the organization of nucleocapsids throughout the final steps of maturation [1,12,18]. Earlier studies using a HCMV UL32-deletion mutant showed that pp150 is critical for virion egress [11]. However, subsequent transmission-electron-microscopic studies of UL32-deletion-mutant (ΔUL32)-infected cells revealed that although nucleocapsids in the nucleus appeared similar in both wild-type and ΔUL32-mutant cells, there were fewer cytoplasmic virus particles in mutant cells compared to WT, indicating that in the absence of pp150, maturing nucleocapsids are unstable in the cytoplasm following nuclear egress [12].

Pp150nt (amino acid residues 1–275) is necessary and sufficient for capsid–tegument interaction, and it consists of conserved regions including CR1 and CR2. Deletion of the amino terminal of pp150 or disruption of either CR1 or CR2 in SCMV blocked the secondary spread of mutant virus, thereby indicating that CR1 and CR2 domains are critical for virus maturation [11]. Further studies by Tandon and Mocarski successfully characterized CR2 phenotypes, where they showed that two independent CR2 point mutants failed to support viral replication. These replication-defective CR2 mutants exhibited a phenotype similar to a virus that carried a complete deletion of UL32 ORF (ΔUL32). The UL32 mutant-virus-infected cells showed defect in the formation of vAC, which was highly vesiculated and contained fewer nucleocapsids or complete virus particles in contrast with the intact, wild-type vAC [12]. The pp150 localization results in HCMV-infected fibroblasts (Figure 6) are consistent with these earlier findings, where the perturbation of pp150-CR2 by peptides (pep-CR2) compromises the formation of vAC, thereby affecting virus maturation.

To further strengthen our hypothesis that CR2 of pp150 can be targeted by pep-CR2, we investigated the inhibitory efficiency of pep-CR2 against MCMV. CR1 and CR2 are the two most highly conserved regions among various CMV species, and it is expected that pep-CR2 will target CR2 of M32 (UL32 homolog of pp150 in MCMV). The results in Figure 7 show that pep-CR2 treatment significantly reduces MCMV growth with no impact on cell viability, which corroborates the results of pep-CR2 treatment on HCMV-infected cells.

Overall, the results in this study show that CR2 of pp150 is amenable to targeting by a sequence-specific peptide inhibitor, which can be developed into an effective antiviral.

Author Contributions: R.T. designed and guided the experiments; D.M., M.H.H., J.T.B. and R.T. performed the experiments and analyzed the data. R.T. and D.M. wrote and edited the manuscript. Conceptualization, R.T., D.M. and G.L.B.III; methodology, R.T., M.H.H., D.M., and J.T.B.; software, D.M. and J.T.B.; validation, R.T., D.M. and G.L.B.III; formal analysis, R.T. and D.M.; investigation, R.T., D.M and G.L.B.III; resources, R.T.; writing—original draft preparation, D.M.; writing—review and editing, R.T. and G.L.B.III; supervision, R.T.; project administration, R.T.; funding acquisition, R.T. All authors have read and agreed to the published version of the manuscript.

Funding: This research was funded by NASA, grant number 80NSSC19K1603 (PI: Tandon).

Institutional Review Board Statement: Not applicable.

Informed Consent Statement: Not applicable.

Data Availability Statement: All data are included in the manuscript itself.

Acknowledgments: The authors thank Lauren A. Fassero for assistance with cell-culture maintenance and Christian Yu for helping with the ATP assay for review comments.

Conflicts of Interest: The authors declare no conflict of interest

References

1. Mocarski, E.S., Jr.; Shenk, T.; Pass, R.F. *Fields Virology*, 5th ed.; Howley, D.M.K.P.M., Ed.; Lippincott Williams & Wilkins: Philadelphia, PA, USA, 2006.
2. Griffiths, P.; Reeves, M. Pathogenesis of human cytomegalovirus in the immunocompromised host. *Nat. Rev. Microbiol.* **2021**, *19*, 759–773. [CrossRef]
3. Britt, W.J. Congenital Human Cytomegalovirus Infection and the Enigma of Maternal Immunity. *J. Virol* **2017**, *91*, 759–773. [CrossRef]
4. Mettenleiter, T.C.; Klupp, B.G.; Granzow, H. Herpesvirus assembly: A tale of two membranes. *Curr. Opin. Microbiol.* **2006**, *9*, 423–429. [CrossRef]
5. Hellberg, T.; Passvogel, L.; Schulz, K.S.; Klupp, B.G.; Mettenleiter, T.C. Nuclear Egress of Herpesviruses: The Prototypic Vesicular Nucleocytoplasmic Transport. *Adv. Virus Res.* **2016**, *94*, 81–140. [CrossRef]
6. Alwine, J.C. The human cytomegalovirus assembly compartment: A masterpiece of viral manipulation of cellular processes that facilitates assembly and egress. *PLoS Pathog* **2012**, *8*, e1002878. [CrossRef] [PubMed]
7. Procter, D.J.; Banerjee, A.; Nukui, M.; Kruse, K.; Gaponenko, V.; Murphy, E.A.; Komarova, Y.; Walsh, D. The HCMV Assembly Compartment Is a Dynamic Golgi-Derived MTOC that Controls Nuclear Rotation and Virus Spread. *Dev. Cell* **2018**, *45*, 83–100. [CrossRef] [PubMed]
8. Tandon, R.; Mocarski, E.S. Viral and host control of cytomegalovirus maturation. *Trends Microbiol.* **2012**, *20*, 392–401. [CrossRef] [PubMed]
9. Britt, B. Maturation and egress. In *Human Herpesviruses: Biology, Therapy, and Immunoprophylaxis*; Arvin, A., Campadelli-Fiume, G., Mocarski, E., Moore, P.S., Roizman, B., Whitley, R., Yamanishi, K., Eds.; Cambridge University Press: Cambridge, UK, 2007.
10. Kalejta, R.F. Tegument proteins of human cytomegalovirus. *Microbiol. Mol. Biol. Rev.* **2008**, *72*, 249–265. [CrossRef] [PubMed]
11. AuCoin, D.P.; Smith, G.B.; Meiering, C.D.; Mocarski, E.S. Betaherpesvirus-conserved cytomegalovirus tegument protein ppUL32 (pp150) controls cytoplasmic events during virion maturation. *J. Virol.* **2006**, *80*, 8199–8210. [CrossRef]
12. Tandon, R.; Mocarski, E.S. Control of cytoplasmic maturation events by cytomegalovirus tegument protein pp150. *J. Virol.* **2008**, *82*, 9433–9444. [CrossRef]
13. Baxter, M.K.; Gibson, W. Cytomegalovirus basic phosphoprotein (pUL32) binds to capsids in vitro through its amino one-third. *J. Virol.* **2001**, *75*, 6865–6873. [CrossRef] [PubMed]
14. Tandon, R.; Mocarski, E.S. Cytomegalovirus pUL96 is critical for the stability of pp150-associated nucleocapsids. *J. Virol.* **2011**, *85*, 7129–7141. [CrossRef] [PubMed]
15. Brechtel, T.M.; Mocarski, E.S.; Tandon, R. Highly acidic C-terminal region of cytomegalovirus pUL96 determines its functions during virus maturation independently of a direct pp150 interaction. *J. Virol.* **2014**, *88*, 4493–4503. [CrossRef] [PubMed]
16. Liu, W.; Dai, X.; Jih, J.; Chan, K.; Trang, P.; Yu, X.; Balogun, R.; Mei, Y.; Liu, F.; Zhou, Z.H. Atomic structures and deletion mutant reveal different capsid-binding patterns and functional significance of tegument protein pp150 in murine and human cytomegaloviruses with implications for therapeutic development. *PLoS Pathog* **2019**, *15*, e1007615. [CrossRef]
17. Bogdanow, B.; Weisbach, H.; von Einem, J.; Straschewski, S.; Voigt, S.; Winkler, M.; Hagemeier, C.; Wiebusch, L. Human cytomegalovirus tegument protein pp150 acts as a cyclin A2-CDK-dependent sensor of the host cell cycle and differentiation state. *Proc. Natl. Acad. Sci. USA* **2013**, *110*, 17510–17515. [CrossRef]
18. Dai, X.; Yu, X.; Gong, H.; Jiang, X.; Abenes, G.; Liu, H.; Shivakoti, S.; Britt, W.J.; Zhu, H.; Liu, F.; et al. The smallest capsid protein mediates binding of the essential tegument protein pp150 to stabilize DNA-containing capsids in human cytomegalovirus. *PLoS Pathog* **2013**, *9*, e1003525. [CrossRef]
19. Indran, S.V.; Ballestas, M.E.; Britt, W.J. Bicaudal D1-dependent trafficking of human cytomegalovirus tegument protein pp150 in virus-infected cells. *J. Virol.* **2010**, *84*, 3162–3177. [CrossRef]

20. Moorman, N.J.; Sharon-Friling, R.; Shenk, T.; Cristea, I.M. A targeted spatial-temporal proteomics approach implicates multiple cellular trafficking pathways in human cytomegalovirus virion maturation. *Mol. Cell Proteom.* **2010**, *9*, 851–860. [CrossRef]
21. Sampaio, K.L.; Cavignac, Y.; Stierhof, Y.D.; Sinzger, C. Human cytomegalovirus labeled with green fluorescent protein for live analysis of intracellular particle movements. *J. Virol.* **2005**, *79*, 2754–2767. [CrossRef]
22. Yu, X.; Jih, J.; Jiang, J.; Zhou, Z.H. Atomic structure of the human cytomegalovirus capsid with its securing tegument layer of pp150. *Science* **2017**, *356*, 759–773. [CrossRef]
23. Yu, X.; Shah, S.; Lee, M.; Dai, W.; Lo, P.; Britt, W.; Zhu, H.; Liu, F.; Zhou, Z.H. Biochemical and structural characterization of the capsid-bound tegument proteins of human cytomegalovirus. *J. Struct. Biol.* **2011**, *174*, 451–460. [CrossRef] [PubMed]
24. Li, Z.; Pang, J.; Dong, L.; Yu, X. Structural basis for genome packaging, retention, and ejection in human cytomegalovirus. *Nat. Commun.* **2021**, *12*, 4538. [CrossRef] [PubMed]
25. Vilas Boas, L.C.P.; Campos, M.L.; Berlanda, R.L.A.; de Carvalho Neves, N.; Franco, O.L. Antiviral peptides as promising therapeutic drugs. *Cell Mol. Life Sci.* **2019**, *76*, 3525–3542. [CrossRef] [PubMed]
26. Agarwal, G.; Gabrani, R. Antiviral Peptides: Identification and Validation. *Int J. Pept. Res. Ther.* **2020**, 1–20. [CrossRef]
27. Tiwari, V.; Liu, J.; Valyi-Nagy, T.; Shukla, D. Anti-heparan sulfate peptides that block herpes simplex virus infection in vivo. *J. Biol. Chem.* **2011**, *286*, 25406–25415. [CrossRef]
28. Jackson, J.W.; Hancock, T.J.; Dogra, P.; Patel, R.; Arav-Boger, R.; Williams, A.D.; Kennel, S.J.; Wall, J.S.; Sparer, T.E. Anticytomegalovirus Peptides Point to New Insights for CMV Entry Mechanisms and the Limitations of In Vitro Screenings. *mSphere* **2019**, *4*, e00586-18. [CrossRef] [PubMed]
29. Dogra, P.; Martin, E.B.; Williams, A.; Richardson, R.L.; Foster, J.S.; Hackenback, N.; Kennel, S.J.; Sparer, T.E.; Wall, J.S. Novel heparan sulfate-binding peptides for blocking herpesvirus entry. *PLoS ONE* **2015**, *10*, e0126239. [CrossRef]
30. Pitt, E.A.; Dogra, P.; Patel, R.S.; Williams, A.; Wall, J.S.; Sparer, T.E. The D-form of a novel heparan binding peptide decreases cytomegalovirus infection in vivo and in vitro. *Antiviral Res.* **2016**, *135*, 15–23. [CrossRef] [PubMed]
31. Mitra, D.; Hasan, M.H.; Bates, J.T.; Bierdeman, M.A.; Ederer, D.R.; Parmar, R.C.; Fassero, L.A.; Liang, Q.; Qiu, H.; Tiwari, V.; et al. The degree of polymerization and sulfation patterns in heparan sulfate are critical determinants of cytomegalovirus entry into host cells. *PLoS Pathog* **2021**, *17*, e1009803. [CrossRef]
32. Hadigal, S.R.; Agelidis, A.M.; Karasneh, G.A.; Antoine, T.E.; Yakoub, A.M.; Ramani, V.C.; Djalilian, A.R.; Sanderson, R.D.; Shukla, D. Heparanase is a host enzyme required for herpes simplex virus-1 release from cells. *Nat. Commun.* **2015**, *6*, 6985. [CrossRef]
33. Mehta, S.K.; Laudenslager, M.L.; Stowe, R.P.; Crucian, B.E.; Sams, C.F.; Pierson, D.L. Multiple latent viruses reactivate in astronauts during Space Shuttle missions. *Brain Behav. Immun.* **2014**, *41*, 210–217. [CrossRef]
34. Goodrum, F. Human Cytomegalovirus Latency: Approaching the Gordian Knot. *Annu. Rev. Virol.* **2016**, *3*, 333–357. [CrossRef] [PubMed]
35. Hasan, M.H.; Davis, L.E.; Bollavarapu, R.K.; Mitra, D.; Parmar, R.; Tandon, R. Dynamin Is Required for Efficient Cytomegalovirus Maturation and Envelopment. *J. Virol.* **2018**, *92*, 759–773. [CrossRef] [PubMed]
36. Li, X.; Kuromi, H.; Briggs, L.; Green, D.B.; Rocha, J.J.; Sweeney, S.T.; Bullock, S.L. Bicaudal-D binds clathrin heavy chain to promote its transport and augments synaptic vesicle recycling. *EMBO J.* **2010**, *29*, 992–1006. [CrossRef]
37. Chen, D.H.; Jiang, H.; Lee, M.; Liu, F.; Zhou, Z.H. Three-dimensional visualization of tegument/capsid interactions in the intact human cytomegalovirus. *Virology* **1999**, *260*, 10–16. [CrossRef]
38. Trus, B.L.; Gibson, W.; Cheng, N.; Steven, A.C. Capsid structure of simian cytomegalovirus from cryoelectron microscopy: Evidence for tegument attachment sites. *J. Virol.* **1999**, *73*, 2181–2192. [CrossRef] [PubMed]

MDPI

St. Alban-Anlage 66

4052 Basel

Switzerland

Tel. +41 61 683 77 34

Fax +41 61 302 89 18

www.mdpi.com

Viruses Editorial Office

E-mail: viruses@mdpi.com

www.mdpi.com/journal/viruses